Entropy Measures for Data Analysis

Entropy Measures for Data Analysis

Theory, Algorithms and Applications

Special Issue Editor

Karsten Keller

MDPI • Basel • Beijing • Wuhan • Barcelona • Belgrade

MDPI

Special Issue Editor
Karsten Keller
Institut für Mathematik,
Universität zu Lübeck
Germany

Editorial Office
MDPI
St. Alban-Anlage 66
4052 Basel, Switzerland

This is a reprint of articles from the Special Issue published online in the open access journal *Entropy* (ISSN 1099-4300) from 2018 to 2019 (available at: https://www.mdpi.com/journal/entropy/special_issues/Entropy_Data).

For citation purposes, cite each article independently as indicated on the article page online and as indicated below:

LastName, A.A.; LastName, B.B.; LastName, C.C. Article Title. *Journal Name* **Year**, *Article Number*, Page Range.

ISBN 978-3-03928-032-2 (Pbk)
ISBN 978-3-03928-033-9 (PDF)

Contents

About the Special Issue Editor

Karsten Keller teaches and conducts research at the University of Luebeck. His work concerns time-discrete nonlinear dynamical systems and time-series analysis with special emphasis on symbolic dynamics. His recent research is mainly devoted to the development and application of entropies and entropy-like measures, especially in terms of concepts related to the recent concept of permutation entropy.

entropy

MDPI

Editorial

Entropy Measures for Data Analysis: Theory, Algorithms and Applications

Karsten Keller

Institut für Mathematik, Universität zu Lübeck, D-23562 Lübeck, Germany; keller@math.uni-luebeck.de

Received: 11 September 2019; Accepted: 22 September 2019; Published: 25 September 2019

Keywords: data analysis; entropy; entropy measures; automatic learning

Entropies and entropy-like quantities are playing an increasing role in modern non-linear data analysis and beyond. Fields that benefit from their application reach from diagnostics in physiology, for instance, electroencephalography (EEG), magnetoencephalography (MEG) and electrocardiography (ECG), to econophysics and engineering. During the last few years, classical concepts as approximate entropy and sample entropy have been supplemented by new entropy measures, like permutation entropy and its variants. Recent developments are focused on multidimensional generalizations of the concepts with a special emphasis on the quantification of coupling and the similarity between time series and system components behind them. One of the main future challenges in the field include finding a better understanding of the nature of the various entropy measures and their relationships, with the aim of adequate application including good parameter choices. The utilization of entropy measures as features in automatic learning and their application to large and complex data for such tasks as classification, discrimination and finding structural changes requires fast and well-founded algorithms. This issue is facing a different aspect of the use of entropy measures for data analysis in a wide sense, including those described.

Papers 1–3 discuss the problem of parameter choice mentioned and aspects related to it. Ahmadi et al. [1] investigate the sensitivity of sample entropy with respect to different parameters like, for example, tolerance size and sampling rate for gait data. Cuesta-Frau et al. [2] study parameter choice for permutation entropy, particularly embedding dimension and time series length, in the context of a lot of synthetic and real data sets. Here special emphasis is put on practical aspects of data analysis. In particular, the authors point out that in many cases permutation entropy can be used for shorter data sets than reported by other authors. In a certain sense complementary to the paper of Cuesta-Frau et al. [2], Piek and Keller [3] investigate parameter choice for permutation entropy and, more generally, ordinal pattern-based entropies from a computational and theoretical viewpoint. Fast algorithms are presented, possibilities and limits of the estimation of the Kolmogorov–Sinai entropy are discussed. A further aspect of the paper is the generation of artificial data for testing ordinal pattern methods.

The main objective of papers 4 and 5 is entropy-based feature extraction. Lu et al. [4] utilize approximate entropy, sample entropy, composite multiscale entropy and fuzzy entropy for identifying auditory object-specific attention from single-trial EEG signals by support vector machine (SVM)-based learning. For circuit fault diagnosis, He et al. [5] propose a new feature extraction method, which is mainly based on a measure called joint cross-wavelet singular entropy and a special dimension reduction technique. The obtained features are entered into a support vector machine classifier in order to locate faults. Besides feature extraction, direct applications of entropy for automatic learning is also addressed in this issue. Bukovsky et al. [6] discuss and further develop the recently introduced concept of learning entropy (LE) as a learning-based information measure, which is targeted at real-time novelty detection based on unusual learning efforts. For assessing the quality of data transformations in machine learning, Valverde-Albacete et al. [7] introduce an information-theoretic tool. They analyze performance of the tool for different types of data transformation, among them principal component analysis and independent component analysis.

Papers 8–10 are devoted to the aspect of coupling and similarity analysis. For studying the Chinese stock market around the 2015 crash, Wang and Hui [8] utilize effective transfer entropy (ETE), which is an adaption of transfer entropy to limited and noisy data. From this base, they discuss and compare dependencies of 10 Chinese stock sectors during four characteristic time periods near the crash. In [9], Craciunescu et al. introduce a new measure for describing coupling in interconnected dynamical systems and test it for different system interactions. Besides such in model systems, real-life system interactions like between the El Niño Southern Oscillation, the Indian Ocean Dipole, and influenza pandemic occurrence are considered. Here, coupling strength is quantified by entropies of adjacency matrices associated to networks constructed. Wang et al. [10] use entropy-based similarity and synchronization indices for relating postural stability and lower-limb muscle activity. Their study is based on two types of signals, one measuring the centre of pressure (COP) in dependence on time and one being an electromyogram (EMG). The authors show high correlation of COP and the low frequency EMG and that the cheaper COP contains much information on the EMG.

The other four papers touch further interesting aspects of entropy measure use. Selvachandran et al. [11] consider complex vague soft sets (CCVS), defined as a hybrid model of vague soft sets and complex fuzzy sets, which is, for example, useful for the description of images. Some distance and entropy measures for CCVSs are axiomatically defined and relations between them are investigated. The work [12] of Pan et al. focuses on Dempster–Shafer evidence theory, which can be considered as a generalization of probability theory. A new belief entropy, measuring uncertainty in this framework, and its performance are discussed on the base of numerical experiments. García-Gutiérrez et al. [13] introduce a new model for the particle size distribution (PSD) of granular media, which relates two models known for a long time. For this purpose, a differential equation involving the information entropy is used. The interesting point is that experimental data can be considered as an initial condition for simulating a PSD. Last but not least, Liu et al. [14] demonstrate that entropy methods also can be helpful in solving nonlinear and multimodal optimization problems. They develop an algorithm based on the firefly algorithm and the cross-entropy method and report its good performance, especially powerful global search capacity precision and robustness for numerical optimization problems.

Acknowledgments: I express my thanks to the authors of the above contributions, and to the journal Entropy and MDPI for their support during this work.

Conflicts of Interest: The authors declare no conflict of interest.

References

1. Ahmadi, S.; Sepehri, N.; Wu, C.; Szturm, T. Sample Entropy of Human Gait Center of Pressure Displacement: A Systematic Methodological Analysis. *Entropy* **2018**, *20*, 579. [CrossRef]
2. Cuesta-Frau, D.; Murillo-Escobar, J.P.; Orrego, D.A.; Delgado-Trejos, E. Embedded Dimension and Time Series Length. Practical Influence on Permutation Entropy and Its Applications. *Entropy* **2019**, *21*, 385. [CrossRef]
3. Piek, A.B.; Stolz, I.; Keller, K. Algorithmics, Possibilities and Limits of Ordinal Pattern Based Entropies. *Entropy* **2019**, *21*, 547. [CrossRef]
4. Lu, Y.; Wang, M.; Zhang, Q.; Han, Y. Identification of Auditory Object-Specific Attention from Single-Trial Electroencephalogram Signals via Entropy Measures and Machine Learning. *Entropy* **2018**, *20*, 386. [CrossRef]
5. He, W.; He, Y.; Li, B.; Zhang, C. Analog Circuit Fault Diagnosis via Joint Cross-Wavelet Singular Entropy and Parametric t-SNE. *Entropy* **2018**, *20*, 604. [CrossRef]
6. Bukovsky, I.; Kinsner, W.; Homma, N. Learning Entropy as a Learning-Based Information Concept. *Entropy* **2019**, *21*, 166. [CrossRef]
7. Valverde-Albacete, F.J.; Peláez-Moreno, C. Assessing Information Transmission in Data Transformations with the Channel Multivariate Entropy Triangle. *Entropy* **2018**, *20*, 498. [CrossRef]

8. Wang, X.; Hui, X. Cross-Sectoral Information Transfer in the Chinese Stock Market around Its Crash in 2015. *Entropy* **2018**, *20*, 663. [CrossRef]

9. Craciunescu, T.; Murari, A.; Gelfusa, M. Improving Entropy Estimates of Complex Network Topology for the Characterization of Coupling in Dynamical Systems. *Entropy* **2018**, *20*, 891. [CrossRef]

10. Wang, C.; Jiang, B.C.; Huang, P. The Relationship between Postural Stability and Lower-Limb Muscle Activity Using an Entropy-Based Similarity Index. *Entropy* **2018**, *20*, 320. [CrossRef]

11. Selvachandran, G.; Garg, H.; Quek, S.G. Vague Entropy Measure for Complex Vague Soft Sets. *Entropy* **2018**, *20*, 403. [CrossRef]

12. Pan, Q.; Zhou, D.; Tang, Y.; Li, X.; Huang, J. A Novel Belief Entropy for Measuring Uncertainty in Dempster-Shafer Evidence Theory Framework Based on Plausibility Transformation and Weighted Hartley Entropy. *Entropy* **2019**, *21*, 163. [CrossRef]

13. García-Gutiérrez, C.; Martín, M.Á.; Pachepsky, Y. On the Information Content of Coarse Data with Respect to the Particle Size Distribution of Complex Granular Media: Rationale Approach and Testing. *Entropy* **2019**, *21*, 601. [CrossRef]

14. Li, G.; Liu, P.; Le, C.; Zhou, B. A Novel Hybrid Meta-Heuristic Algorithm Based on the Cross-Entropy Method and Firefly Algorithm for Global Optimization. *Entropy* **2019**, *21*, 494. [CrossRef]

entropy

MDPI

Article

Sample Entropy of Human Gait Center of Pressure Displacement: A Systematic Methodological Analysis

Samira Ahmadi [1], Nariman Sepehri [1,*], Christine Wu [1] and Tony Szturm [2]

[1]   Department of Mechanical Engineering, University of Manitoba, Winnipeg, MB R3T 5V6, Canada;
    ahmadis3@myumanitoba.ca (S.A.); Christine.Wu@umanitoba.ca (C.W.)
[2]   Department of Physical Therapy, College of Rehabilitation Sciences, University of Manitoba, Winnipeg,
    MB R3E 0T6, Canada; Tony.Szturm@umanitoba.ca
*   Correspondence: nariman.sepehri@umanitoba.ca; Tel.: +1-204-474-6834

Received: 18 June 2018; Accepted: 2 August 2018; Published: 6 August 2018

Abstract: Sample entropy (SampEn) has been used to quantify the regularity or predictability of human gait signals. There are studies on the appropriate use of this measure for inter-stride spatio-temporal gait variables. However, the sensitivity of this measure to preprocessing of the signal and to variant values of template size (m), tolerance size (r), and sampling rate has not been studied when applied to "whole" gait signals. Whole gait signals are the entire time series data obtained from force or inertial sensors. This study systematically investigates the sensitivity of SampEn of the center of pressure displacement in the mediolateral direction (ML COP-D) to variant parameter values and two pre-processing methods. These two methods are filtering the high-frequency components and resampling the signals to have the same average number of data points per stride. The discriminatory ability of SampEn is studied by comparing treadmill walk only (WO) to dual-task (DT) condition. The results suggest that SampEn maintains the directional difference between two walking conditions across variant parameter values, showing a significant increase from WO to DT condition, especially when signals are low-pass filtered. Moreover, when gait speed is different between test conditions, signals should be low-pass filtered and resampled to have the same average number of data points per stride.

Keywords: sample entropy; treadmill walking; center of pressure displacement; dual-tasking

---

1. Introduction

Entropy measures quantify the regularity or predictability of a time series [1–5]. Larger entropy values indicate less regularity or predictability in a time series. These measures have been used in gait analysis, and have been shown to discriminate between fallers and non-fallers [6], older and younger adults [7,8] and walk only (WO) and dual-task (DT) walking condition [7]. Various entropy measures have been proposed based on Shannon's entropy [9] and its successor method, Approximate Entropy [10]. One of the most commonly methods used to study gait function is Sample Entropy (SampEn) [3].

Earlier application of SampEn was used to examine various inter-stride spatio-temporal gait variables, derived from endpoints of the gait cycle (heel strikes), for example, stride time and step length signals [11]. It has also been shown that there may be temporal scales in changes that occur in spatio-temporal gait variables [12]. However, these signals lack intra-stride information, which represents important passive and active gait control process [7]. Recent gait studies have examined "whole" gait signals (entire time series as opposed to stride-to-stride gait variables), such as trunk linear acceleration [1,6,13,14], joint angular positions [15] and center of pressure of feet displacement [7,16].

Center of pressure displacement in the mediolateral direction (ML COP-D) has been extensively used to examine the balance performance during standing conditions (base of support stationary) [17–19]. However, its usage in gait analysis is limited to a few studies [1,7,16,20]. This might be due to the limitation of the facilities that collect data on a treadmill or during overground walking. ML COP-D is capable of representing both single and double support phases of the gait cycle. It has also been shown that SampEn of ML COP-D better distinguishes between different treadmill walking conditions as compared to the SampEn of other commonly used signals (e.g., trunk linear acceleration) [7].

A few studies have investigated how variant template size (m), tolerance size (r), and data length (N) would affect SampEn of short [11] and long [21] time series when using inter-stride spatio-temporal gait variables. It was shown that SampEn values are dependent on the combination of m and r, and not on N [11,21]. However, no study has investigated the effect of parameter selection on SampEn of the human gait whole signals, such as ML COP-D or trunk linear acceleration, over an appropriate amount of continuous strides. During continuous, steady-state gait, these signals are similar in nature with a few dominant frequencies and have consistent fluctuations from stride to stride. Considering the increasing use of SampEn in analyzing human gait whole signals, it is essential to investigate how parameter selection would affect the outcomes. The importance of this investigation stems from the fact that parameter selection for calculating SampEn of whole gait signals, in many studies, is based on those that have analyzed inter-stride gait variables [8,13].

Most studies which have examined the effect of aging or dual-tasking on gait function, use self-paced walking and do not control for gait speed. Self-paced walking results in different walking speeds and, therefore, each walking condition will have a different average number of data points per stride. It has been shown that gait speed is significantly reduced during dual-task walking compared to walk only trials [22] and it has been reported that speed has a significant effect on measures of dynamical systems, such as the largest Lyapunov exponent [23–25]. Moreover, researchers have used different sampling frequencies when collecting target whole signals, which in turn have resulted in a different average number of data points per stride. It is unknown whether a different average number of data points per stride caused by varying walking speed or sampling rate would affect SampEn. Furthermore, many researchers have opted to apply SampEn, or other entropy measures, to raw unfiltered signals [8,16,26] to avoid losing or altering information due to filtering. While others have filtered the high-frequency components of trunk linear acceleration signal using a cut-off frequency of 20 Hz [14,27]. Therefore, investigating the effect of filtering would also be beneficial.

The first objective of this study is to systematically examine the sensitivity of SampEn of ML COP-D signals, obtained during treadmill walking, to variant m, r, and sampling rate values. The second objective is to determine the effect of the choice of low-pass filtering and data resampling, to have the same average number of data points per stride, on the SampEn of ML COP-D signals. Discriminatory ability of SampEn will be examined through comparing walk only condition to dual-task walking, which has been shown to adversely affect gait performance [28,29].

2. Materials and Methods

2.1. Experimental Procedure

A convenience sample of 29 healthy young participants (eight females, 28.3 $\pm$ 2.7 years, 173.4 $\pm$ 8.8 cm, 69.7 $\pm$ 14.2 kg, mean $\pm$ standard deviation (SD)) was recruited. They were screened to ensure that no participant had any illnesses, neuromuscular injuries or previous surgeries that might affect their balance and gait. The University of Manitoba Human Research Ethics Committee approved the study and all participants signed the informed consent form prior to the tests.

Participants were asked to walk on an instrumented Bertec treadmill (Bertec Corporation, Columbus, OH, USA) under three different walking conditions:

(a)   Walk only (WO) trial of 1 min at a speed of 1.0 m/s, and

(b)   Dual-task (DT) walking trial of 1 min at a speed of 1.0 m/s, which is described below, and

(c)   Walk only trial of 1 min at a speed of 1.3 m/s (WO-1.3).

Center of pressure displacements in the mediolateral (ML COP-D) and anteroposterior (AP COP-D) directions (Figure 1) were calculated from the force and moment components, which were sampled at 1000 Hz. Forty seconds of each signal, which contained at least 30 strides [16], were used after discarding approximately the first 4 strides.

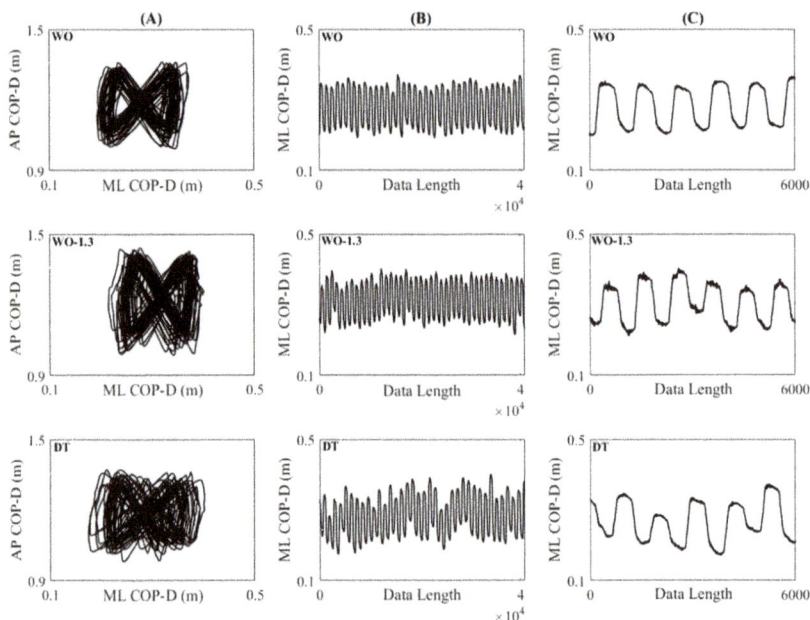

Figure 1. Trajectory of the center of pressure under WO, WO-1.3, and DT conditions: (**A**) The low-pass filtered trajectory of center of pressure displayed as AP COP-D vs. ML COP-D, (**B**) Filtered ML COP-D, (**C**) Several strides of unfiltered ML COP-D.

During all walking trials, participants viewed an 80 cm computer monitor positioned 1 meter away at eye level. During the WO trials, participants watched a scenery video to maintain gaze and head position relative to the monitor. For the purpose of hands-free interaction with game activities, a commercial inertial-based wireless mouse (Elite mouse, SMK Electronics, Chula Vista, CA, USA) was mounted on a plastic headband worn by each participant. Therefore, during walking, the head rotation was used to control the motion of the computer cursor. The goal of the game was to move a game paddle horizontally to interact with moving game objects. The game objects were categorized as designated targets or designated distractors, with the shape of a soccer ball and dotted sphere, respectively. They appeared at random locations at the top of the display every 2 s and moved diagonally toward the bottom of the display. In response to each "game event" (target appearance), the participant produced a head rotation (i.e., rotation of the motion-sense mouse) to move the game paddle (left/right) to catch the target objects and avoid the distractors. For a full description of the interactive cognitive computer game, see Szturm et al. [29].

2.2. Sample Entropy

SampEn (m, r, N) [3] of a dataset of length N is the negative natural logarithm of the conditional probability of two successive counts of similar pairs (Chebyshev distance less than a tolerance size of r) of template size m and $m + 1$ without allowing self-matches. SampEn is calculated as follows [7]; consider a time series of length N given below:

$$u = u(1), u(2), \ldots, u(N) \text{ or } u = \{u(j) : 1 \leq j \leq N\} \tag{1}$$

First, m value is chosen to construct series of pairs, size m as:

$$X_m(i) = \{u(i+k) : 0 \leq k \leq m-1\}, \ 1 \leq i \leq N - m + 1 \tag{2}$$

Next, matching templates are found by comparing their chebyshev distance to a pre-determined r value while excluding self-comparison. Next, a variable called B_i is built which is the number of pairs satisfying the aforementioned criteria:

$$B_i^m(r) = \frac{1}{N-m-1} \ (\# \text{ of } d|X_m(i) - X_m(j)| \ \leq r, \text{ where } j = 1 : N - m \ \& \ i \neq j) \tag{3}$$

$$d|X_m(i) - X_m(j)| = \max\{|u(i+k) - u(j+k)| : 0 \leq k \leq m-1\} \tag{4}$$

Next, $B^m(r)$ is defined as:

$$B^m(r) = \frac{1}{N-m} \sum_{i=1}^{N-m} B_i^m(r) \tag{5}$$

This process is repeated for $m+1$ and r to form $A^m(r)$:

$$A_i^m(r) = \frac{1}{N-m-1} \ (\# \text{ of } d|X_{m+1}(i) - X_{m+1}(j)| \ \leq r, \text{ where } j = 1 : N - m \ \& \ i \neq j) \tag{6}$$

$$A^m(r) = \frac{1}{N-m} \sum_{i=1}^{N-m} A_i^m(r) \tag{7}$$

Lastly, SampEn is calculated based on $B^m(r)$ and $A^m(r)$ as

$$\text{SampEn}\,(m, r, N) = -\ln \frac{A^m(r)}{B^m(r)} \tag{8}$$

2.3. Data Analysis

This study consists of two parts. In the first part, the sensitivity of SampEn to changing m, r, and sampling rate was investigated when comparing WO to DT. Two methods were used to downsample signals from 1000 Hz to lower sampling rates (Table 1). The goal was to downsample signals by factors of 1, 2, 4, 8, 16 and 32. The first method, decimation (D) by a factor of f, used an eighth-order low-pass Chebyshev Type I filter, which filtered the signal in forward and reverse directions to remove phase distortions and then selected every fth point (MATLAB command *decimate*). The filter had a normalized cut-off frequency of $0.8/f$. This method was chosen to avoid aliasing distortion that might occur by simply downsampling a signal.

Table 1. Summary of downsampling factors (f), sampling rates and cut-off frequency for decimation and filtering-and-downsampling methods.

f	Sampling Rate (Hz)	Cut-Off Frequency (Hz)	
		Decimation	Filtering-and-Downsampling
1	1000	800	30
2	500	400	30
4	250	200	30
8	125	100	30
16	62	50	30
32	31	25	30

The second method, filtering-and-downsampling (FD) by a factor of f, used a second-order Butterworth low-pass filter with a cut-off frequency of 30 Hz, and then downsampled the signal by a factor of f (MATLAB command *downsample*). Butterworth low-pass filter is the most common filter used in the literature to reduce the effect of noise [30] along with maintaining the variability in the lower range frequencies where the musculoskeletal motion occurs [31]. A nonparametric PSD estimator, Welch's algorithm, was used to obtain the cut-off frequency. The dominant peak was at 0.89 ± 0.06 Hz (mean $\pm$ SD) for WO, 0.91 ± 0.06 for DT, and 0.99 ± 0.06 for WO-1.3. The last peak before noise floor occurred in the 8–15 Hz frequency range. Therefore, 15 Hz was considered as the highest frequency component and 30 Hz was used as the cut-off frequency.

The two methods yielded approximately the same results with respect to the low-pass filtering for $f = 32$. Therefore, the first five f values could shed light on the effect of low-pass filtering prior to the calculation of SampEn.

SampEn was calculated using all combinations of parameter values, $m = 2, 4, 6, 8, 10$ and $r = 0.2$ and $0.3 \times$ standard deviation (SD) of all the time series, and for all downsampling factors $f = 1, 2, 4, 8, 16, 32$, and for both decimated and filtered-and-downsampled signals of WO and DT condition. The present investigation was based on more m and f values in the selected ranges. However, the necessity for statistical analysis with the purpose of studying the discriminatory ability of SampEn, led to choosing fewer parameter values (levels within a factor); e.g., five levels versus nine levels for template size ($m = 2$~10). In a previous study [11], $m = 2, 3, 4$ were tested when SampEn was applied to the inter-stride spatio-temporal gait variables. The present work included more m values to study the SampEn of the entire gait signals and not just times at heel strike or step distances. It was hypothesized that larger m values could better discern changes when there is a much greater number of data points per gait cycle or stride. Additionally, unlike ApEn, SampEn decreases almost monotonically with increasing r value [3,11] and 0.1–0.3 times the standard deviation has been suggested for inter-stride spatio-temporal gait variables [11]. The current analysis was based on $r = 0.1 \times$ SD, $r = 0.2 \times$ SD and $r = 0.3 \times$ SD. However, when the parameter value $r = 0.1 \times$ SD was used, many SampEn values converged to infinity. Therefore this level was not included in the results. Large r values were not included because they result in much smaller SampEn values for each condition, i.e., more matched templates, which diminish the discriminatory ability of SampEn.

In the second part, the effect of low-pass filtering and resampling, to have the same average number of data points per stride, was investigated. SampEn of ML COP-D signal of WO, DT, and WO-1.3 was calculated using $m = 4$, $r = 0.2 \times$ SD, and $f = 8$ (based on the results of the first part). Four methods of preprocessing were used for each condition;

- decimation (D),
- decimation-and-resampling (D-R),
- filtering-and-downsampling (FD) and,
- filtering-and-downsampling-and-resampling (FD-R).

The average number of data points per stride for WO, DT and WO-1.3 were 142, 140 and 128, respectively. Therefore, 30 strides of each time series were resampled (MATLAB command *resample*) so that all of the signals would have an average of 142 data points per stride.

2.4. Statistical Analysis

In the first part of this study, there were 4 factors of within-subject repeated measures, which are 2 levels of walking condition (WO and DT), 2 levels of r, 6 levels of f, and 5 levels of m. The following steps were taken to perform the statistical analysis separately for both decimated and filtered-and-downsampled signals.

i.   A two-factor repeated measure ANOVA (walking condition*m) was performed at each *f* level while considering the first tolerance level.

ii.   A two-factor repeated measure ANOVA (walking condition*m) was performed at each *f* level while considering the second tolerance level.

iii.   A two-factor repeated measure ANOVA (walking condition*f) was performed at each *m* level while considering the first tolerance level.

iv.   A two-factor repeated measure ANOVA (walking condition*f) was performed at each *m* level while considering the second tolerance level.

v.   Post hoc pairwise comparisons with Bonferroni correction were performed to examine the effect of dual-tasking at each level.

vi.   Finally, a two-factor (walking condition*r) repeated measure ANOVA was performed at fixed $m = 4$ and $f = 8$ values, which were chosen based on the previous step's statistical results.

In the second part of this study, two two-factor within-subject ANOVA were used to examine the main and interaction effects of the following factors on SampEn;

- walking condition (WO versus DT) and preprocessing method (D, D-R, FD, FD-R)
- gait speed (1.0 m/s versus 1.3 m/s) and preprocessing method (D, D-R, FD, FD-R)

Normality of all dependent variables was checked using the Shapiro-Wilk normality test. Results confirmed that the data was normally distributed. Statistical analyses were carried out using SPSS software version 24. In all the tests, a *p*-value less than 0.05 was considered significant. A Bonferroni correction was used in the software for multiple comparisons.

3. Results

For the vast majority of combinations of parameter values, SampEn of ML COP-D during dual-task walking was significantly larger than that of walk only. In general, SampEn decreased as *m* increased, as *r* increased, and as *f* factor decreased, i.e., as sampling rate increased or as the number of points per stride increased. However, there were a few exceptions, which will be discussed further. The results of the main and interaction effects of walking condition (WO and DT), *m*, and *f* at each *r* value are presented in Tables 2 and 3. In addition, the results of pairwise comparisons of the significant main effects of walking condition are presented in Tables A1 and A2. The detailed results for each downsampling method are presented in the following subsections followed by the results of the effects of the preprocessing methods.

Table 2. Main and interaction effects (*p*-values) of "walking condition (W-C)*f" at each *r* and *m* value for FD (Filtered-and-Downsampled) and D (Decimated) ML COP-D. The two conditions are WO and DT. *p*-values in bold indicate a significant difference.

		W-C	*f*	W-C*f			W-C	*f*	W-C*f
FD $r = 0.2 \times SD$	$m = 2$	<0.001	<0.001	<0.001	D $r = 0.2 \times SD$	$m = 2$	0.009	<0.001	<0.001
	$m = 4$	<0.001	<0.001	0.006		$m = 4$	0.009	<0.001	<0.001
	$m = 6$	0.001	<0.001	0.123		$m = 6$	0.021	<0.001	0.023
	$m = 8$	0.004	<0.001	0.334		$m = 8$	0.037	<0.001	0.070
	$m = 10$	0.020	<0.001	0.421		$m = 10$	0.075	<0.001	0.211
		W-C	*f*	**W-C**f**			**W-C**	*f*	**W-C**f**
FD $r = 0.3 \times SD$	$m = 2$	<0.001	<0.001	<0.001	D $r = 0.3 \times SD$	$m = 2$	0.008	<0.001	<0.001
	$m = 4$	<0.001	<0.001	<0.001		$m = 4$	0.001	<0.001	<0.001
	$m = 6$	0.002	<0.001	0.080		$m = 6$	0.015	<0.001	0.034
	$m = 8$	0.006	<0.001	0.232		$m = 8$	0.032	<0.001	0.154
	$m = 10$	0.020	<0.001	0.387		$m = 10$	0.068	<0.001	0.279

Table 3. Main and interaction effects (*p*-values) of "walking condition (W-C)*m" at each *r* and *f* value for FD (Filtered-and-Downsampled) and D (Decimated) ML COP-D. The two conditions are WO and DT. *p*-values in bold indicate a significant difference.

		W-C	*m*	W-C*m			W-C	*m*	W-C*m
	$f = 1$	0.007	<0.001	0.001		$f = 1$	0.094	<0.001	0.004
	$f = 2$	0.002	<0.001	0.007		$f = 2$	0.806	<0.001	0.000
FD	$f = 4$	<0.001	<0.001	0.128	D	$f = 4$	0.049	<0.001	0.176
$r = 0.2 \times$ SD	$f = 8$	<0.001	<0.001	0.015	$r = 0.2 \times$ SD	$f = 8$	0.008	<0.001	0.145
	$f = 16$	<0.001	<0.001	0.002		$f = 16$	0.001	<0.001	0.003
	$f = 32$	0.007	<0.001	0.070		$f = 32$	0.005	<0.001	0.114
		W-C	***m***	**W-C**m***			**W-C**	***m***	**W-C**m***
	$f = 1$	0.011	<0.001	0.010		$f = 1$	0.329	<0.001	0.044
	$f = 2$	0.006	<0.001	0.000		$f = 2$	0.938	<0.001	<0.001
FD	$f = 4$	0.001	<0.001	0.170	D	$f = 4$	0.049	<0.001	0.020
$r = 0.3 \times$ SD	$f = 8$	<0.001	<0.001	0.062	$r = 0.3 \times$ SD	$f = 8$	0.006	<0.001	0.178
	$f = 16$	<0.001	<0.001	0.002		$f = 16$	0.001	<0.001	0.004
	$f = 32$	0.005	<0.001	0.006		$f = 32$	0.005	<0.001	0.010

3.1. Sensitivity of SampEn to Variant Parameter Values When Using Filtering-and-Downsampling

Figures 2 and 3 show the effect of changing *f* and *m* values on SampEn of the filtered-and-downsampled signals at $r = 0.2 \times$ SD. The figures are virtually the same as those of $r = 0.3 \times$ SD. The statistical results are also reported in Tables 2, 3, A1 and A2. A statistically significant interaction was found between walking condition and *f* at $m = 2$ and $m = 4$ for both tolerance values. Since the direction of the changes of SampEn with increasing *f* was increasing for both walking conditions, the interaction effect would signify a difference in the rate of the changes between levels. A statistically significant interaction was found between walking condition and *m* at all *f* values except for; (a) $f = 4$ and $f = 32$ for $r = 0.2 \times$ SD and (b) $f = 4$ and $f = 8$ for $r = 0.3 \times$ SD. At each *f* value, the direction of the changes of SampEn of WO, with respect to *m*, was similar to those of DT. The two-factor repeated measures of ANOVA of walking condition *r showed that there was no significant interaction between the walking condition and the *r* value ($p = 0.813$). However, there was a statistically significant decrease of SampEn with increasing the *r* value ($p < 0.001$). At each *r* value, SampEn of DT was statistically significantly larger than that of WO ($p < 0.001$).

For all *m* values, there was a statistically significant main effect of walking condition and *f* value on SampEn. Similarly, for all *f* values, there was a statistically significant main effect of walking condition and *m* value on SampEn. SampEn significantly increased from WO to DT for most combinations of *r*, *m*, and *f* values except for; (a) 4 out of 30 combinations of *f* and *m* values for $r = 0.2 \times$ SD and (b) 3 out of 30 combinations for $r = 0.3 \times$ SD. Nevertheless, for these exceptions, there was a trend of increased SampEn from WO to DT. Based on the statistical analysis, the increasing effect of dual-tasking on SampEn of ML COP-D signal can be captured for most combinations of *r*, *m*, and *f* values. The only exceptions are $m = 10$ at $f = 16$ and $m = 6, 8, 10$ at $f = 32$.

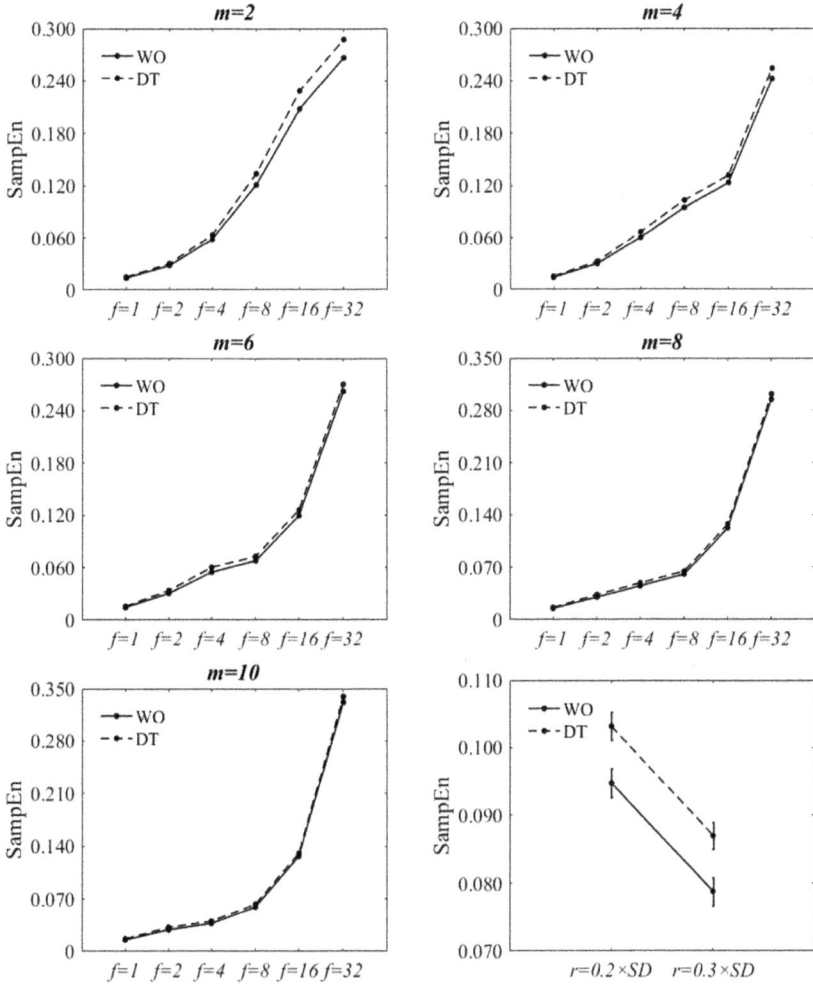

Figure 2. Effect of changing f on SampEn of WO and DT at each m value at $r = 0.2 \times$ SD. for filtered-and-downsampled ML COP-D. Bottom-right: Effect of r on SampEn at $f = 8$ and $m = 4$. The error bars reflect the standard error of the means.

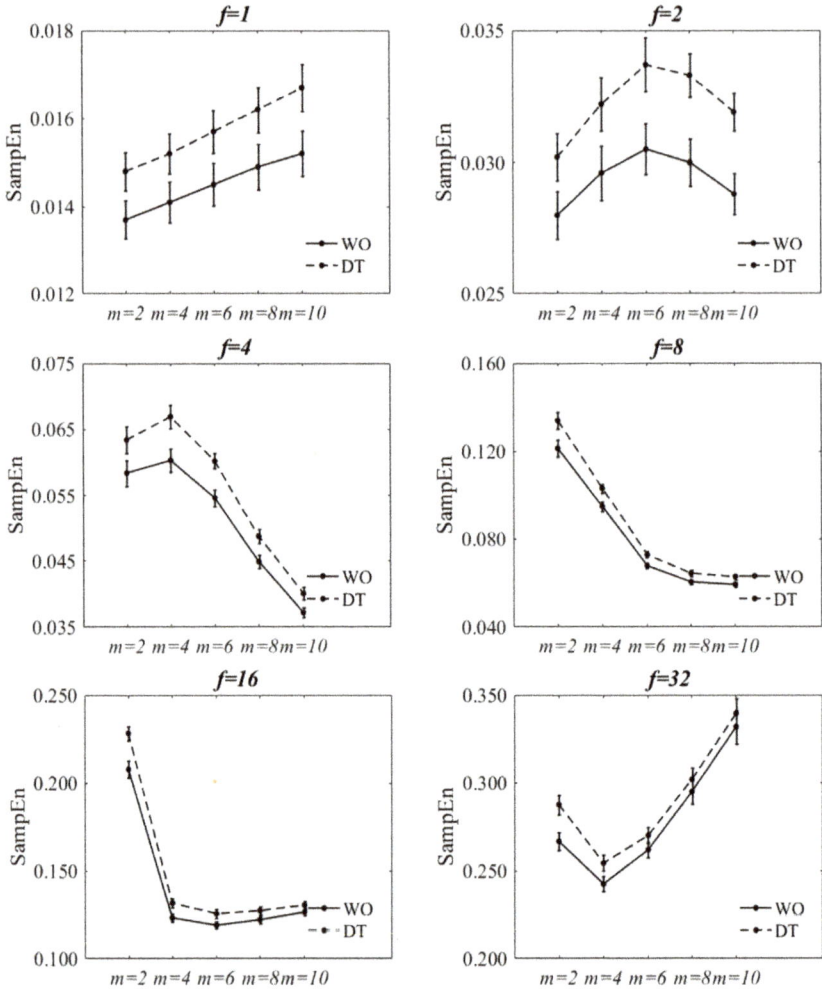

Figure 3. Effect of changing m on SampEn of WO and DT at each f value at $r = 0.2 \times SD$ for filtered-and-downsampled ML COP-D. The error bars reflect the standard error of the means.

3.2. Sensitivity of SampEn to Variant Parameter Values When Using Decimation

Figures 4 and 5 show the effect of changing f and m values on SampEn of the decimated signal at $r = 0.2 \times SD$. The figures are virtually the same as those of $r = 0.3 \times SD$. The statistical results are also reported in Tables 2, 3, A1 and A2. A statistically significant interaction was found between walking condition and f at $m = 2$, $m = 4$, and $m = 6$ for both tolerance values. The direction of the changes of SampEn with increasing f was increasing for both tasks. A statistically significant interaction was found between walking condition and m at all f values except for; (a) $f = 4$, $f = 8$ and $f = 32$ for $r = 0.2 \times SD$ and (b) $f = 8$ for $r = 0.3 \times SD$. At each f value, the direction of changes of SampEn of WO, with respect to m, was the same to those of DT. The two-factor repeated measure of ANOVA of walking condition-r showed that there was no significant interaction between the walking condition and the r value ($p = 0.980$). However, there was a statistically significant decrease of SampEn with increasing the *r* value (*p* = 0.001). At each *r* value, SampEn of DT was statistically significantly larger than that of WO (*p* < 0.001).

For all *m* values, there was a statistically significant main effect of walking condition (except for *m* = 10) and *f* values on SampEn. And for all *f* values, there was a statistically significant main effect of walking condition (except for *f* = 1 and *f* = 2) and *m* values on SampEn. SampEn significantly increased from WO to DT for most combinations of *m*, *r*, and *f* (4, 8, 16, and 32) values with a few exceptions; smaller *m* values at *f* = 4 and larger *m* values at *f* = 16 and *f* = 32. Nevertheless, SampEn was seen to increase from WO to DT for those exceptions.

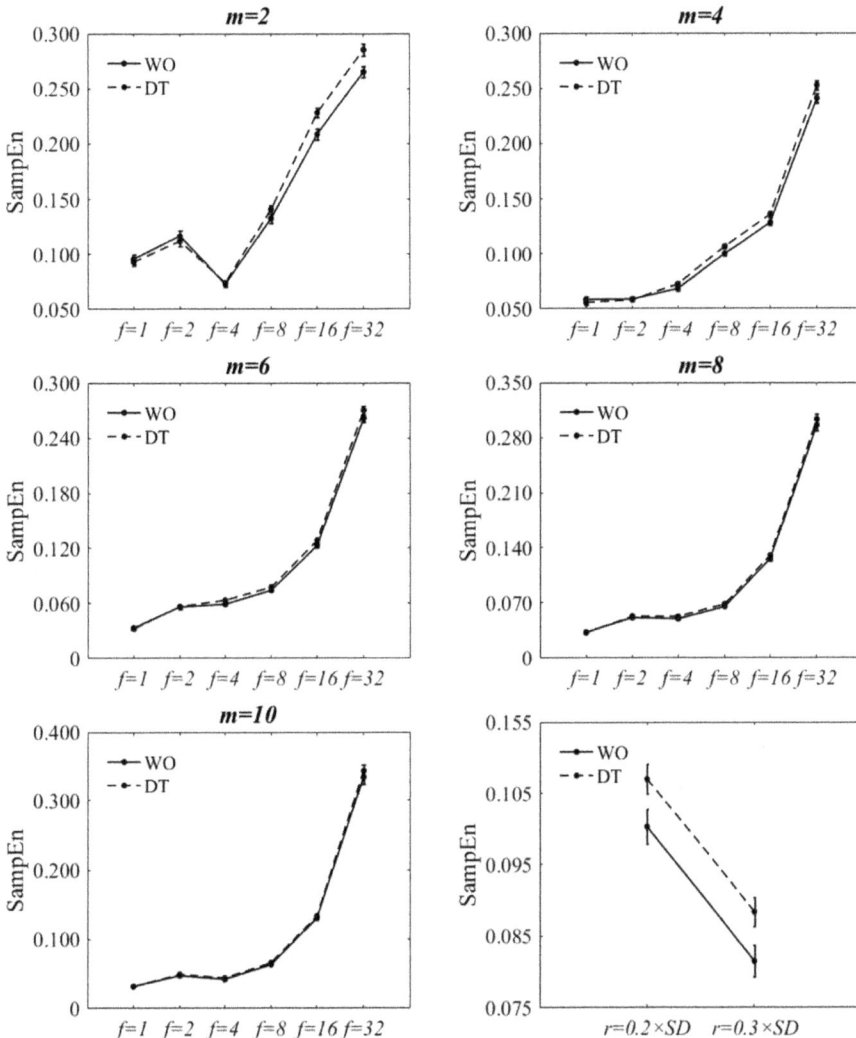

Figure 4. Effect of changing *f* on SampEn of WO and DT at each *m* value at *r* = 0.2 × SD for decimated ML COP-D. Bottom-right: Effect of *r* on SampEn at *f* = 8 and *m* = 4. The error bars reflect the standard error of the means.

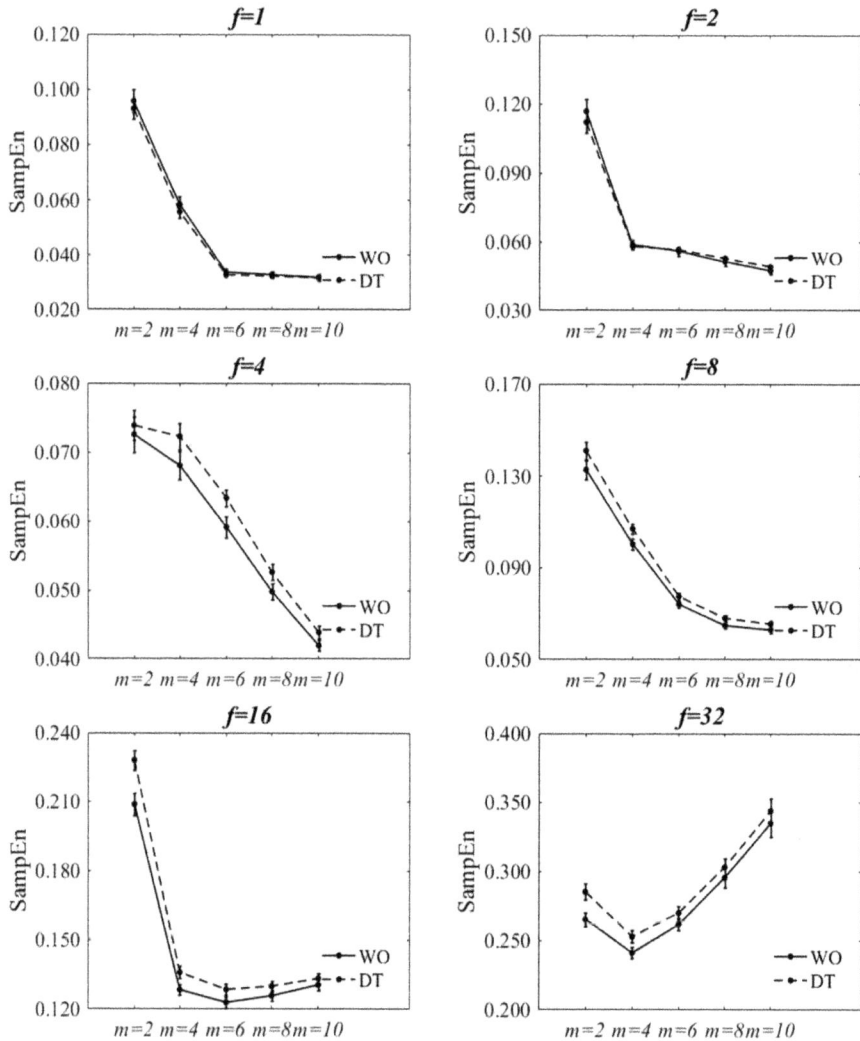

Figure 5. Effect of changing m on SampEn of WO and DT at each f value at $r = 0.2 \times$ SD for decimated ML COP-D. The error bars reflect the standard error of the means.

3.3. Effects of Preprocessing Methods

The descriptive and statistical results of the walking condition (WO versus DT), walking speed (1.0 m/s versus 1.3 m/s) and preprocessing method (D, D-R, FD, FD-R) on SampEn for the combination of $f = 8$, $m = 4$ and $r = 0.2 \times$ SD are presented in Figure 6 and Table 4.

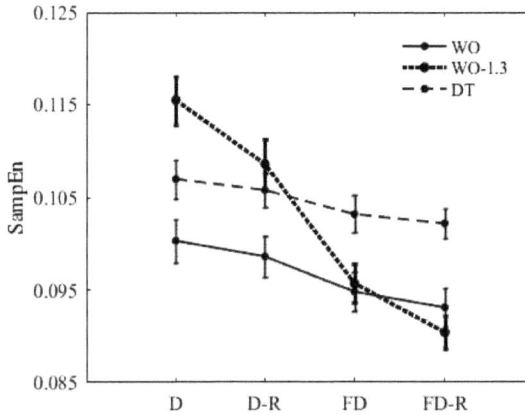

Figure 6. Effects of different preprocessing methods on SampEn of the ML COP-D signal of WO, DT, and WO-1.3 conditions for the combination of $f = 8$, $m = 4$, and $r = 0.2 \times SD$; The error bars reflect the standard error of the means. D: Decimation; D-R: Decimation-and-Resampling; FD: Filtering-and-Downsampling; and FD-R: Filtering-and-Downsampling-and-Resampling.

Table 4. Statistical results of the walking condition (WO and DT), walking speed (1.0 m/s and 1.3 m/s) and preprocessing method (D, D-R, FD, and FD-R) on SampEn of ML COP-D for the combination of $f = 8$, $m = 4$, and $r = 0.2 \times SD$; The top section presents main and interaction effects. The middle section presents the pairwise comparisons between preprocessing methods for each walking condition (WO, WO-1.3, and DT). The bottom section presents the pairwise comparisons between walking conditions for each preprocessing method. *p*-values in bold indicate a significant difference. D: Decimation; D-R: Decimation-and-Resampling; FD: Filtering-and-Downsampling; and FD-R: Filtering-and-Downsampling-and-Resampling.

	Main and Interaction Effects (*p*-Value)		
	Condition/Speed	Method	Interaction
WO vs. WO-1.3	**0.017**	**<0.001**	**<0.001**
WO vs. DT	**0.002**	**<0.001**	0.057
	Pairwise Comparisons (*p*-Value)		
	D vs. D-R	D vs. FD	FD vs. FD-R
WO	0.104	**<0.001**	**0.042**
WO-1.3	**<0.001**	**<0.001**	**<0.001**
DT	0.981	**<0.001**	1.000
Method	WO vs. WO-1.3	WO vs. DT	
D	**<0.001**	**0.013**	
D-R	**0.001**	**0.006**	
FD	0.701	**0.001**	
FD-R	0.225	**<0.001**	

There was a significant interaction effect of walking speed and method, while no significant interaction of walking condition and method was found. In addition, all three walking condition, speed, and method had a significant main effect on SampEn. Results revealed that resampling signals to have a larger average number of data points per stride had a decreasing effect on SampEn of WO-1.3 signals. However, there was no significant effect of resampling on SampEn of DT and WO signals. The only exception was the significant decrease of SampEn of WO from FD to FD-R. In addition, SampEn significantly decreased by filtering the high-frequency components. Furthermore, SampEn increased significantly from WO to DT when using all four methods. Finally, there was a significant increase in SampEn with increasing walking speed (WO to WO-1.3) only when using decimation or decimation-and-resampling.

4. Discussion

The goal of this investigation was to identify the sensitivity of SampEn to variant values of parameters (m and r, and sampling rate) and two preprocessing methods when applied to ML COP-D signal obtained during treadmill walking under WO and DT conditions. There were three main observations in this study. First, SampEn of ML COP-D consistently increased significantly from WO to DT for most of the combinations of the parameter values and methods of pre-processing. This finding is in agreement with the previous studies [7,16,32]. The only exceptions were $m = 10$ at $f = 16$ and $m = 6$, 8, 10 at $f = 32$ for filtered-and-downsampled ML COP-D. And for decimated ML COP-D, all m values at $f = 1$ and $f = 2$, and smaller m values at $f = 4$ and larger m values at $f = 16$ and $f = 32$. Second, the results demonstrated that walking speed should be controlled when studying the effect of another factor, such as adding a cognitive task. Finally, using a low-pass filter to eliminate high-frequency signal components and smoothing a time series improved the consistency (a significant increase from WO to DT) of SampEn analysis.

The results of this study showed that SampEn of ML COP-D signal increases as participants perform a concurrent cognitive task while walking. Steady-state gait signals are intrinsically periodic with consistent fluctuation from stride to stride. It can be argued that when human gait is disrupted by performing a secondary task or negatively affected by aging or diseases, these fluctuations would change. The unplanned fluctuations might increase during walking under challenging conditions known to cause gait disturbances and stumbles [33]. Another study has reported that SampEn of trunk linear acceleration signal was smaller in older adults who had reported a fall versus non-fallers when tested during overground walking [6]. However, in that study, speed was not controlled among participants or between groups. Reducing one's walking speed is a consistent strategy to manage threats to balance and when attending to concurrent cognitive tasks. In this regard, a main finding of the present study was that walking speed had a significant effect on SampEn of ML COP-D signal. One reason is an increase in vibrational noise of the treadmill-mounted force plate at higher walking speeds. Increased noise will cause an increase in SampEn. This problem can be solved by preprocessing the signal using a low-pass filter to eliminate high-frequency vibrational noise. Another issue when comparing signals collected at different speeds is that stride time will be reduced as speed increases, and therefore the number of data points per stride will be reduced. The present results of the effect of different f values on SampEn demonstrated that SampEn increased as the number of data points per stride decreased. One method to deal with this issue is to resample the signals so that the average number of data points per stride would be the same across different walking speeds. Based on these findings, it is important to control the walking speed when comparing SampEn of gait signal between WO and DT conditions; since people, especially older adults, slow down when they perform a secondary task [34,35].

The results of this study suggest that SampEn benefits from a relative consistency (a significant increase from WO to DT) across different combinations of the variant values of m, r, and f. For chaotic signals like Mackey-Glass system, which resemble periodic time series, entropy values decrease with increasing the m values [36]. In the present study, there was a decreasing trend of SampEn of ML COP-D signal with increasing m. However, there were exceptions and SampEn values plateaued only at $m = 4$ for $f = 16$. Nevertheless, for most combinations, there was a statistically significant increase of SampEn from WO to DT. With respect to changes in template size at higher sampling rates, SampEn showed an increasing trend with increasing m. A possible reason for this behavior may be related to the strong periodicity of the signal. To overcome this, larger template sizes might be chosen for signals collected at higher sampling rates. A similar issue exists for lower sampling rates where SampEn values increased with increasing m, but only after a specific m value. This suggests that smaller m

values should be chosen for lower sampling rates. The two tolerance values performed similarly since the smaller one was already larger than the noise level of the signal.

There are three limitations to this study. First, the sample size of the current study (29 participants) which led to performing several 2-factor repeated measure ANOVA instead of for example two 3-factor repeated measure ANOVA. A much larger sample size would be more appropriate to study the effect of four factors each with many levels. Second, this study examined the discriminatory ability of SampEn by only comparing WO to DT conditions. Several other factors, such as aging, should be considered to generalize the results of this study. Finally, only SampEn, which is a single-scale entropy measure, was studied. Multi-scale SampEn analysis [12] or modified SampEn analysis (incorporating a time delay greater than one) [37] would likely yield important findings.

For future studies using the ML COP-D signal, it is recommended to use a low-pass filter prior to the calculation of SampEn. In addition, a sampling rate of 125 Hz or 62 Hz with $m = 2 \sim 6$ and $r = 0.2 \times SD$ would be the preferred combinations. For studies testing overground walking where speed is difficult to control, an integer number of strides should be resampled so that the average number of data points per stride remains the same. For research investigations using other gait signals, such as trunk/pelvis linear acceleration, a similar approach should be performed to select the best combinations of m, r, and f values. This study was not designed to investigate the effect of data length on SampEn. Nevertheless, SampEn of whole signals plateaus after a few strides [7] and it has been shown that 30 strides are sufficient to calculate SampEn of whole gait signals [16].

Author Contributions: S.A., T.S., and C.W. conceived and designed this work; S.A. collected and analyzed the data; S.A. and T.S. interpreted the results; S.A. drafted the manuscript; S.A., N.S., T.S. and C.W. revised and finalized the manuscript. All authors read and approved the final manuscript.

Funding: This research was funded by the Natural Science and Engineering Research Council of Canada grant number RGPIN-314032, the Canadian Institute of Health Research grant number CEH-134155, and the University of Manitoba Graduate Fellowship UMGF-7765611.

Conflicts of Interest: The authors declare no conflict of interest.

Appendix

Table A1. Descriptive results (mean ± SD) and pairwise comparisons of "walking condition*m" at $r = 0.2 \times SD$ and each f value for FD (Filtered-and-Downsampled) and D (Decimated) ML COP-D. The two conditions are WO and DT. p-values in bold indicate a significant difference.

$r = 0.2 \times SD$		FD			D		
		WO	DT	p-Value	WO	DT	p-Value
	$m = 2$	0.014 ± 0.002	0.015 ± 0.002	**0.008**	0.096 ± 0.023	0.093 ± 0.020	
	$m = 4$	0.014 ± 0.002	0.015 ± 0.002	**0.008**	0.058 ± 0.015	0.056 ± 0.012	
$f = 1$	$m = 6$	0.015 ± 0.003	0.016 ± 0.003	**0.008**	0.034 ± 0.007	0.033 ± 0.005	-
	$m = 8$	0.015 ± 0.003	0.016 ± 0.003	**0.008**	0.033 ± 0.007	0.032 ± 0.005	
	$m = 10$	0.015 ± 0.003	0.017 ± 0.003	**0.004**	0.032 ± 0.007	0.032 ± 0.004	
	$m = 2$	0.028 ± 0.005	0.030 ± 0.005	**0.008**	0.117 ± 0.029	0.112 ± 0.024	
	$m = 4$	0.030 ± 0.006	0.032 ± 0.005	**0.008**	0.059 ± 0.011	0.058 ± 0.008	
$f = 2$	$m = 6$	0.031 ± 0.005	0.034 ± 0.006	**0.002**	0.056 ± 0.011	0.057 ± 0.008	-
	$m = 8$	0.030 ± 0.005	0.033 ± 0.004	**0.001**	0.051 ± 0.009	0.053 ± 0.006	
	$m = 10$	0.029 ± 0.004	0.032 ± 0.004	**<0.001**	0.047 ± 0.008	0.049 ± 0.005	
	$m = 2$	0.058 ± 0.011	0.063 ± 0.011	**0.008**	0.073 ± 0.014	0.074 ± 0.012	0.568
	$m = 4$	0.060 ± 0.010	0.067 ± 0.009	**0.001**	0.068 ± 0.011	0.072 ± 0.010	**0.042**
$f = 4$	$m = 6$	0.055 ± 0.007	0.060 ± 0.006	**<0.001**	0.059 ± 0.008	0.063 ± 0.007	**0.011**
	$m = 8$	0.045 ± 0.006	0.049 ± 0.006	**0.002**	0.050 ± 0.006	0.053 ± 0.006	**0.035**
	$m = 10$	0.037 ± 0.004	0.040 ± 0.005	**0.002**	0.042 ± 0.004	0.044 ± 0.005	0.058

Table A1. *Cont.*

$r = 0.2 \times$ SD		FD			D		
		WO	DT	*p*-Value	WO	DT	*p*-Value
$f = 8$	$m = 2$	0.121 ± 0.020	0.134 ± 0.021	**0.002**	0.133 ± 0.022	0.141 ± 0.021	**0.042**
	$m = 4$	0.095 ± 0.012	0.103 ± 0.011	**0.001**	0.100 ± 0.013	0.107 ± 0.011	**0.013**
	$m = 6$	0.068 ± 0.007	0.073 ± 0.008	**0.002**	0.074 ± 0.007	0.078 ± 0.009	**0.041**
	$m = 8$	0.061 ± 0.006	0.064 ± 0.007	**0.001**	0.065 ± 0.006	0.068 ± 0.007	**0.008**
	$m = 10$	0.059 ± 0.006	0.063 ± 0.007	**0.001**	0.063 ± 0.007	0.066 ± 0.007	**0.015**
$f = 16$	$m = 2$	0.208 ± 0.025	0.228 ± 0.022	**<0.001**	0.209 ± 0.026	0.228 ± 0.022	**0.001**
	$m = 4$	0.123 ± 0.012	0.132 ± 0.014	**0.001**	0.128 ± 0.012	0.136 ± 0.015	**0.002**
	$m = 6$	0.119 ± 0.012	0.126 ± 0.013	**0.001**	0.123 ± 0.012	0.128 ± 0.013	**0.007**
	$m = 8$	0.122 ± 0.011	0.128 ± 0.012	**0.010**	0.126 ± 0.012	0.130 ± 0.012	**0.049**
	$m = 10$	0.127 ± 0.011	0.131 ± 0.012	0.052	0.130 ± 0.012	0.133 ± 0.012	0.181
$f = 32$	$m = 2$	0.267 ± 0.027	0.288 ± 0.030	**<0.001**	0.265 ± 0.026	0.285 ± 0.030	**0.001**
	$m = 4$	0.243 ± 0.023	0.254 ± 0.024	**0.004**	0.241 ± 0.023	0.253 ± 0.023	**0.004**
	$m = 6$	0.262 ± 0.023	0.270 ± 0.025	0.051	0.262 ± 0.023	0.270 ± 0.025	**0.037**
	$m = 8$	0.295 ± 0.037	0.302 ± 0.036	0.138	0.296 ± 0.037	0.303 ± 0.036	0.105
	$m = 10$	0.332 ± 0.052	0.340 ± 0.045	0.224	0.335 ± 0.053	0.344 ± 0.047	0.154

Table A2. Descriptive results (mean ± SD) and pairwise comparisons of "walking condition*$*m*$" at $r = 0.3 \times$ SD and each f value for FD (Filtered-and-Downsampled) and D (Decimated) ML COP-D. The two conditions are WO and DT. *p*-values in bold indicate a significant difference.

$r = 0.3 \times$ SD		FD			D		
		WO	DT	*p*-Value	WO	DT	*p*-Value
$f = 1$	$m = 2$	0.009 ± 0.002	0.010 ± 0.002	**0.012**	0.058 ± 0.014	0.057 ± 0.011	
	$m = 4$	0.009 ± 0.002	0.010 ± 0.002	**0.012**	0.034 ± 0.008	0.032 ± 0.006	
	$m = 6$	0.010 ± 0.002	0.010 ± 0.002	**0.011**	0.019 ± 0.003	0.019 ± 0.002	-
	$m = 8$	0.010 ± 0.002	0.011 ± 0.002	**0.011**	0.018 ± 0.003	0.018 ± 0.002	
	$m = 10$	0.010 ± 0.002	0.011 ± 0.002	**0.011**	0.018 ± 0.003	0.018 ± 0.002	
$f = 2$	$m = 2$	0.019 ± 0.003	0.020 ± 0.003	**0.012**	0.069 ± 0.016	0.067 ± 0.013	
	$m = 4$	0.019 ± 0.003	0.021 ± 0.003	**0.011**	0.034 ± 0.006	0.034 ± 0.004	
	$m = 6$	0.020 ± 0.004	0.022 ± 0.004	**0.011**	0.032 ± 0.006	0.032 ± 0.004	-
	$m = 8$	0.021 ± 0.004	0.023 ± 0.004	**0.005**	0.031 ± 0.006	0.032 ± 0.004	
	$m = 10$	0.021 ± 0.003	0.023 ± 0.003	**0.002**	0.030 ± 0.005	0.031 ± 0.004	
$f = 4$	$m = 2$	0.038 ± 0.007	0.041 ± 0.007	**0.011**	0.047 ± 0.009	0.048 ± 0.007	0.578
	$m = 4$	0.041 ± 0.008	0.045 + 0.008	**0.006**	0.046 ± 0.009	0.048 ± 0.008	0.130
	$m = 6$	0.041 ± 0.006	0.045 ± 0.006	**0.001**	0.044 ± 0.007	0.047 ± 0.007	**0.018**
	$m = 8$	0.039 ± 0.005	0.043 ± 0.005	**0.001**	0.041 ± 0.006	0.044 ± 0.005	**0.007**
	$m = 10$	0.035 ± 0.004	0.038 ± 0.004	**0.001**	0.037 ± 0.005	0.040 ± 0.004	**0.008**
$f = 8$	$m = 2$	0.081 ± 0.015	0.088 ± 0.015	**0.008**	0.089 ± 0.017	0.093 ± 0.016	0.127
	$m = 4$	0.079 ± 0.011	0.087 ± 0.010	**0.001**	0.081 ± 0.012	0.088 ± 0.011	**0.005**
	$m = 6$	0.065 ± 0.007	0.070 ± 0.007	**0.001**	0.068 ± 0.008	0.072 ± 0.007	**0.010**
	$m = 8$	0.054 ± 0.004	0.058 ± 0.005	**<0.001**	0.057 ± 0.005	0.060 ± 0.006	**0.003**
	$m = 10$	0.050 ± 0.004	0.053 ± 0.005	**<0.001**	0.053 ± 0.004	0.055 ± 0.005	**0.002**
$f = 16$	$m = 2$	0.161 ± 0.024	0.178 ± 0.024	**0.001**	0.163 ± 0.025	0.178 ± 0.024	**0.002**
	$m = 4$	0.112 ± 0.010	0.120 ± 0.011	**<0.001**	0.115 ± 0.010	0.122 ± 0.012	**0.001**
	$m = 6$	0.100 ± 0.008	0.105 ± 0.010	**<0.001**	0.102 ± 0.009	0.107 ± 0.010	**0.002**
	$m = 8$	0.102 ± 0.009	0.106 ± 0.010	**0.004**	0.104 ± 0.010	0.108 ± 0.011	**0.014**
	$m = 10$	0.106 ± 0.010	0.109 ± 0.010	**0.042**	0.108 ± 0.010	0.111 ± 0.011	0.102
$f = 32$	$m = 2$	0.247 ± 0.024	0.268 ± 0.024	**<0.001**	0.246 ± 0.024	0.265 ± 0.024	**<0.001**
	$m = 4$	0.202 ± 0.018	0.212 ± 0.021	**0.001**	0.201 ± 0.018	0.211 ± 0.020	**0.001**
	$m = 6$	0.219 ± 0.021	0.225 ± 0.022	0.069	0.218 ± 0.021	0.224 ± 0.022	0.066
	$m = 8$	0.240 ± 0.029	0.245 ± 0.030	0.143	0.241 ± 0.030	0.246 ± 0.030	0.156
	$m = 10$	0.255 ± 0.035	0.260 ± 0.034	0.220	0.257 ± 0.036	0.263 ± 0.035	0.211

References

1. Kaptein, R.G.; Wezenberg, D.; IJmker, T.; Houdijk, H.; Beek, P.J.; Lamoth, C.J.; Daffertshofer, A. Shotgun approaches to gait analysis: Insights & limitations. *J. Neuroeng. Rehabil.* **2014**, *11*, 120. [PubMed]
2. Pincus, S.M.; Goldberger, A.L. Physiological time-series analysis: What does regularity quantify? *Am. J. Physiol.* **1994**, *266*, H1643–H1656. [CrossRef] [PubMed]
3. Richman, J.S.; Moorman, J.R. Physiological time-series analysis using approximate entropy and sample entropy. *Am. J. Physiol. Heart Circ. Physiol.* **2000**, *278*, H2039–H2049. [CrossRef] [PubMed]
4. Gow, B.; Peng, C.-K.; Wayne, P.; Ahn, A. Multiscale Entropy Analysis of Center-of-Pressure Dynamics in Human Postural Control: Methodological Considerations. *Entropy* **2015**, *17*, 7926–7947. [CrossRef]
5. Lake, D.E.; Richman, J.S.; Griffin, M.P.; Moorman, J.R. Sample entropy analysis of neonatal heart rate variability. *Am. J. Physiol. Regul. Integr. Comp. Physiol.* **2002**, *283*, R789–R797. [CrossRef] [PubMed]
6. Ihlen, E.A.F.; Weiss, A.; Bourke, A.; Helbostad, J.L.; Hausdorff, J.M. The complexity of daily life walking in older adult community-dwelling fallers and non-fallers. *J. Biomech.* **2016**, 1–9. [CrossRef] [PubMed]
7. Ahmadi, S.; Wu, C.; Sepehri, N.; Kantikar, A.; Nankar, M.; Szturm, T. The Effects of Aging and Dual Tasking on Human Gait Complexity During Treadmill Walking: A Comparative Study Using Quantized Dynamical Entropy and Sample Entropy. *J. Biomech. Eng.* **2018**, *140*, 1–10. [CrossRef] [PubMed]
8. Bisi, M.; Riva, F.; Stagni, R. Measures of gait stability: Performance on adults and toddlers at the beginning of independent walking. *J. Neuroeng. Rehabil.* **2014**, *11*, 131. [CrossRef] [PubMed]
9. Shannon, C.E. A mathematical theory of communication. *Bell Syst. Tech. J.* **1948**, *27*, 379–423. [CrossRef]
10. Pincus, S. Approximate entropy (ApEn) as a complexity measure. *Chaos* **1995**, *5*, 110–117. [CrossRef] [PubMed]
11. Yentes, J.M.; Hunt, N.; Schmid, K.K.; Kaipust, J.P.; McGrath, D.; Stergiou, N. The appropriate use of approximate entropy and sample entropy with short data sets. *Ann. Biomed. Eng.* **2013**, *41*, 349–365. [CrossRef] [PubMed]
12. Costa, M.; Peng, C.K.; Goldberger, A.L.; Hausdorff, J.M. Multiscale entropy analysis of human gait dynamics. *Phys. A Stat. Mech. Appl.* **2003**, *330*, 53–60. [CrossRef]
13. Van Schooten, K.S.; Pijnappels, M.; Rispens, S.M.; Elders, P.J.M.; Lips, P.; Daffertshofer, A.; Beek, P.J.; Van Dieën, J.H. Daily-life gait quality as predictor of falls in older people: A 1-year prospective cohort study. *PLoS ONE* **2016**, *11*, 1–13. [CrossRef] [PubMed]
14. Lamoth, C.J.; van Deudekom, F.J.; van Campen, J.P.; Appels, B.A.; de Vries, O.J.; Pijnappels, M. Gait stability and variability measures show effects of impaired cognition and dual tasking in frail people. *J. Neuroeng. Rehabil.* **2011**, *8*, 2. [CrossRef] [PubMed]
15. Kurz, M.J.; Hou, J.G. Levodopa influences the regularity of the ankle joint kinematics in individuals with Parkinson's disease. *J. Comput. Neurosci.* **2010**, *28*, 131–136. [CrossRef] [PubMed]
16. Leverick, G.; Szturm, T.; Wu, C.Q. Using Entropy Measures to Characterize Human Locomotion. *J. Biomech. Eng.* **2014**, *136*, 121002. [CrossRef] [PubMed]
17. Fino, P.C.; Mojdehi, A.R.; Adjerid, K.; Habibi, M.; Lockhart, T.E.; Ross, S.D. Comparing Postural Stability Entropy Analyses to Differentiate Fallers and Non-fallers. *Ann. Biomed. Eng.* **2016**, *44*, 1636–1645. [CrossRef] [PubMed]
18. Rhea, C.K.; Kiefer, A.W.; Wright, W.G.; Raisbeck, L.D.; Haran, F.J. Interpretation of postural control may change due to data processing techniques. *Gait Posture* **2015**, *41*, 731–735. [CrossRef] [PubMed]
19. Ramdani, S.; Seigle, B.; Lagarde, J.; Bouchara, F.; Bernard, P.L. On the use of sample entropy to analyze human postural sway data. *Med. Eng. Phys.* **2009**, *31*, 1023–1031. [CrossRef] [PubMed]
20. Terrier, P.; Dériaz, O. Non-linear dynamics of human locomotion: Effects of rhythmic auditory cueing on local dynamic stability. *Front. Physiol.* **2013**, *4*, 1–13. [CrossRef] [PubMed]
21. Yentes, J.M.; Denton, W.; Mccamley, J.; Ra, P.C.; Schmid, K.K. Effect of parameter selection on entropy calculation for long walking trials. *Gait Posture* **2018**, *60*, 128–134. [CrossRef] [PubMed]
22. Howcroft, J.; Kofman, J.; Lemaire, E.D.; McIlroy, W.E. Analysis of Dual-Task Elderly Gait in Fallers and Non-Fallers using Wearable Sensors. *J. Biomech.* **2016**, *49*, 992–1001. [CrossRef] [PubMed]
23. Bruijn, S.M.; van Dieën, J.H.; Meijer, O.G.; Beek, P.J. Is slow walking more stable? *J. Biomech.* **2009**, *42*, 1506–1512. [CrossRef] [PubMed]

24. Kang, H.G.; Dingwell, J.B. Effects of walking speed, strength and range of motion on gait stability in healthy older adults. *J. Biomech.* **2008**, *41*, 2899–2905. [CrossRef] [PubMed]
25. Rispens, S.M.; Dieën, J.H. Van; Schooten, K.S. Van; Lizama, L.E.C.; Daffertshofer, A.; Beek, P.J.; Pijnappels, M. Fall-related gait characteristics on the treadmill and in daily life. *J. Neuroeng. Rehabil.* **2016**, *13*, 12. [CrossRef] [PubMed]
26. Katsavelis, D.; Mukherjee, M.; Decker, L.; Stergiou, N. Variability of lower extremity joint kinematics during backward walking in a virtual environment. *Nonlinear Dyn. Psychol. Life Sci.* **2010**, *14*, 165–178.
27. Kavanagh, J.; Barrett, R.; Morrison, S. The role of the neck and trunk in facilitating head stability during walking. *Exp. Brain Res.* **2006**, *172*, 454–463. [CrossRef] [PubMed]
28. Dingwell, J.B.; Robb, R.T.; Troy, K.L.; Grabiner, M.D. Effects of an attention demanding task on dynamic stability during treadmill walking. *J. Neuroeng. Rehabil.* **2008**, *5*, 12. [CrossRef] [PubMed]
29. Szturm, T.; Maharjan, P.; Marotta, J.J.; Shay, B.; Shrestha, S.; Sakhalkar, V. The interacting effect of cognitive and motor task demands on performance of gait, balance and cognition in young adults. *Gait Posture* **2013**, *38*, 596–602. [CrossRef] [PubMed]
30. Sloot, L.H.; Houdijk, H.; Harlaar, J. Technical note A comprehensive protocol to test instrumented treadmills. *Med. Eng. Phys.* **2015**, *37*, 610–616. [CrossRef] [PubMed]
31. England, S.A.; Granata, K.P. The influence of gait speed on local dynamic stability of walking. *Gait Posture* **2007**, *25*, 172–178. [CrossRef] [PubMed]
32. Riva, F.; Toebes, M.J.P.; Pijnappels, M.; Stagni, R.; van Dieën, J.H. Estimating fall risk with inertial sensors using gait stability measures that do not require step detection. *Gait Posture* **2013**, *38*, 170–174. [CrossRef] [PubMed]
33. Vaillancourt, D.E.; Newell, K.M. Changing complexity in human behavior and physiology through aging and disease. *Neurobiol. Aging* **2002**, *23*, 1–11. [CrossRef]
34. Dorfman, M.; Herman, T.; Brozgol, M.; Shema, S.; Weiss, A.; Hausdorff, J.; Mirelman, A. Dual-Task Training on a Treadmill to Improve Gait and Cognitive Function in Elderly Idiopathic Fallers. *J. Neurol. Phys. Ther.* **2014**, *38*, 246–253. [CrossRef] [PubMed]
35. Lamoth, C.J.C.; Ainsworth, E.; Polomski, W.; Houdijk, H. Variability and stability analysis of walking of transfemoral amputees. *Med. Eng. Phys.* **2010**, *32*, 1009–1014. [CrossRef] [PubMed]
36. Restrepo, J.F.; Schlotthauer, G.; Torres, M.E. Maximum approximate entropy and threshold: A new approach for regularity changes detection. *Phys. A Stat. Mech. Appl.* **2014**, *409*, 97–109. [CrossRef]
37. Govindan, R.B.; Wilson, J.D.; Eswaran, H.; Lowery, C.L.; Preissl, H. Revisiting sample entropy analysis. *Phys. A Stat. Mech. Appl.* **2007**, *376*, 158–164. [CrossRef]

entropy

MDPI

Article

Embedded Dimension and Time Series Length. Practical Influence on Permutation Entropy and Its Applications

David Cuesta-Frau [1],*, Juan Pablo Murillo-Escobar [2], Diana Alexandra Orrego [2] and Edilson Delgado-Trejos [3]

[1] Technological Institute of Informatics, Universitat Politècnica de València, Alcoi Campus, 03801 Alcoi, Spain
[2] Grupo de Investigación e Innovación Biomédica (GI2B), Instituto Tecnológico Metropolitano (ITM), Medellín, Colombia; juanmurillo@itmeduco.onmicrosoft.com (J.P.M.-E.); dianaorrego@itm.edu.co (D.A.O.)
[3] CM&P, Instituto Tecnológico Metropolitano (ITM), Medellín, Colombia; edilsondelgado@itm.edu.co
* Correspondence: dcuesta@disca.upv.es; Tel.: +34-966528505

Received: 13 February 2019; Accepted: 8 April 2019; Published: 9 April 2019

Abstract: Permutation Entropy (PE) is a time series complexity measure commonly used in a variety of contexts, with medicine being the prime example. In its general form, it requires three input parameters for its calculation: time series length N, embedded dimension m, and embedded delay τ. Inappropriate choices of these parameters may potentially lead to incorrect interpretations. However, there are no specific guidelines for an optimal selection of N, m, or τ, only general recommendations such as $N >> m!$, $\tau = 1$, or $m = 3, \ldots, 7$. This paper deals specifically with the study of the practical implications of $N >> m!$, since long time series are often not available, or non-stationary, and other preliminary results suggest that low N values do not necessarily invalidate PE usefulness. Our study analyses the PE variation as a function of the series length N and embedded dimension m in the context of a diverse experimental set, both synthetic (random, spikes, or logistic model time series) and real–world (climatology, seismic, financial, or biomedical time series), and the classification performance achieved with varying N and m. The results seem to indicate that shorter lengths than those suggested by $N >> m!$ are sufficient for a stable PE calculation, and even very short time series can be robustly classified based on PE measurements before the stability point is reached. This may be due to the fact that there are forbidden patterns in chaotic time series, not all the patterns are equally informative, and differences among classes are already apparent at very short lengths.

Keywords: permutation entropy; embedded dimension; short time records; signal classification; relevance analysis

1. Introduction

The influence of input parameters on the performance of entropy statistics is a well known issue. If the selected values do not match the intended purpose or application, the results can be completely meaningless. Since the first widely used methods, such as Approximate Entropy (ApEn) [1], or Sample Entropy (SampEn) [2], the characterization of this influence has become a topic of intense research. For example, ref [3] proposed the computation of all the ApEn results with the tolerance threshold varying from 0 to 1 in order to find its maximum, which leads to a more correct complexity assessment. The authors also proposed a method to reduce the computational cost of this approach. For SampEn, works such as [4] have focused on optimizing the input parameters for a specific field of application, the estimation of atrial fibrillation organisation. In [5], an analysis of ApEn and SampEn performance with changing parameters, using short length spatio–temporal gait time series was researched. According to their results, SampEn is more stable than ApEn, and the required minimum length should be at least 200 samples. They also noticed that longer series can have a detrimental effect due to non-stationarities and drifts, and therefore these issues should always be checked in advance.

The research into this parameter has been extended to other entropy statistics. The study in [6], addresses the problem of parameter configuration for ApEn, SampEn, Fuzzy (FuzzyEn) [7], and Fuzzy Measure (FuzzyMEn) [8] entropies in the framework of heart rate variability. These methods require from 3 up to 6 parameters. FuzzyEn and FuzzyMEn are apparently quite insensitive to r values, whereas ApEn exhibits the flip–flop effect (depending on r, the entropy values of two signals under comparison may swap order [9]). Although this work acknowledges the extreme difficulty of studying the effect of up to 6 degrees of freedom, and the need for more studies, they were able to conclude that length N should be at least 200 samples for $r = 0.2\sigma$. Another important conclusion of [6], strongly related to the present work, is that length has an almost negligible effect on the ability of the entropy measurements to classify records. PE parameters have been addressed in works such as in [10]. The authors explored the effect of $m = 3$–7 and $\tau = 1$–5 on anaesthetic depth assessment, based on the electroencephalogram. Their conclusion was that PE performed best for $m = 3$, and $\tau = 2, 3$, and proposed to combine those two cases in a single index. However, as far as we know, there is no study that quantifies the effect of N and its relationship with m on PE applications.

Since PE conception [11], the length N of a time series under analysis using PE has been recommended to be significantly greater than the number of possible order permutations [12–15], given by the factorial of the embedded dimension m, that is, $m! << N$, or some of its variants, such as $5m! \leq N$ [16]. For example, in [12], the authors describe the choice of algorithmic parameters based on a survey of many PE studies. They also performed a PE study using synthetic records of length $N = 6025$: Lorenz system, Van–der–Pol oscillator, the logistic map, and an autoregressive model, varying τ and m, and from an absolute point of view (no classification analysis). The main conclusions of these works were to recommend $\tau = 1$ and m the highest possible value, with $N > 5m!$. The study in [16] is devoted to distinguishing white noise from noisy deterministic time series. They look for forbidden patterns to ensure determinism, and therefore have to use long enough synthetic records (Hénon maps), since the probability that any existing pattern remains undetected tend towards 0 exponentially as N grows. Their recommendation is also $N > 5m!$. The PE proposers [11] worked with logistic map records of $N = 10^6$ to obtain accurate PE results for $m \leq 15$, but they also found that PE could be reliably estimated in this case with $N = 1000$.

The rationale of the $m! << N$ recommendation, as for other entropy metrics [5,7,17–19], is to ensure a high number of matches for a confident estimation of the probability ratios [20,21] and also ensure that all possible patterns become visible [16]. An original recipe for m [11] was choosing the embedding dimension from within the range $3, \ldots, 7$, from which a suitable N value can be inferred.

However, in some contexts, it is not possible to obtain long time series [22], or for decisions have to be made as quickly as possible, once a few samples are already available for analysis [21] in a real time system. In addition, long records are more likely to exhibit changes in the underlying dynamics. In other words, the required stationarity for a stable PE measurement cannot be assured [23]. As a consequence, N is sometimes out of the researcher's control, and short records are often unavoidable. Therefore, only relatively small values of the embedded dimension m should be used, in accordance with the recommendation stated above. Unfortunately, high values of m usually provide better signal classification performance [24–26], and this fact leads to an antagonistic and counterproductive relationship between PE stability, and its segmentation power. For example, in reference [24], the classification performance of PE using electroencephalogram records of 4096 samples, temperature records of 480 samples, RR records of some 1000 samples, and continuous glucose monitoring records of 280 samples was analysed. Using m values from 3 up to 9, classification performance was highest for $m = 9$ for all the signal types, even the shortest ones, which is in high contrast to the recommendation assessed.

Thus, there are studies where, despite analysing short time series with high m values that did not fulfil the relationship $m! << N$, the classification achieved using PE was very good [24,26,27].

This led to the hypothesis that PE probably achieves stability before it was initially thought, especially for larger m values, and additionally, such stability is not required to attain a significant classification accuracy. The stability criterion proposed is based on the step response of a first order system: the time needed to achieve a steady state response or its final value. This settling time is defined as the time required for that response to reach and stay within a percentage of its final value, typically between 2% and 5% [28]. Thus, we consider PE reaches stability when that measurement stays within a 2% error band of the PE value obtained for the entire record, and instead of time, the independent variable is the number of samples. This is the same criterion used in similar works, such as in [1]. If this error band is not satisfied for the maximum length available, we consider stability is not reached for that m and N.

Furthermore, entropy values are relative, they cannot be correctly interpreted if they are analyzed in isolation, without a comparison between a control and an experimental group [5]. This has already been demonstrated in previous studies [24], where PE differences in relative terms were key to obtaining a significant classification, not the absolute PE values that were influenced by the presence of ties in the sub–sequences.

In this paper, we try to fine–tune the general recommendation $m! << N$ by computing exactly what is the required length for a stable PE calculation using different m values, from 3 to 7, and in a few cases even 9. A classification analysis using short records and PE as the distinctive feature is also included. The experimental dataset will be composed of a miscellaneous set of records from different scientific and technical fields, including synthetic and real–world time series.

2. Materials and Methods

2.1. Permutation Entropy

Given an input time series $\{x_t : t = 0, \ldots, N - 1\}$, and an embedding dimension $m > 1$, for each extracted subsequence at time s, $(s) \longmapsto \left(x_{s-(m-1)}, x_{s-(m-2)}, \ldots, x_{s-1}, x_s\right)$, an ordinal pattern π related to s is obtained as $\pi = (r_0, r_1, \ldots, r_{m-1})$, defined by $x_{s-r_{m-1}} \leq x_{s-r_{m-2}} \leq \cdots \leq x_{s-r_1} \leq x_{s-r_0}$ [15]. For all the possible $m!$ permutations, each probability $p(\pi)$ is estimated as the relative frequency of each different π pattern found. Once all these probabilities have been obtained, the final value of PE is given by [11]:

$$\text{PE} = -\sum_{j=0}^{m!-1} p(\pi_j)\log_2(p(\pi_j)), \text{if } p(\pi_j) > 0 \tag{1}$$

More details of the PE algorithm, including examples, can be found in [11]. The implicit input parameters for PE are:

1. The embedded dimension m. The recommended range for this parameter is $3, \ldots, 7$ [11], but other greater values have been used successfully [12,24,26,27]. Since this parameter is also part of the inequality under analysis in this work, m will be varied in the experiments, taking values from within the recommended range, and in some cases beyond that.
2. The embedded delay τ. The influence of the embedded delay has been studied in several previous publications [10,29] for specific applications. This parameter is not directly involved in the $m! << N$ relationship, and therefore it will not be assessed in this work. Moreover, this parameter contributes to a reduction in the amount of data available when $\tau > 1$ in practical terms [30], and therefore might have a detrimental effect on the analysis. Thus, τ will be considered as $\tau = 1$ in all the experiments except a few cases for illustrative purposes.
3. The length of the time series N. As stated before, the recommended relationship $m! << N$ is commonplace in practically all the publications related to PE, but no study so far has quantified this relationship as planned in the present paper. N will be varied in the experiments to obtain a representative set of PE curve points accounting for increasing time series lengths, from 10

samples up to the maximum length available. Each time series was run at different lengths and m values.

2.2. Experimental Dataset

The experimental data contains a varied and diverse set of real–world time series, in terms of length and frequency content and distribution, from scientific frameworks where PE or other similar methods have proven to be a useful tool [14,31–34]. Synthetic time series are also included for a more controlled analysis. These synthetic time series enable a fine tuning of their parameters to elicit the desired effects, such as exhibiting a random, chaotic, or more deterministic behaviour. All the records were normalised before computing PE (zero mean, unit variance). The key specific features of each dataset utilized are described in Sections 2.2.1 and 2.2.2.

2.2.1. Synthetic Dataset

The main goal of this synthetic dataset was to test the effect of randomness on the rate of PE stabilisation. In principle, 100 random realisations of each case were created, and all the records contained 1000 samples to study the evolution for low m values. Most of them were also generated with 5000 data points to study the effect of greater m values, as described in Section 3. In the specific case of the logistic map, the resulting records were also used for classification tests since their chaotic behaviour can be parametrically controlled. This dataset, along with the key features and abbreviations, is described below. Examples of some synthetic records are shown in Figure 1.

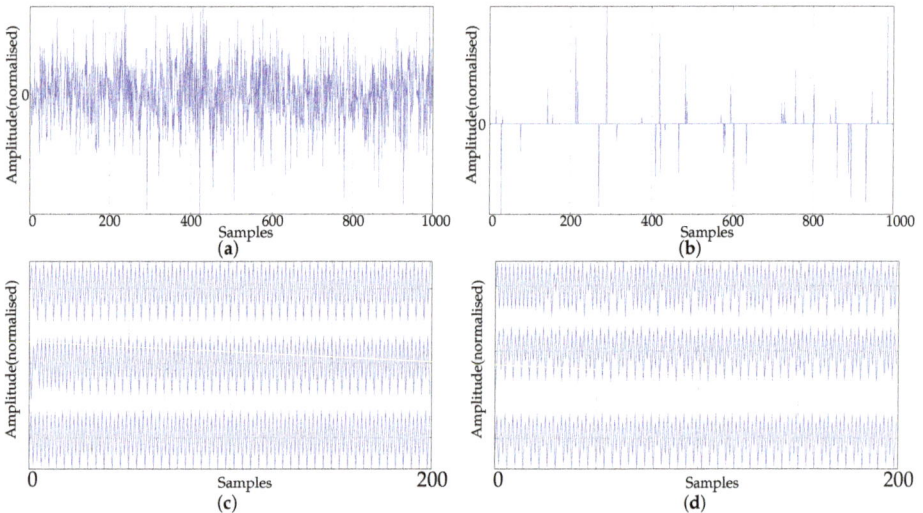

Figure 1. Synthetic data experimental dataset examples. (**a**) Example of a synthetic random sequence from the RAND experimental dataset; (**b**) Example of a synthetic spikes sequence from the SPIKES experimental dataset; (**c**) Example of a synthetic logistic map periodic sequence from the LMAP experimental dataset. The three records correspond to $R = 3.50, 3.51$, and 3.52. Only the first 200 samples are shown for resolution purposes; (**d**) Example of a synthetic logistic map chaotic sequence from the LMAP experimental dataset. The three records correspond to $R = 3.57, 3.58$, and 3.59. Only the first 200 samples are shown for resolution purposes.

- RAND. A sequence of random numbers following a normal distribution (Figure 1a).
- SPIKES. A sequence of zeros including random spikes generated by a binomial distribution with probability 0.05, and whose amplitude follows a normal distribution (Figure 1b). This sequence is generated as in [35].
- LMAP. A sequence of numbers computed from the logistic map equation $x_{t+1} = R \cdot x_t(1 - x_t)$. This dataset really corresponds to 2 subsets obtained by changing the value of the parameter R: 100 random initialisations of x_0 with $x_0 \in]0, 1[$, and with $R = 3.50, 3.51$, and 3.52 to create 3 classes of 100 periodic records each (Figure 1c), and 3x100 randomly initialised records with $R = 3.57, 3.58$, and 3.59 to create 3 classes of 100 more chaotic records each (Figure 1d).
- SIN. A sequence of values from a sinusoid with random phase variations. Used specifically to study the number of patterns found in deterministic records.

The logistic map has been used in several previous similar studies. In [1], records of this type were analysed using ApEn, and lengths of 300, 1000, and 3000 samples. Random values are also a reference dataset in many works, such as in [36], where sequences of 2000 uniform random numbers were used in some experiments. Spikes have been used in studies such as [22,35], with $N = 1000$.

2.2.2. Real Dataset

The real–world dataset was chosen from different contexts where time series are processed using PE. This dataset, along with the key features and abbreviations, is described below. Examples of some of these records are shown in Figure 2.

- CLIMATOLOGY. Symbolic dynamics have a place in the study of climatology [33], with many time series databases publicly available nowadays [37–39]. This group includes time series of temperature anomalies from the Global Historic Climatology Network temperature database available through the National Oceanic and Atmospheric Administration [39]. The data correspond to monthly global surface temperature anomaly readings dating back from 1880 to the present. The temperature anomaly corresponds to the difference between the long–term average temperature, and the actual temperature. In this case, anomalies are based on the climatology from 1971 to 2000, with a total of 1662 samples for each record. These time series exhibit a clear growing trend from year 2000, probably due to the global warming effect, as illustrated in Figure 2a. In [36], average daily temperatures in Mexico City and New York City were used, with more than 2000 samples. Other works have also used climate data, such as in [40], where surface temperature anomaly data in Central Europe were analysed using Multi-scale entropy, with $N = 2000$.
- SEISMIC. Seismic data have also been successfully analysed using PE [41], and these time series are a very promising field of research using PE. The data included in this paper was drawn from the Seismic data database, US Geological Survey Earthquake Hazards Program [42]. The time series correspond to worldwide earthquakes whose magnitude is greater than 2.5, detected each month, from January to July 2018. The lengths of these time series are not uniform, since they depend on the number of earthquakes detected each month. It ranges from 2104 up to 9090 samples. An example of these records is show in Figure 2b.
- FINANCIAL. This set of financial time series was included as an additional representative field of application of PE [43]. Specifically, data corresponding to daily simple returns of Apple, American Express, and IBM, from 2001 to 2010 [44] were included, with a total length of 2519 samples. One of these time series are shown in Figure 2c. There is a good review of entropy applications to financial data in [45].
- Biomedical time series. This is probably the most thoroughly studied group of records using PE [14]. Three subsets have been included:

  1. EMG. Three (healthy, myopathy, neuropathy) very extensive records corresponding to electromyographic data (Examples of electromyograms [46]). The data were acquired at

50 kHz and downsampled to 4 kHz, and band–pass filtered during the recording process between 20 Hz and 5 kHz. All three records contain more than 50,000 samples. These records were later split into consecutive non-overlapping sequences of 5000 samples to create three corresponding groups for classification analysis (10 healthy, 22 myopathy, and 29 neuropathy resulting records).

2. PAF. The PAF (Paroxysmal Atrial Fibrillation) prediction challenge database is also publicly available at Physionet [46], and is described in [47]. The PAF records used correspond to 50 time series of short duration (5 minute records), coming from subjects with PAF. Even–numbered records contain an episode of PAF, whereas odd–numbered records are PAF–free (Figure 2e). This database was selected because the two classes are easily distinguishable, and the short duration of the records (some 400–500 samples) can be challenging for PE, even at low m values.

3. PORTLAND. Very long time series (more than 1,000,000 samples) from Portland State University corresponding to traumatic brain injury data. Arterial blood, central venous, and intracranial pressure, sampled at 125 Hz during 6 h (Figure 2f) from a single paediatric patient, are available in this public database [48]. Time series of this length enable the study of the influence of great m values on PE, and are also very likely to exhibit non-stationarities or drifts [5].

4. EEG. Electroencephalograph records with 4097 samples from the Department of Epileptology, University of Bonn [49], publicly available at http://epileptologie-bonn.de. This database is included in the present paper because it has been used in a myriad of classification studies using different feature extraction methods [50–54], including PE [55], and whose results make an interesting comparison here. Records correspond to the 100 EEGs of this database from epilepsy patients, but with no seizures included, and 100 EEGs including seizures. More details of this database can be found in the references included and in many other papers.

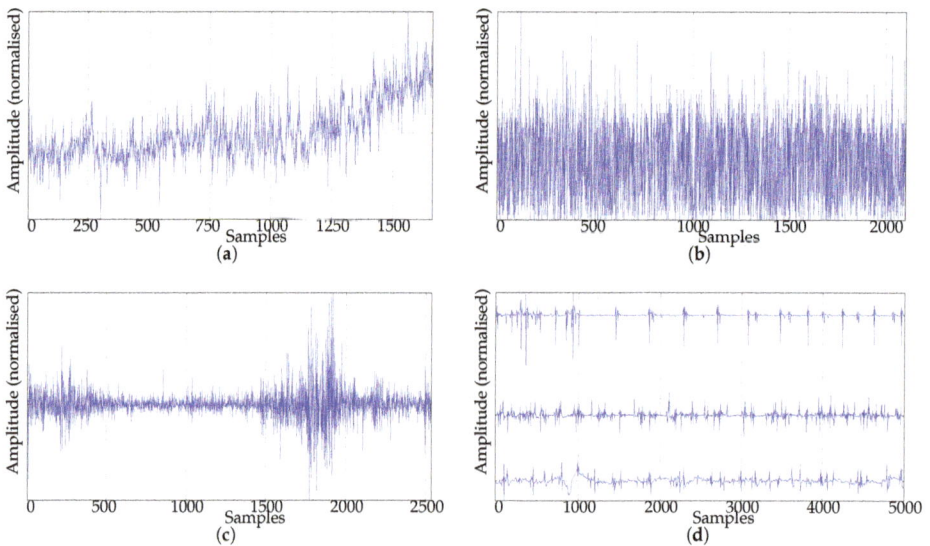

Figure 2. *Cont.*

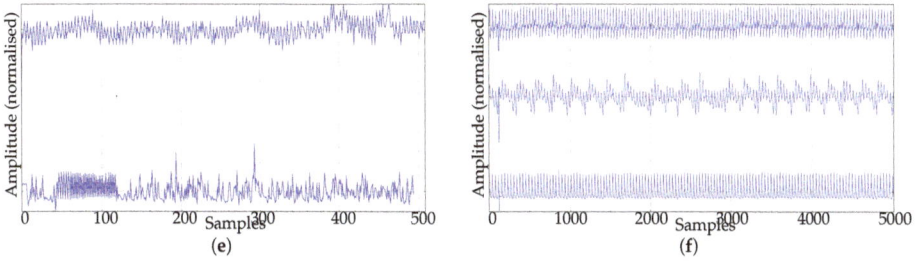

Figure 2. Real data experimental dataset examples. (**a**) Example of temperature anomaly data from the CLIMATOLOGY subset. Record comprises from 1880 to 2018, with 1662 readings (12 per year), and a growing trend in recent years. (**b**) Example of seismic data from the SEISMIC subset. Record comprises worldwide earthquakes of greater intensity than 2.5, registered during May 2018. (**c**) Example of financial time series from the FINANCIAL subset. (**d**) EMG records included in the dataset (top: Neuropathy, center: Myopathy, bottom: Healthy). Only the first 5000 samples out of more than 50,000 are shown for clarity. (**e**) Examples of the records in the two groups of the PAF dataset included in the experiments. (**f**) Examples of the records in the PORTLAND dataset: arterial, central venous, and intracranial pressure. Only the first 5000 samples are shown for clarity.

To analyse the real–world records using PE, the minimum length should be that stated in Table 1. This length, given by $10m!$ according to our interpretation of $m! << N$, is an even more conservative approach than those used in other studies [16]. Therefore, the hypothesis of this work is that PE reaches stability at that length, and that will be the reference used in the experiments.

Table 1. Records in the real-world experimental database and their agreement with the recommendation $N >> m!$ for m in the usual range. Initially, N is considered to be much greater than $m!$ when it is at least equal to 10 times $m!$. Data length is included in brackets under the database name.

m	$m!$	$10m!$	CLIMATOLOGY (1662)	SEISMIC (2104–9090)	FINANCIAL (2519)	EMG (>50,000)	EEG (4097)	PAF (400–500)	PORTLAND ($1 \cdot 10^6$)
3	6	60	✓	✓	✓	✓	✓	✓	✓
4	24	240	✓	✓	✓	✓	✓	✓	✓
5	120	1200	✓	✓	✓	✓	✓	–	✓
6	720	7200	–	✓	–	✓	–	–	✓
7	5040	50,400	–	–	–	✓	–	–	✓
8	40,320	403,200	–	–	–	–	–	–	✓
9	362,880	3,628,800	–	–	–	–	–	–	–

3. Experiments and Results

The experiments addressed the influence of time series length on PE computation from two standpoints: absolute and relative. The absolute case corresponds to the stable value that PE reaches if a sufficient number of samples is provided (see the analysis in Section 3.1). This is considered the true PE value for that time series. The relative standpoint studies the PE variations for different classes, in order to assess whether, despite PE not being constant with N, the curve for each class can at least still be distinguished significantly from the others. If that is the case, that would certainly relax the requirements in terms of N for signal classification purposes. This issue is addressed in the experiments in Section 3.2.

In the absolute case, all the datasets described in Sections 2.2.1 and 2.2.2 were tested. The PE was computed for all the records in each dataset and for an equally distributed set of lengths, to obtain the points of a PE–N plot from the mean PE(m, N) value. In an ideal scenario, the resulting plot should be a constant value, that is, PE would be independent of N. However, in practice, PE will exhibit a transient response before it stabilises, if the time series under analysis is stationary and has enough samples. This number of samples is usually considered as that length that ensures all the ordinal patterns can be found. That is why the possible relationship between PE stability and the number of ordinal patterns for each length was also studied in this case.

The classification analysis used only those datasets that at least contain two different record classes. This analysis used first the complete records for PE computation, from which the classification performance was obtained. Then, this classification analysis was repeated using a set of lengths well below the baseline N length in order to assess the possible detrimental effect on the performance. Additional experiments were conducted in order to justify why that detrimental effect was found to be negligible, based on three hypotheses raised by the authors: PE–N curves are somehow divergent among classes, not all the ordinal patterns are necessary to find differences, and some ordinal patterns carry more discriminant information than others.

3.1. Length Analysis

When the results of PE are plotted against different time series lengths, a two-phase curve is obtained: a parabolic–like region and a saturation region. For very short lengths, PE increases as the number of samples also increases. At a certain length value, the rate of PE evolution levels off, and no further length increases cause a significant variation of the PE value. This behaviour is the same for all the datasets studied, except those with a strong prevalence of drifts, or markedly non-stationary. There are no guidelines to quantitatively define this point of stabilisation. We used the approach applied in [1], where stability was considered to be reached when the relative error was smaller than 2%. The ground truth with regard to the real PE value was that obtained at a certain length beyond which further PE variations were smaller than 2%.

The length analysis graphic results of the synthetic dataset (RAND, SPIKES, chaotic LMAP, and periodic LMAP records of length 1000) are shown in Figure 3, with $m = 3, 4, 5, 6, 7$. RAND records exhibit the most frequently found behaviour in real–world records, a kind of first–order system step response, with stability achieved at 50 samples for $m = 3$, 200 for $m = 4$ and at 500 for $m = 5$. Other lengths are not shown in the plot, but the experiments yielded a stabilisation length of 20,000 samples for $m = 6$, and 55,000 samples for $m = 7$, approximately. This can be considered in accordance with the $m! \ll N$ recommendation. The remaining synthetic records exhibited a different behaviour. The PE results for the SPIKES dataset were quite unstable, there was no clear stabilisation point. This can be due to the fact that PE is hypothetically sensitive to the presence of spikes, since it has been used as a spike detector [30,56]. Both LMAP datasets displayed the same behaviour. A PE maximum at very short lengths, and a very fast stabilisation for any m value, around 400 samples. Both datasets are very deterministic, even the chaotic one, and it can arguably be hypothesized that a relative low value of patterns suffice to estimate PE in these cases.

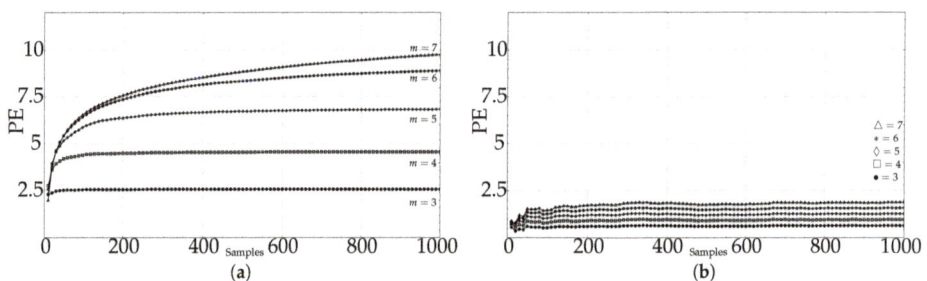

Figure 3. *Cont.*

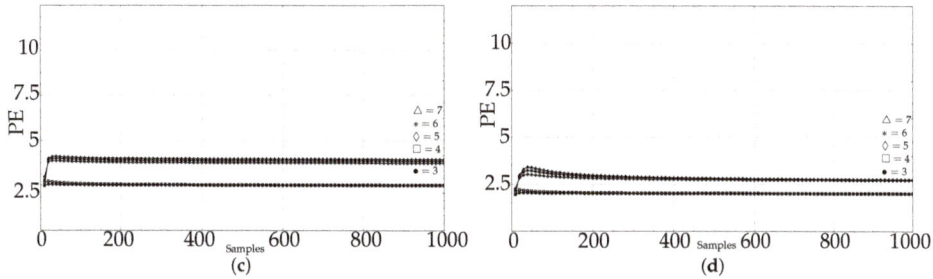

Figure 3. PE evolution for synthetic time series as a function of length *N*. Average PE results for the 100 time series generated in each dataset when *N* was varied from 10 up to 1000. ($\triangle$)m = 7, ($*$)m = 6, ($\lozenge$)m = 5, ($\square$)m = 4, ($\bullet$)m = 3. (**a**) Length analysis of the synthetic RAND dataset. (**b**) Length analysis of the synthetic SPIKES dataset. (**c**) Length analysis of the synthetic chaotic LMAP dataset (average of the three seeds). (**d**) Length analysis of the synthetic periodic LMAP dataset (average of the three seeds).

As for the real datasets: RAND, CLIMATOLOGY, SEISMIC, FINANCIAL, and EMG (only the first 5000 samples for EMG records), they exhibit the same behaviour depicted in Figure 3a, as shown individually in Figure 4a–d: An initial fast growing trend that later converges asymptotically to the supposedly true PE value.

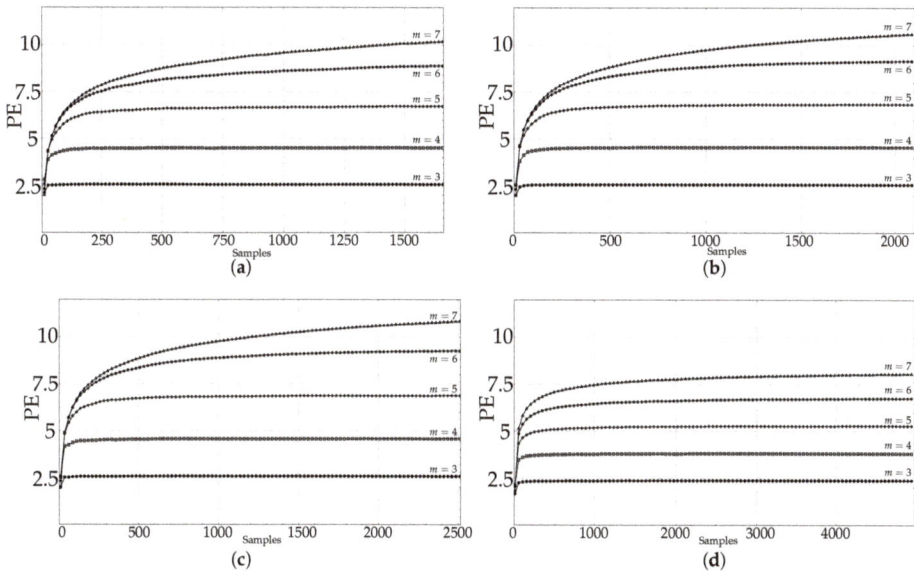

Figure 4. Average PE evolution for real–world time series as a function of length *N*. ($\triangle$)m = 7, ($*$)m = 6, ($\lozenge$)m = 5, ($\square$)m = 4, ($\bullet$)m = 3. (**a**) Average PE evolution for all the records in the CLIMATOLOGY database, with *m* from 3 to 7. Maximum length was 1500 samples. (**b**) Average PE evolution for all the records in the SEISMIC database, with *m* from 3 to 7. Maximum length was 2000 samples. (**c**) Average PE evolution for all the records in the FINANCIAL database, with *m* from 3 to 7. Maximum length was 2500 samples. (**d**) Average PE evolution for all the records in the EMG database (healthy, myopathy, neuropathy), with *m* from 3 to 7. Maximum length was 5000 samples.

Figure 5 shows in more detail the results corresponding to averaged PE values at 100 different lengths for all the PAF records, with *m* ranging from 3 up to 7. For the *m* values 3, 4, and 5, it is clear that PE becomes stable at the 200 samples mark at latest, which is before the recommended number. However, stability is not achieved for the maximum length available, less than 300 samples, for $m = 6$ and $m = 7$. According to Table 1, lengths around 7200 and 50,400 samples would be necessary, but such lengths are not available.

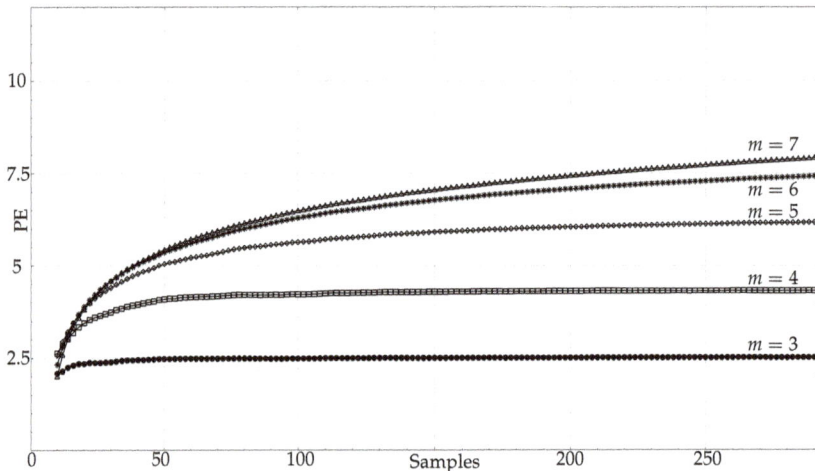

Figure 5. Detailed. average PE evolution for all the real records in the PAF database, with m from 3 to 7. Maximum length is taken from the shortest record, approximately 290 samples.

For lengths in the range 10,000–50,000 samples, the full–length EMG records were used for characterisation. The results for the healthy EMG record are shown in Figure 6, including those for very high m values of up to 9. As anticipated, there is a clear trend towards later stabilisation with increasing m, but not as demanding as $m! \ll N$ entails. Approximately, PE reaches stability at 40,000 samples for $m = 9$, at 20,000 samples for $m = 8$, and at 10,000 samples for $m = 7$ (for smaller m values, see Figure 4d). According to the general recommendation, around 3,600,000, 400,000, or 50,000 samples would have been required respectively instead (Table 1). With other less demanding recommendations such as $5m! \leq N$ [16], the real difference is still very significant.

Although PE is very robust against non-stationarities [57], they can also pose a problem as signal length increases. To illustrate this point, Figure 7 shows the PE results for the very long signals from the PORTLAND database. In this specific case, even for low m values, there is not a clear stabilisation at any point. These results suggest that a prior stationarity analysis would be required in case of very long time series.

Since PE measurements are related to the ordinal patterns found, we also analysed the evolution of the number of patterns with a relative frequency greater than 0, as a function of N. The results are shown in Figure 8. The trend is similar to that of PE itself, a fast growing curve for short lengths that later stabilises to the maximum number of patterns that can be found (this number can be smaller than $m!$ due to the presence of forbidden patterns). However, the stabilisation takes place far later than for PE, which seems to indicate that PE values do not depend equally on all the patterns, as will be further demonstrated in Section 3.2.

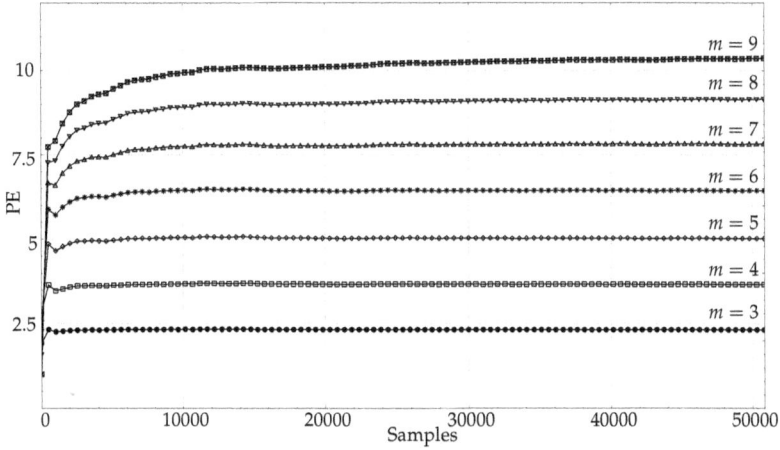

Figure 6. Average. PE evolution using the entire length of the healthy EMG. (⊠)$m = 9$, (∇)$m = 8$, (△)$m = 7$, (∗)$m = 6$, (◊)$m = 5$, (□)$m = 4$, (•)$m = 3$. This figure complements Figure 4d, where EMG short–term evolution was depicted instead of this long–term evolution. The availability of very long records enabled the analysis using greater m values.

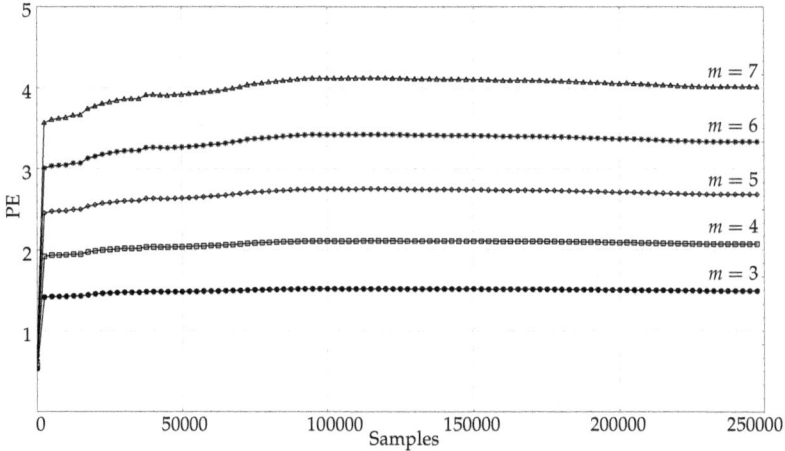

Figure 7. Average. PE evolution using the records from the PORTLAND database. Contrary to the previous cases, PE does not become stable even for very high values of N and low m values, probably due to non-stationarities or changes in record dynamics that impact on PE results. (△)$m = 7$, (∗)$m = 6$, (◊)$m = 5$, (□)$m = 4$, (•)$m = 3$.

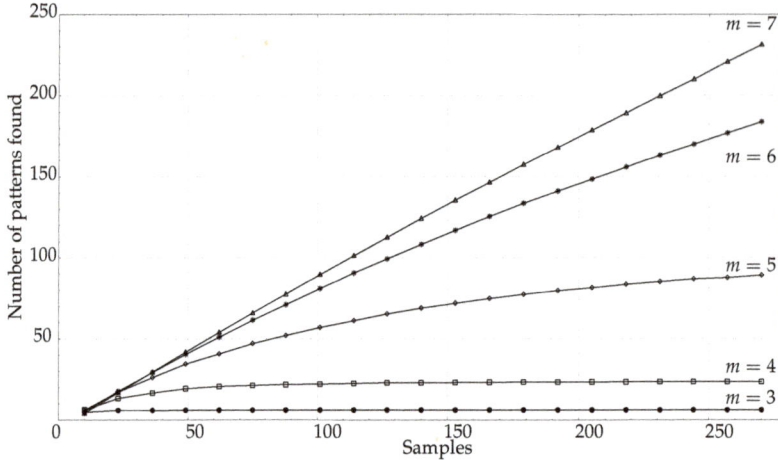

Figure 8. Average. number of ordinal patterns found for all the PAF records as a function of the length N for m between 3 and 7. $(\triangle)m = 7, (*)m = 6, (\Diamond)m = 5, (\square)m = 4, (\bullet)m = 3$.

3.2. Classification Analysis

There is a clear dependence of PE on the record length, mainly for very short records and large m values. However, as other previous studies have already demonstrated [24], PE might be able to capture the differences between signal groups even under unfavourable conditions, provided these conditions are the same for all the classes. Along these lines, it was hypothesised that well before PE reaches stability, differences become apparent. This hypothesis was developed following observations in previous class segmentation studies using PE and short records [24,26,27], as a generalisation of the PE capability to distinguish among record classes despite not satisfying the $m! << N$ condition.

The present classification analysis used records from the datasets that included several groups that were presumably separable. Specifically, from the synthetic database, the LMAP records were in principle separable since 3 different R coefficient values were used (3.50, 3.51, 3.52). This initial separability was first confirmed with a classification analysis whose results are listed in Table 2. This analysis took place using the entire 100 sequences of 1000 samples each, and the classes were termed 0, 1, and 2 respectively. The embedded dimension was varied from 3 up to 7, the usual range, but cases $m = 8$ and $m = 9$ were analysed too, which would require very long time series according to the recommendation under assessment (403,200 and 3,628,800 samples respectively). Classification performance was measured in terms of Sensitivity, Specificity, ROC Area Under Curve (AUC), and statistical significance, quantified using an unpaired Wilcoxon–Mann–Whitney test. This is the same scheme used in previous works [22]. The classes became significantly separable in all cases for $N = 1000$ and $m > 5$, which seems counter–intuitive in terms of the recommendation stated: better classification accuracy for worse $m! << N$ agreement.

Table 2. Baseline average classification results for synthetic LMAP periodic records using all the samples (1000) and different m values. For $m = 3$, the standard deviation is included in brackets.. The classes were studied in pairs, 01, 02, and 12. Very significant differences were found between classes 0 and 1, and 0 and 2. For classes 1 and 2, higher m values were required, although for less significant differences.

m	Sensitivity			Specificity			p			AUC		
	Se_{01}	Se_{02}	Se_{12}	Sp_{01}	Sp_{02}	Sp_{12}	p_{01}	p_{02}	p_{12}	01	02	12
3	0.67(0.06)	0.66(0.05)	0.66(0.13)	0.38(0.04)	0.35(0.04)	0.38(0.14)	0.7837	0.8990	0.6981	0.51(0.01)	0.50(0.01)	0.51(0.02)
4	0.49	0.68	0.67	0.55	0.41	0.41	0.6891	0.4214	0.6681	0.51	0.53	0.51
5	1	1	0.58	1	1	0.5	<0.0001	<0.0001	0.5807	1	1	0.52
6	1	1	0.61	1	1	0.65	<0.0001	<0.0001	0.0006	1	1	0.64
7	1	1	0.56	1	1	0.66	<0.0001	<0.0001	0.0193	1	1	0.59
8	1	1	0.64	1	1	0.66	<0.0001	<0.0001	<0.0001	1	1	0.66
9	1	1	0.64	1	1	0.76	<0.0001	<0.0001	<0.0001	1	1	0.73

The experiments in Table 2 were repeated for other lengths of the LMAP periodic records. These new results are shown in Table 3. The goal of this analysis was to find out if the entire length of the records was necessary to achieve the same classification results. As can be seen, the same classification performance can be obtained using only the initial 200–300 samples out of the complete time series of 1000 samples. The performance also improves when m is greater, contrary to what $m! << N$ would suggest.

Table 3. Classification results for synthetic LMAP periodic records for different N and m values. The classes were studied in pairs, 01, 02, and 12. These results should be compared to results in Table 2, where the same dataset was used, but using the entire length. With lengths as short as 200 samples, results are almost the same achieved with the complete records. More difficulties were found to separate groups 1 and 2, also in line with the results using $N = 1000$.

m	N	Sensitivity			Specificity			p			AUC		
		Se_{01}	Se_{02}	Se_{12}	Sp_{01}	Sp_{02}	Sp_{12}	p_{01}	p_{02}	p_{12}	01	02	12
3	100	0.59	0.53	0.59	0.49	0.49	0.47	0.0959	0.4840	0.3219	0.56	0.52	0.54
3	200	0.51	0.74	0.69	0.54	0.34	0.40	0.6228	0.4599	0.1891	0.51	0.53	0.55
4	100	0.33	0.35	0.44	0.71	0.71	0.58	0.9359	0.5087	0.5919	0.50	0.52	0.52
4	200	0.46	0.50	0.48	0.69	0.56	0.60	0.0965	0.4465	0.3909	0.56	0.53	0.53
5	100	1	1	0.52	1	1	0.59	<0.0001	<0.0001	0.7850	1	1	0.51
5	200	1	1	0.52	1	1	0.53	<0.0001	<0.0001	0.9414	1	1	0.50
6	100	0.86	0.83	0.46	0.98	1	0.69	<0.0001	<0.0001	0.0075	0.95	0.92	0.61
6	200	1	1	0.61	1	1	0.54	<0.0001	<0.0001	0.1867	1	1	0.55
7	100	0.44	0.44	0.67	1	0.84	0.54	0.0001	0.0424	0.0074	0.65	0.58	0.61
7	200	0.98	0.98	0.63	1	1	0.54	<0.0001	<0.0001	0.1212	0.99	0.99	0.55
8	100	0.67	0.52	0.66	0.72	0.82	0.72	<0.0001	0.0012	0.0025	0.71	0.63	0.62
8	200	0.98	0.94	0.66	0.95	1	0.6	<0.0001	<0.0001	0.0087	0.99	0.98	0.60
9	100	0.94	0.92	0.61	0.94	0.99	0.66	<0.0001	<0.0001	0.0053	0.97	0.97	0.61
9	200	1	1	0.5	1	1	0.78	<0.0001	<0.0001	0.0899	1	1	0.57
9	300	1	1	0.5	1	1	0.83	<0.0001	<0.0001,	<0.0001	1	1	0.66

The classification analysis using real–world signals was based on PAF, EMG, and EEG records from the biomedical database. Table 4 shows the results for the classification of the two groups in the PAF database (fibrillation episode and no–episode) for the lengths available in each 5 minutes record, and for m between 3 and 7. These classes were significantly distinguishable in all cases studied, although the approximately 400 samples available fell well below the amount recommended, mainly for $m \geq 5$.

Table 4. Baseline classification results for PAF records using all the samples of each 5 minutes record and different m values. Sensitivity improves with greater m values, but the opposite for Specificity. Maximum AUC is obtained for $m = 5$. Anyway, the dataset is separable for any m value.

m	Sensitivity	Specificity	p	AUC
3	0.76	0.88	<0.0001	0.8560
($\tau = 2$)	0.92	0.72	<0.0001	0.8560
($\tau = 4$)	0.84	0.72	0.0002	0.8016
4	0.80	0.84	<0.0001	0.8608
5	0.80	0.80	<0.0001	0.8688
6	0.92	0.72	<0.0001	0.8672
7	0.96	0.68	<0.0001	0.8432

The experiments in Table 4 were repeated using only a subset of the samples located at the beginning of the time series. These additional results are shown in Table 5. Although there is a detrimental effect on the classification performance, significant results are achieved with even very short time series of some 45 ($m = 3$) or 50 samples ($m = 4, 5$).

Table 5. PAF records classification results for different values of N and m. These results should be compared with those in Table 4, where the same dataset was used, but the complete time series instead. For lengths around 50 samples, classification performance is very similar to that achieved with the entire records.

m	N	Sensitivity	Specificity	p	AUC
3	10	0.52	0.68	1.0000	0.5000
3	25	0.68	0.56	0.0857	0.6416
3	40	0.68	0.72	0.0045	0.7336
3	45	0.76	0.84	0.0002	0.8048
3	50	0.80	0.80	0.0002	0.7984
3	60	0.84	0.72	0.0003	0.7920
3	75	0.76	0.76	0.0004	0.7904
3	100	0.92	0.60	0.0003	0.7920
4	10	0.64	0.52	0.1278	0.6184
4	25	0.52	0.68	0.2169	0.6016
4	50	0.72	0.76	0.0004	0.7904
4	75	0.80	0.72	0.0003	0.7936
4	100	0.88	0.68	0.0001	0.8096
4	150	0.92	0.68	<0.0001	0.8496
5	10	0.00	1.00	0.8083	0.5200
5	25	0.52	0.60	0.2192	0.5984
5	50	0.68	0.84	0.0012	0.7664
5	75	0.60	0.84	0.0007	0.7784
5	100	0.76	0.72	0.0017	0.7584
5	200	0.88	0.64	0.0001	0.8208

Table 6 shows the classification results for the EMG records of length 5000 samples. Each class is termed 0, 1, or 2 healthy, myopathy, and neuropathy, respectively. Pairs 01 and 12 were easily distinguishable for any m value, but pair 02 could not be significantly segmented.

Table 6. Baseline classification results for the three classes of. EMG records using all 5000 samples and different m values. Groups 0 and 2 were not distinguishable in any case.

m	Sensitivity			Specificity			p			AUC		
	Se_{01}	Se_{02}	Se_{12}	Sp_{01}	Sp_{02}	Sp_{12}	p_{01}	p_{02}	p_{12}	01	02	12
3	1	1	0.51	1	0.62	0.81	<0.0001	0.2602	0.0203	1	0.6206	0.6912
4	1	1	1	1	0.62	1	<0.0001	0.2602	<0.0001	1	0.6209	1
5	1	1	1	1	0.62	1	<0.0001	0.2602	<0.0001	1	0.6209	1
6	1	0.9	1	1	0.62	1	<0.0001	0.3033	<0.0001	1	0.6103	1
7	1	0.9	1	1	0.55	1	<0.0001	0.3678	<0.0001	1	0.5965	1

As with the LMAP and PAF data, the EMG experiments were repeated using only a subset of the samples at the beginning of each record. These results are shown in Table 7. As with the entire records, pairs 01 and 12 can be separated even using very short records (200 samples for $m = 3$, 100 for $m = 4, 5$). As can be seen, the classification performance improves more with m than with N, probably because longer patterns provide more information about the signal dynamics [12]. Pair 02 could not be separated, but that was also the case when the entire records were processed using PE.

Table 7. EMG classification results for different values of N and m using the subset of 5000 samples extracted from each of the three EMG records as described in Section 2.2.2. These results should be compared with those in Table 6, where the same dataset was used, but with $N = 5000$. Similar were indeed achieved for lengths as short as 300 samples.

m	N	Sensitivity			Specificity			p			AUC		
		Se_{01}	Se_{02}	Se_{12}	Sp_{01}	Sp_{02}	Sp_{12}	p_{01}	p_{02}	p_{12}	01	02	12
3	100	0.40	0.55	0.51	1	0.6	0.91	0.6843	0.8469	0.2699	0.5454	0.5206	0.5909
3	200	0.80	0.80	0.76	0.81	0.58	0.59	0.0009	0.2602	0.0236	0.8681	0.6206	0.6865
3	300	0.80	0.70	0.72	0.91	0.58	0.63	0.0008	0.7722	0.0034	0.8727	0.5310	0.7413
3	400	0.90	0.90	0.58	0.91	0.55	0.77	<0.0001	0.4994	0.0036	0.9409	0.5724	0.7398
3	500	0.90	0.90	0.51	1	0.62	0.68	<0.0001	0.2340	0.0347	0.9636	0.6275	0.6739
4	100	0.7	0.41	0.58	0.86	0.8	0.81	0.0064	0.8976	0.0034	0.8045	0.5137	0.7413
4	200	1	0.80	0.86	0.95	0.51	0.86	<0.0001	0.4594	<0.0001	0.9863	0.5793	0.9090
4	400	1	0.70	0.89	1	0.62	1	<0.0001	0.5200	<0.0001	1	0.5689	0.9623
4	600	1	0.90	0.93	1	0.58	1	<0.0001	0.3678	<0.0001	1	0.5965	0.9890
4	800	1	1	1	1	0.58	0.95	<0.0001	0.2216	<0.0001	1	0.6310	0.9968
5	100	0.8	0.48	0.82	0.91	0.80	0.81	0.0008	0.6758	<0.0001	0.8727	0.5448	0.8463
5	200	1	0.60	0.89	0.95	0.51	0.95	<0.0001	1	<0.0001	0.9954	0.5	0.9502
5	500	1	0.80	1	1	0.62	0.95	<0.0001	0.4594	<0.0001	1	0.5793	0.9952
5	750	1	0.80	1	1	0.58	1	<0.0001	0.3851	<0.0001	1	0.5931	1
5	1000	1	0.80	1	1	0.58	1	<0.0001	0.3678	<0.0001	1	0.5965	1

Finally, the EEG records were also analysed, in order to provide a similar scheme to compare the results to those achieved in other works [55], although the experimental dataset and the specific conditions may vary across studies. The quantitative results are shown in Tables 8 and 9.

Table 8. Baseline classification results for EEG records using all 4097 samples and different m values. For any m value, the classification performance was very significant.

m	Sensitivity	Specificity	p	AUC
3	0.93	0.90	<0.0001	0.9619
($\tau = 2$)	0.72	0.64	<0.0001	0.7186
($\tau = 4$)	0.62	0.56	0.2569	0.5464
4	0.93	0.89	<0.0001	0.9579
5	0.92	0.89	<0.0001	0.9563
6	0.91	0.89	<0.0001	0.9526
7	0.93	0.85	<0.0001	0.9443

Table 9. EEG classification results for different values of N and m. These results should be compared with those of Table 8, where the same dataset was used, but with all the 4097 samples.

m	N	Sensitivity	Specificity	p	AUC
3	100	0.76	0.86	<0.0001	0.8604
3	200	0.83	0.83	<0.0001	0.8966
3	300	0.85	0.84	<0.0001	0.9183
3	400	0.86	0.86	<0.0001	0.9241
3	500	0.89	0.83	<0.0001	0.9336
3	1000	0.87	0.87	<0.0001	0.9362
4	100	0.75	0.85	<0.0001	0.8531
4	200	0.86	0.81	<0.0001	0.8898
4	300	0.86	0.80	<0.0001	0.9086
4	400	0.87	0.83	<0.0001	0.9167
4	500	0.83	0.88	<0.0001	0.9264
4	1000	0.86	0.87	<0.0001	0.9307
5	100	0.74	0.84	<0.0001	0.8441
5	200	0.82	0.82	<0.0001	0.8746
5	300	0.84	0.80	<0.0001	0.8963
5	400	0.85	0.83	<0.0001	0.8999
5	500	0.86	0.84	<0.0001	0.9132
5	1000	0.87	0.85	<0.0001	0.9260
6	100	0.73	0.83	<0.0001	0.8239
6	200	0.81	0.79	<0.0001	0.8513
6	300	0.82	0.79	<0.0001	0.8729
6	400	0.85	0.81	<0.0001	0.8800
6	500	0.86	0.81	<0.0001	0.8940
6	1000	0.89	0.81	<0.0001	0.9146
7	100	0.71	0.79	<0.0001	0.7991
7	200	0.78	0.79	<0.0001	0.8283
7	300	0.75	0.81	<0.0001	0.8461
7	400	0.87	0.82	<0.0001	0.8533
7	500	0.85	0.78	<0.0001	0.8700
7	1000	0.89	0.78	<0.0001	0.8942

3.3. Justification Analysis

All the classification results hint that the necessary length N to achieve a significant performance is far shorter than that stated by the recommendation $m! << N$. This may be due to several factors:

- Firstly, the possible differences among classes in terms of PE may become apparent before stability is reached. As occurred with ties [24], artefacts, including lack of samples, exert an equal impact on all the classes under analysis, and therefore, PE results are skewed, but differences remain almost constant. In other words, the curves corresponding to the evolution of PE with N remain parallel even for very small N values. An example of this relationship is shown in Figure 9 for PAF records using $m = 3$ and $m = 5$. Analytically, PE reaches stability at 45 samples for $m = 3$, but at 30 samples, both classes become significantly separable, which is confirmed by numerical results in Table 5. For $m = 5$ there are not enough samples to reach stability, as defined in Section 3.1, but class separability can be achieved with less than 50 samples. Shorter lengths may have a detrimental effect on classification accuracy, but such accuracy is still very significant. This behaviour is quite common (Tables 3 and 5).
- Secondly, the recommendation $m! << N$ was devised to ensure that all patterns could be found with high probability [16]. However, this is a very restrictive limitation, since this is only achievable for random time series. More deterministic time series, even chaotic time series like the ones included in the experimental dataset, have forbidden patterns that cannot be found whatever the length is [58]. Therefore, all the possible different patterns involved in a chaotic time series can be found with shorter records than the recommendation suggests. This is very well illustrated in Table 10, where random sequences (RANDOM, SEISMIC) exhibit more different

patterns than chaotic ones (EMG, PAF) per length unit. Thus, for most real–world signals that recommendation could arguably be softened.

- Third, and finally, not all the patterns, in terms of estimated probability, have the same impact, positive or negative, on PE calculation. Indirectly, this impact will also have an influence on the discriminative power of PE. In other words, a subset of the patterns can be more beneficial than the entire set. To assess this point, we modified the PE algorithm to sort the estimated non-zero probabilities in ascending order, and remove the k–smallest ones from the final computation. The approximated PE value was used in the classification analysis instead. Some experiments were carried out to quantify the possible loss incurred by this removal in cases previously studied. The corresponding results are shown in Table 11, for records with a significant number of patterns as per the data in Table 10.

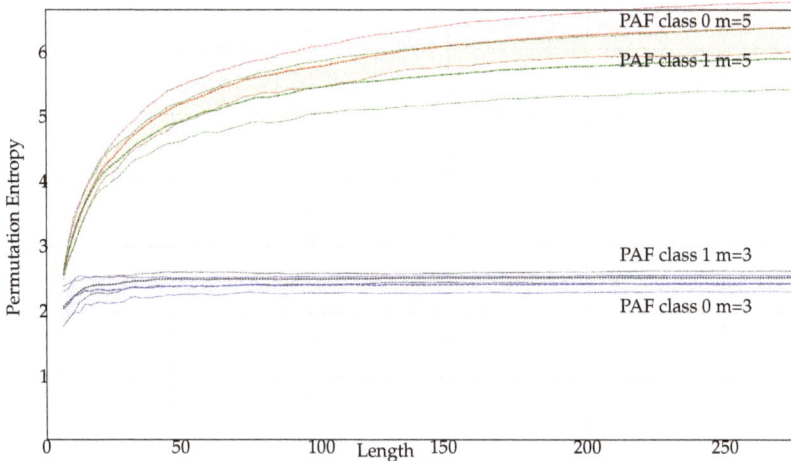

Figure 9. PE evolution with N for PAF records and $m = 3$ and $m = 5$. In contrast to previous results, not only average values are shown, but also one standard deviation interval to illustrate the possible overlapping between classes.

Table 10. Average number of patterns found in several datasets compared to the maximum number of patterns that $m!$ implies (found/expected). Randomness and determinism are related to the number of patterns found per length unit, and the number of forbidden patterns.

	N	$m = 3$	$m = 4$	$m = 5$	$m = 6$	$m = 7$
RANDOM	5000	6/6	24/24	120/120	719.37/720	3176.74/5040
EMG	5000	6/6	24/24	115.213/120	455.82/720	1053.31/5040
SINUS	5000	4/6	6/24	8/120	10/720	10/5040
LMAP (Periodic)	5000	4.366/6	5.01/24	9.367/120	11.28/720	12.80/5040
LMAP (Chaotic)	5000	4.31/6	4.94/24	9.31/120	10.45/720	11.21/5040
PAF	400	6/6	23.76/24	97.84/120	249.98/720	346.9/5040
SEISMIC	2000–9000	6/6	24/24	120/120	699.714/720	2602.29/5040

Table 11. Influence of number of patterns used for PE computation on classification performance. The first column corresponds to the normal case of no–pattern–restriction, the other ones account for the performance when the smallest PE relative frequencies were discarded, and only the reported number of patterns remained in the calculation.

	m	Remaining patterns (Sensitivity)(Specificity)					
PAF	3	6 (0.76)(0.88)	5 (0.76)(0.92)	4 (0.8)(0.8)	3 (0.8)(0.8)	2 (0.72)(0.8)	1 (0.76)(0.68)
	4	24 (0.80)(0.84)	20 (0.72)(0.88)	16 (0.8)(0.88)	12 (0.8)(0.84)	8 (0.8)(0.84)	4 (0.84)(0.8)
	5	120 (0.8)(0.8)	100 (0.8)(0.8)	80 (0.84)(0.76)	60 (0.92)(0.76)	40 (0.88)(0.8)	20 (0.88)(0.8)
EMG	3	6 (1,1,0.51)(1,0.62,0.81)	5 (1,1,0.51)(1,0.62,0.81)	4 (1,1,0.86)(1,0.62,0.44)	3 (1,1,0.41)(1,0.62,1)	2 (1,1,0.62)(1,0.62,0.8)	1 (1,1,0.62)(1,0.62,0.81)
	4	24 (1,1,1)(1,0.62,1)	20 (1,1,1)(1,0.62,1)	16 (1,0.7,1)(1,0.65,1)	12 (1,0.7,1)(1,0.62,1)	8 (1,0.38,1)(1,0.8,1)	4 (1,0.51,1)(1,0.8,1)
	5	120 (1,1,1)(1,0.62,1)	100 (1,0.8,1)(1,0.48,1)	80 (1,0.9,1)(1,0.44,1)	60 (1,0.7,1)(1,0.55,1)	40 (1,0.8,1)(1,0.58,1)	20 (1,0.9,1)(1,0.62,0.91)

3.3.1. Relevance Analysis

The results in Table 11 show that only a few patterns suffice to find differences between classes. For PAF records and $m = 3$, with only 3 patterns it is possible to achieve a sensitivity and specificity as high as 0.8. For $m = 5$, a subset of patterns can be better for classification, since only 40 or 20 achieve more accuracy than 120 or 100. This is also the case for other m values or other signals. Probably, a more careful selection of the remaining patterns could yield even better results.

Since not only the quantity of attributes may play an important role, but also their quality, a relevance analysis to the ordinal patterns for $m = 3$ (6 patterns) obtained when processing the PAF database was applied. Relevance analysis aims to reduce the complexity in a representation space, removing redundant and/or irrelevant information according to an objective function, in order to improve classification performance and discover the intrinsic information for decision support purposes [59]. In this paper, a relevance analysis routine based on the RELIEF-F algorithm was used to highlight the most discriminant patterns [60].

RELIEF-F is an inductive learning procedure, which gives a weight to every feature, where a higher weight means that the feature is more relevant for the classification [61]. For selecting relevant ordinal patterns the RELIEF-F algorithm shown in Algorithm 1.

Algorithm 1: RELIEF-F for ordinal patterns selection

Inputs: Π, k and n
Outputs: W
for $j \in \{1, m!\}$ **do**
$\quad W[\pi_j] := 0$
for $i \in \{1, n\}$ **do**
$\quad$ Randomly select an instance R
$\quad$ Find the k nearest Hits $H_{1,2\ldots k}$
$\quad$ Find the k nearest Misses $M_{1,2\ldots k}$
$\quad$ **for** $l \in \{1, k\}$ **do**
$\quad\quad$ **for** $j \in \{1, m!\}$ **do**
$\quad\quad\quad W[\pi_j] := W[\pi_j] - \dfrac{diff(P(\pi_j), R, H_l) - diff(P(\pi_j), R, M_l)}{k}$

The nearest Hits makes reference to its nearest neighbours in the same class, while the nearest Misses refers to the nearest neighbours of a different class. Likewise, $diff(P(\pi_j), A, B)$ function expresses the normalized difference, i.e., $[0, 1]$ range, for the relative frequency of the ordinal pattern π_j, between the instances A and B.

The results in Table 12 confirm this hypothesis. As the number and content of the patterns in PE is known in advance, this could become a field of intense study in future works due to its potential as a tool to improve the segmentation capability of PE or any related method.

Table 12. Results of the relevance analysis for the patterns obtained using the PAF records and $m = 3$.

	Ordinal Pattern					
	123	**132**	**213**	**231**	**321**	**312**
Rank	1	3	5	6	2	4
Weight	0.02	0.01	−0.005	−0.0077	0.013	0.0074
p-value	0.0002	0.0170	0.0270	0.1510	0.0123	0.0681

Additionally, according to Figure 10, in the boxplots of relative frequencies for the six ordinal patterns assessed, the discriminant effectiveness is different for each pattern. E.g., pattern 123 is the one which offers the best classification capability (Figure 10a), while pattern 231 is not recommended (Figure 10f). These results suggest that for classification purposes it may not be necessary to compute the relative frequency for all patterns, which means a reduction in the computational cost, a very important issue for real time systems.

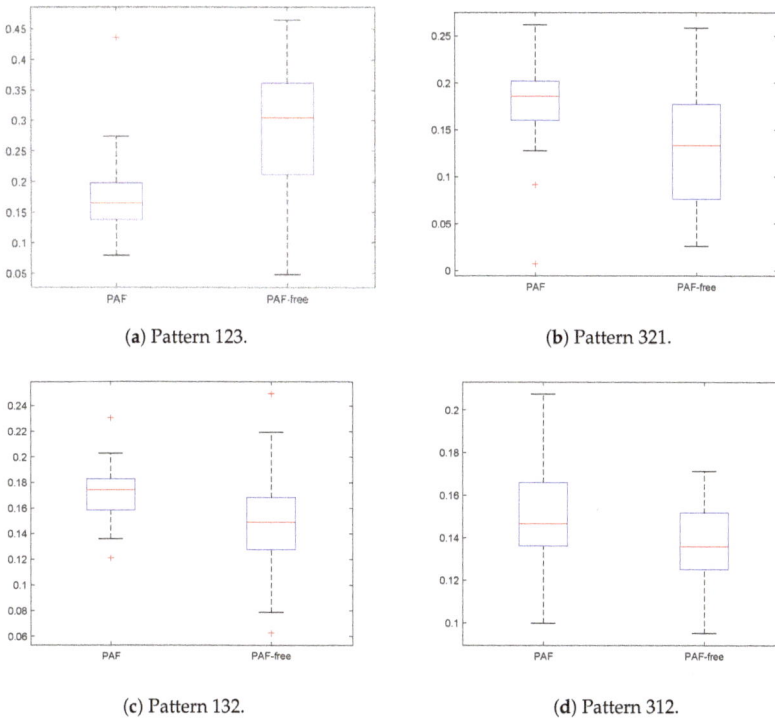

(a) Pattern 123.

(b) Pattern 321.

(c) Pattern 132.

(d) Pattern 312.

Figure 10. *Cont.*

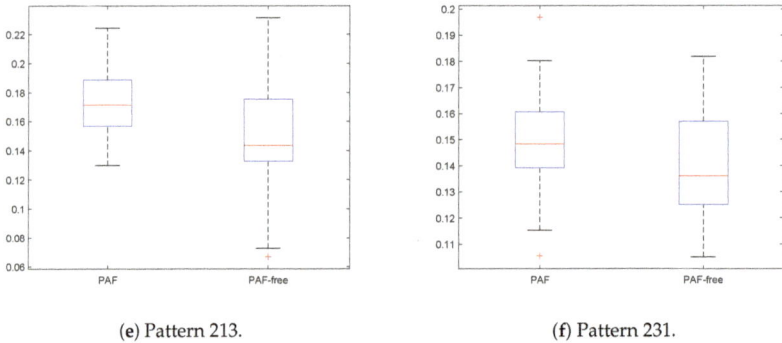

(e) Pattern 213.   (f) Pattern 231.

Figure 10. Boxplots of relative frequencies of ordinal patterns over time series with PAF and PAF–free.

4. Discussion

The recommendation $m! << N$ is aimed at ensuring that all possible patterns become visible [16], even those with low probability. This is a sure and safe bet, and is clearly true for random time series, where any pattern can appear [32].

For both synthetic and real signals, there is a clear dependence of PE on N, which is depicted in Figures 3 and 4, with the exception of the SPIKES and LMAP datasets. PE initially grows very fast, which can be interpreted as a complexity increase due to the addition of new patterns π_j since more terms $p(\pi_j)$ become greater than 0. PE tends to quickly stabilise once all the allowed patterns have been found [58], and at some point, more samples only contribute to increasing the counters of the already occupied probability bins, but PE remains almost constant. However, PE stabilises before the number of patterns found does (Figure 8), probably because not all the patterns are equally significant when computing PE. SPIKES are not very well suited for PE since most of the subsequences will be composed of 0 values, yielding a very biased distribution, but they have been included since there are quite many works where PE was used to detect spikes, and to illustrate this anomalous behaviour (Figure 3b).

Numerically, there is a great variability of the point where PE stabilises in each case. The RAND dataset is probably the one that best follows the $m! << N$ recommendation, with approximate real stabilisation points at 50, 200, 500, 20000, and 55000 samples (for $m = 3, \ldots, 7$), compared with the estimated values of 60, 240, 1200, 7200, and 50,400.

For the PAF database, PE becomes stable at 50 samples for $m = 3$, 150 for $m = 4$, and 250 for $m = 5$. There were not enough data to study greater m values. However, the lengths available seem to suggest that shorter lengths suffice to compute PE, and the greater the m value, the greater the difference between the real length needed, and the length suggested.

The other real signals yielded very similar results. The CLIMATOLOGY database stabilised PE at lengths shorter than 100 samples for $m = 3$, at 250 for $m = 4$, and at 750 samples approximately for $m = 5$. Using the SEISMIC records, the lengths were 50, 300, and 900 for $m = 3, 4, 5$. The FINANCIAL database needed 80, 450, and 850 samples for the same embedded dimensions. The EMG records of length 5000 became stable at 100, 400, and 950 respectively. All these signals were not long enough for $m = 6$ and $m = 7$.

These values of m were tested with the full–length EMG records (Figure 6), along with the long records of the PORTLAND database (Figure 7). In the EMG case, stability was reached for $m = 6$ at length 16,000, and at 30,000 for $m = 7$. It was also possible to see that the length required for $m = 8$ was 35,000, and 50,000 for $m = 9$. The PORTLAND records did not yield any stabilisation point as defined in this study, probably because such great lengths are counterproductive in terms of stationarity. This case was included in order to illustrate the detrimental effect that longer records may also have.

The classification analysis reported in Tables 2 and 3 suggests length is far less important to find differences among classes using PE. In Table 2, the results for LMAP records using 1000 samples seem to show that for a significant classification, it is necessary to have $m > 4$, and maximum classification performance is achieved for $m = 9$, which would imply, according to $m! << N$, a length in the vicinity of $1 \cdot 10^6$ samples at least, 1000 times more samples. These results are supported by an analysis based on Sensitivity, Specificity, statistical significance, and AUC, from $m = 5$, where $m! << N$ is still fulfilled, up to $m = 9$. There is also a clear direct correlation between m and classification accuracy. With regard to the effect of τ, as hypothesized, it has a detrimental impact on the classification performance due to the information loss that it entails, which is not compensated by a clear multi-scale structure of the data analysed. This parameter does not only imply a length reduction, as others analyses in this study do, but also a sub-sampling effect.

The analysis using shorter versions of LMAP records in Table 3 confirms differences can be found using a subset of an already short time series. With as few as 100 samples, clear differences can be found even at $m = 9$, with a performance level very close to that achieved with the complete records.

Using real signals, as in Tables 4 and 5, the trend is exactly the same. The classification performance for PAF records reaches its maximum at $m = 5$, being significant all the tests for $m = 3$ up to $m = 7$, despite not having enough samples for $m > 5$. Again, with as few as 100 samples (Table 5), the classification is very similar to that in Table 4. The same occurs with the EMG records of length 5000, where best classification is achieved at $m = 5$, with good results for $m > 4$ (Table 7).

The classification of the EEG records from a very well known database by the scientific community working on this field follows the same pattern. Although the experiments are not exactly the same, the results achieved for the full length records (4097 samples) are very similar to those in [55], and in [54], among other papers, with AUCs in the 0.95 range for the specific classes compared. However, as demonstrated in Table 9, a significant separability is achieved for as few as 100 samples, and for any m between 3 and 7. This length is still within the limits suggested by $m! << N$ if $m = 3$, but that relationship is not satisfied for $m > 3$, with $m = 7$ being very far from doing so (some 50,000 samples required, see Table 1). In fact, m seems to have an almost negligible effect on the classification performance. In terms of AUC, a length of 3000 samples seems to suffice to achieve the maximum class separability, with a 0.1 AUC difference between $N = 3000$ and $N = 100$, except for $m = 7$, with a slightly greater AUC difference. Although length has a positive correlation (very small) with classification performance, once again records can be much shorter than $m! << N$ entails.

Signal differences become apparent well before PE stabilisation is reached (Figure 9) and even for very short records and great m values [26,27]. Some patterns have more influence than others (Table 11), and some do not show up at all (Table 10). All these facts may arguably explain why classification can be successfully performed even with as few as 100 samples. A short pattern relevance exploratory analysis (Table 12) seemed to additionally confirm some patterns have a greater contribution to the class differences than others, as is the case in many feature selection applications [62].

5. Conclusions

The well known recommendation of $N >> m!$ for robust PE computation is included in almost any study related to this measurement. However, this recommendation can be too vague and subject to a disparity of interpretations. In addition, it may cast doubt on PE results for short time series despite statistical significance or high classification accuracy.

This study was aimed at shedding some light on this issue from two viewpoints: the stability of the absolute value of PE, and its power as a distinguishing feature for signal classification. A varied and diverse experimental dataset was analysed, trying to include representative time series from different contexts and exhibiting different properties from a chaos point of view. Sinusoidal signals were included for deterministic behaviour, logistic maps also for deterministic and chaotic behaviour. Spike records to account for typical disturbances in many biological records and semi-periodic records. Random records for truly random time series and white noise. The real set included climatology data, non-stationary stochastic data, seismic geographically dispersed data that can be considered random, and stochastic financial data. EMG aimed to characterise the behaviour for very long semiperiodic signals and noise. PAF records are short non-stationary records that have been used in other classification studies previously, and EEG records are broadband records also used in other works. In total, 12 signal types were used in the experiments.

In absolute terms, PE values seem to reach a reasonable stability with 100 samples for $m = 3$, 500 samples for $m = 4$, and 1000 samples for $m = 5$. This can be arguably considered in agreement with the $m! << N$ recommendation, but it is far more specific, and can be further relaxed if the records under analysis are more deterministic. In other words, they can be considered an upper limit. For greater m values, we very much doubt that stationarity could be assured for real–world signals and for the lengths required, and further studies are necessary.

When comparing PE values in relative terms, $N >> m!$ becomes almost meaningless. Results in Tables 5 and 7 already demonstrate this, in agreement with other PE classification studies [26,27]. In all cases analysed, 200 samples seem to suffice to find differences among time series using PE, if not less. This seems to be due to three main factors: length is equally detrimental to all the classes, there is no need to "wait" for all the patterns to appear, since some of them never will, and not all the patterns are balanced in terms of relevance. In fact, considering the ordinal patterns relative frequencies as the features of a classifier, a relevance analysis could arguably improve the results achieved so far using PE, and this is probably a very promising field of research in the coming years. The recommendations are summarised in Table 13.

Table 13. Summary of the conclusions of the paper and the supporting information.

	Recommendation	Supporting Data	Justification
PE (absolute value)	$N >> m!$	Figures 3–7	Pattern probability estimation in other works.
PE (relative value) For classification	$N = 200$	Figures 8–10. Very similar results in other studies ([24,26,27]). Tables 2–7, 9–12. Very similar results for 10 datasets exhibiting a varied and diverse set of features and properties.	Class differences are present at any length in stationary records. Long records are usually non-stationary. There are forbidden patterns. No need to look for them. Not all the ordinal patterns are representative of the differences. Real signals are mostly chaotic.

As far as we know, there is no similar study that analysed quantitatively the $N >> m!$ recommendation. It is based on a conservative assumption to ensure that all ordinal patterns can be found with certain probability. Once that recommendation was proposed, all the subsequent works followed that recommendation in most cases without questioning it. In this work we have provided evidence that for PE absolute value computation that recommendation is reasonable, but it might be completely wrong for classification purposes (relative PE values). In the classification case we have proposed to use specific lengths of some 200 samples, but there is no formula that could mathematically provide an explicit value.

Furthermore, large m values should not be prevented from being used in classification studies based on PE due to the recommendation $N >> m!$. Similar works [24] have already demonstrated that higher m values frequently capture the dynamics of the underlying signal better, as is the case in the present study, and only computational resources should limit the highest m value available. Even for very short records, m values beyond the recommendation seem to perform better than those within $m! << N$.

Our main goal was to make a first step in the direction of questioning the $m! << N$ recommendation, overcome that barrier, and foster the development of other studies with more freedom to choose N. The preliminary relevance analysis introduced should be extended to more signals and cases, even using synthetic records where the probability density function of each order pattern is known and controlled in order to enable to use more analytic calculations.

Author Contributions: D.C.-F. conceived the presented idea, arranged the experimental dataset, and designed the experiments. J.P.M.-E., D.A.O., and E.D.-T. carried out the experiments and introduced the concept of relevance analysis. All authors discussed the results and contributed to the final manuscript. D.C.-F. wrote the paper. All authors have given final approval of the version submitted.

Acknowledgments: No funding was received to support this research work.

Conflicts of Interest: The authors declare no conflict of interest.

References

1. Pincus, S.M. Approximate entropy as a measure of system complexity. *Proc. Natl. Acad. Sci. USA* **1991**, *88*, 2297–2301. [CrossRef] [PubMed]
2. Lake, D.E.; Richman, J.S.; Griffin, M.P.; Moorman, J.R. Sample entropy analysis of neonatal heart rate variability. *Am. J. Physiol.-Regul. Integr. Comp. Physiol.* **2002**, *283*, R789–R797. [CrossRef] [PubMed]
3. Lu, S.; Chen, X.; Kanters, J.K.; Solomon, I.C.; Chon, K.H. Automatic Selection of the Threshold Value *r* for Approximate Entropy. *IEEE Trans. Biomed. Eng.* **2008**, *55*, 1966–1972.
4. Alcaraz, R.; Abásolo, D.; Hornero, R.; Rieta, J. Study of Sample Entropy ideal computational parameters in the estimation of atrial fibrillation organization from the ECG. In Proceedings of the 2010 Computing in Cardiology, Belfast, UK, 26–29 September 2010; pp. 1027–1030.
5. Yentes, J.M.; Hunt, N.; Schmid, K.K.; Kaipust, J.P.; McGrath, D.; Stergiou, N. The Appropriate Use of Approximate Entropy and Sample Entropy with Short Data Sets. *Ann. Biomed. Eng.* **2013**, *41*, 349–365. [CrossRef]
6. Mayer, C.C.; Bachler, M.; Hörtenhuber, M.; Stocker, C.; Holzinger, A.; Wassertheurer, S. Selection of entropy-measure parameters for knowledge discovery in heart rate variability data. *BMC Bioinform.* **2014**, *15*, S2. [CrossRef]
7. Chen, W.; Zhuang, J.; Yu, W.; Wang, Z. Measuring complexity using FuzzyEn, ApEn, and SampEn. *Med. Eng. Phys.* **2009**, *31*, 61–68. [CrossRef] [PubMed]
8. Liu, C.; Li, K.; Zhao, L.; Liu, F.; Zheng, D.; Liu, C.; Liu, S. Analysis of heart rate variability using fuzzy measure entropy. *Comput. Biol. Med.* **2013**, *43*, 100–108. [CrossRef]
9. Bošković, A.; Lončar-Turukalo, T.; Japundžić-Žigon, N.; Bajić, D. The flip-flop effect in entropy estimation. In Proceedings of the 2011 IEEE 9th International Symposium on Intelligent Systems and Informatics, Subotica, Serbia, 8–10 September 2011; pp. 227–230.
10. Li, D.; Liang, Z.; Wang, Y.; Hagihira, S.; Sleigh, J.W.; Li, X. Parameter selection in permutation entropy for an electroencephalographic measure of isoflurane anesthetic drug effect. *J. Clin. Monit. Comput.* **2013**, *27*, 113–123. [CrossRef] [PubMed]
11. Bandt, C.; Pompe, B. Permutation Entropy: A Natural Complexity Measure for Time Series. *Phys. Rev. Lett.* **2002**, *88*, 174102. [CrossRef] [PubMed]
12. Riedl, M.; Müller, A.; Wessel, N. Practical considerations of permutation entropy. *Eur. Phys. J. Spec. Top.* **2013**, *222*, 249–262. [CrossRef]
13. Amigó, J.M.; Zambrano, S.; Sanjuán, M.A.F. True and false forbidden patterns in deterministic and random dynamics. *Europhys. Lett. (EPL)* **2007**, *79*, 50001. [CrossRef]
14. Zanin, M.; Zunino, L.; Rosso, O.A.; Papo, D. Permutation Entropy and Its Main Biomedical and Econophysics Applications: A Review. *Entropy* **2012**, *14*, 1553–1577. [CrossRef]
15. Rosso, O.; Larrondo, H.; Martin, M.; Plastino, A.; Fuentes, M. Distinguishing Noise from Chaos. *Phys. Rev. Lett.* **2007**, *99*, 154102. [CrossRef]
16. Amigó, J.M.; Zambrano, S.; Sanjuán, M.A.F. Combinatorial detection of determinism in noisy time series. *EPL* **2008**, *83*, 60005. [CrossRef]
17. Yang, A.C.; Tsai, S.J.; Lin, C.P.; Peng, C.K. A Strategy to Reduce Bias of Entropy Estimates in Resting-State fMRI Signals. *Front. Neurosci.* **2018**, *12*, 398. [CrossRef]
18. Shi, B.; Zhang, Y.; Yuan, C.; Wang, S.; Li, P. Entropy Analysis of Short-Term Heartbeat Interval Time Series during Regular Walking. *Entropy* **2017**, *19*. doi:10.3390/e19100568. [CrossRef]
19. Karmakar, C.; Udhayakumar, R.K.; Li, P.; Venkatesh, S.; Palaniswami, M. Stability, Consistency and Performance of Distribution Entropy in Analysing Short Length Heart Rate Variability (HRV) Signal. *Front. Physiol.* **2017**, *8*, 720. [CrossRef]

20. Cirugeda-Roldán, E.; Cuesta-Frau, D.; Miró-Martínez, P.; Oltra-Crespo, S.; Vigil-Medina, L.; Varela-Entrecanales, M. A new algorithm for quadratic sample entropy optimization for very short biomedical signals: Application to blood pressure records. *Comput. Methods Programs Biomed.* **2014**, *114*, 231–239. [CrossRef]

21. Lake, D.E.; Moorman, J.R. Accurate estimation of entropy in very short physiological time series: The problem of atrial fibrillation detection in implanted ventricular devices. *Am. J. Physiol.-Heart Circ. Physiol.* **2011**, *300*, H319–H325. [CrossRef]

22. Cuesta-Frau, D.; Novák, D.; Burda, V.; Molina-Picó, A.; Vargas, B.; Mraz, M.; Kavalkova, P.; Benes, M.; Haluzik, M. Characterization of Artifact Influence on the Classification of Glucose Time Series Using Sample Entropy Statistics. *Entropy* **2018**, *20*, doi:10.3390/e20110871. [CrossRef]

23. Costa, M.; Goldberger, A.L.; Peng, C.K. Multiscale entropy analysis of biological signals. *Phys. Rev. E* **2005**, *71*, 021906. [CrossRef]

24. Cuesta–Frau, D.; Varela-Entrecanales, M.; Molina-Picó, A.; Vargas, B. Patterns with Equal Values in Permutation Entropy: Do They Really Matter for Biosignal Classification? *Complexity* **2018**, *2018*, 1–15. [CrossRef]

25. Keller, K.; Unakafov, A.M.; Unakafova, V.A. Ordinal Patterns, Entropy, and EEG. *Entropy* **2014**, *16*, 6212–6239. [CrossRef]

26. Cuesta-Frau, D.; Miró-Martínez, P.; Oltra-Crespo, S.; Jordán-Núñez, J.; Vargas, B.; Vigil, L. Classification of glucose records from patients at diabetes risk using a combined permutation entropy algorithm. *Comput. Methods Programs Biomed.* **2018**, *165*, 197–204. [CrossRef]

27. Cuesta-Frau, D.; Miró-Martínez, P.; Oltra-Crespo, S.; Jordán-Núñez, J.; Vargas, B.; González, P.; Varela-Entrecanales, M. Model Selection for Body Temperature Signal Classification Using Both Amplitude and Ordinality-Based Entropy Measures. *Entropy* **2018**, *20*, 853. doi:10.3390/e20110853. [CrossRef]

28. Tay, T.-T.; Moore, J.B.; Mareels, I. *High Performance Control*; Springer: Berlin, Germany, 1997.

29. Little, D.J.; Kane, D.M. Permutation entropy with vector embedding delays. *Phys. Rev. E* **2017**, *96*, 062205. [CrossRef]

30. Azami, H.; Escudero, J. Amplitude-aware permutation entropy: Illustration in spike detection and signal segmentation. *Comput. Methods Programs Biomed.* **2016**, *128*, 40–51. [CrossRef]

31. Naranjo, C.C.; Sanchez-Rodriguez, L.M.; Martínez, M.B.; Báez, M.E.; García, A.M. Permutation entropy analysis of heart rate variability for the assessment of cardiovascular autonomic neuropathy in type 1 diabetes mellitus. *Comput. Biol. Med.* **2017**, *86*, 90–97. [CrossRef]

32. Zunino, L.; Zanin, M.; Tabak, B.M.; Pérez, D.G.; Rosso, O.A. Forbidden patterns, permutation entropy and stock market inefficiency. *Phys. A Stat. Mech. Appl.* **2009**, *388*, 2854–2864. [CrossRef]

33. Saco, P.M.; Carpi, L.C.; Figliola, A.; Serrano, E.; Rosso, O.A. Entropy analysis of the dynamics of El Niño/Southern Oscillation during the Holocene. *Phys. A Stat. Mech. Appl.* **2010**, *389*, 5022–5027. [CrossRef]

34. Konstantinou, K.; Glynn, C. Temporal variations of randomness in seismic noise during the 2009 Redoubt volcano eruption, Cook Inlet, Alaska. In Proceedings of the EGU General Assembly Conference Abstracts, Vienna, Austria, 23–28 April 2017; Volume 19, p. 4771.

35. Molina-Picó, A.; Cuesta-Frau, D.; Aboy, M.; Crespo, C.; Miró-Martínez, P.; Oltra-Crespo, S. Comparative Study of Approximate Entropy and Sample Entropy Robustness to Spikes. *Artif. Intell. Med.* **2011**, *53*, 97–106. [CrossRef]

36. DeFord, D.; Moore, K. Random Walk Null Models for Time Series Data. *Entropy* **2017**, *19*. [CrossRef]

37. Chirigati, F. Weather Dataset. 2016. Available online: https://doi.org/10.7910/DVN/DXQ8ZP (accessed on 1 August 2018).

38. Thornton, P.; Thornton, M.; Mayer, B.; Wilhelmi, N.; Wei, Y.; Devarakonda, R.; Cook, R. *Daymet: Daily Surface Weather Data on a 1-km Grid for North America, Version 2*; ORNL DAAC: Oak Ridge, TN, USA, 2014.

39. Zhang, H.; Huang, B.; Lawrimore, J.; Menne, M.; Smith, T.M. NOAA Global Surface Temperature Dataset (NOAAGlobalTemp, ftp.ncdc.noaa.gov), Version 4.0, August 2018. Available online: https://doi.org/10.7289/V5FN144H (accessed on 1 August 2018).

40. Balzter, H.; Tate, N.J.; Kaduk, J.; Harper, D.; Page, S.; Morrison, R.; Muskulus, M.; Jones, P. Multi-Scale Entropy Analysis as a Method for Time-Series Analysis of Climate Data. *Climate* **2015**, *3*, 227–240. [CrossRef]

41. Glynn, C.C.; Konstantinou, K.I. Reduction of randomness in seismic noise as a short-term precursor to a volcanic eruption. *Nat. Sci. Rep.* **2016**, *6*, 37733. [CrossRef]

42. Search Earthquake Catalog, National Earthquake Hazards Reduction Program (NEHRP). 2018. Available online: https://earthquake.usgs.gov/earthquakes/search/ (accessed on 1 August 2018).

43. Zhang, Y.; Shang, P. Permutation entropy analysis of financial time series based on Hill's diversity number. *Commun. Nonlinear Sci. Numer. Simul.* **2017**, *53*, 288–298. [CrossRef]

44. Wharton Research Data Services (WRDS), 1993–2018. Available online: https://wrds-web.wharton.upenn.edu/wrds/ (accessed on 1 August 2018).

45. Zhou, R.; Cai, R.; Tong, G. Applications of Entropy in Finance: A Review. *Entropy* **2013**, *15*, 4909–4931. [CrossRef]

46. Goldberger, A.L.; Amaral, L.A.N.; Glass, L.; Hausdorff, J.M.; Ivanov, P.C.; Mark, R.G.; Mietus, J.E.; Moody, G.B.; Peng, C.K.; Stanley, H.E. PhysioBank, PhysioToolkit, and PhysioNet: Components of a New Research Resource for Complex Physiologic Signals. *Circulation* **2000**, *101*, 215–220. [CrossRef]

47. Moody, G.B.; Goldberger, A.L.; McClennen, S.; Swiryn, S. Predicting the Onset of Paroxysmal Atrial Fibrillation: The Computers in Cardiology Challenge 2001. *Comput. Cardiol.* **2001**, *28*, 113–116.

48. Aboy, M.; McNames, J.; Thong, T.; Tsunami, D.; Ellenby, M.S.; Goldstein, B. An automatic beat detection algorithm for pressure signals. *IEEE Trans. Biomed. Eng.* **2005**, *52*, 1662–1670. [CrossRef]

49. Andrzejak, R.G.; Lehnertz, K.; Mormann, F.; Rieke, C.; David, P.; Elger, C.E. Indications of nonlinear deterministic and finite-dimensional structures in time series of brain electrical activity: Dependence on recording region and brain state. *Phys. Rev. E* **2001**, *64*, 061907. [CrossRef]

50. Polat, K.; Güneş, S. Classification of epileptiform EEG using a hybrid system based on decision tree classifier and fast Fourier transform. *Appl. Math. Comput.* **2007**, *187*, 1017–1026. [CrossRef]

51. Subasi, A. EEG signal classification using wavelet feature extraction and a mixture of expert model. *Expert Syst. Appl.* **2007**, *32*, 1084–1093. [CrossRef]

52. İnan Güler.; Übeyli, E.D. Adaptive neuro-fuzzy inference system for classification of EEG signals using wavelet coefficients. *J. Neurosci. Methods* **2005**, *148*, 113–121.

53. Lu, Y.; Ma, Y.; Chen, C.; Wang, Y. Classification of single-channel EEG signals for epileptic seizures detection based on hybrid features. *Technol. Health Care* **2018**, *26*, 1–10. [CrossRef]

54. Cuesta-Frau, D.; Miró-Martínez, P.; Núñez, J.J.; Oltra-Crespo, S.; Picó, A.M. Noisy EEG signals classification based on entropy metrics. Performance assessment using first and second generation statistics. *Comput. Biol. Med.* **2017**, *87*, 141–151. [CrossRef]

55. Redelico, F.O.; Traversaro, F.; García, M.D.C.; Silva, W.; Rosso, O.A.; Risk, M. Classification of Normal and Pre-Ictal EEG Signals Using Permutation Entropies and a Generalized Linear Model as a Classifier. *Entropy* **2017**, *19*, 72. [CrossRef]

56. Fadlallah, B.; Chen, B.; Keil, A.; Príncipe, J. Weighted-permutation entropy: A complexity measure for time series incorporating amplitude information. *Phys. Rev. E* **2013**, *87*, 022911. [CrossRef]

57. Zunino, L.; Pérez, D.; Martín, M.; Garavaglia, M.; Plastino, A.; Rosso, O. Permutation entropy of fractional Brownian motion and fractional Gaussian noise. *Phys. Lett.* **2008**, *372*, 4768–4774. [CrossRef]

58. Zanin, M. Forbidden patterns in financial time series. *Chaos: Interdiscip. J. Nonlinear Sci.* **2008**, *18*, 013119. [CrossRef]

59. Vallejo, M.; Gallego, C.J.; Duque-Muñoz, L.; Delgado-Trejos, E. Neuromuscular disease detection by neural networks and fuzzy entropy on time-frequency analysis of electromyography signals. *Expert Syst.* **2018**, *35*, 1–10. [CrossRef]

60. Robnik-Šikonja, M.; Kononenko, I. Theoretical and Empirical Analysis of ReliefF and RReliefF. *Mach. Learn.* **2003**, *53*, 23–69. [CrossRef]

61. Kononenko, I.; Šimec, E.; Robnik-Šikonja, M. Overcoming the Myopia of Inductive Learning Algorithms with RELIEFF. *Appl. Intell.* **1997**, *7*, 39–55. [CrossRef]

62. Rodríguez-Sotelo, J.; Peluffo-Ordoñez, D.; Cuesta-Frau, D.; Castellanos-Domínguez, G. Unsupervised feature relevance analysis applied to improve ECG heartbeat clustering. *Comput. Methods Programs Biomed.* **2012**, *108*, 250–261. [CrossRef]

Article

Algorithmics, Possibilities and Limits of Ordinal Pattern Based Entropies

Albert B. Piek [1,2,*], **Inga Stolz** [3] and **Karsten Keller** [1]

[1] Institute of Mathematics, University of Lübeck, Lübeck D-23562, Germany; keller@math.uni-luebeck.de
[2] Graduate School for Computing in Medicine and Life Sciences, University of Lübeck, Lübeck D-23562, Germany
[3] Department of Mathematics, The University of Flensburg, Flensburg D-24943, Germany; inga.stolz@uni-flensburg.de
* Correspondence: piek@math.uni-luebeck.de; Tel.: +49-451-3101-6025; Fax: +49-451-3101-6004

Received: 15 May 2019; Accepted: 27 May 2019; Published: 29 May 2019

Abstract: The study of nonlinear and possibly chaotic time-dependent systems involves long-term data acquisition or high sample rates. The resulting big data is valuable in order to provide useful insights into long-term dynamics. However, efficient and robust algorithms are required that can analyze long time series without decomposing the data into smaller series. Here symbolic-based analysis techniques that regard the dependence of data points are of some special interest. Such techniques are often prone to capacity or, on the contrary, to undersampling problems if the chosen parameters are too large. In this paper we present and apply algorithms of the relatively new ordinal symbolic approach. These algorithms use overlapping information and binary number representation, whilst being fast in the sense of algorithmic complexity, and allow, to the best of our knowledge, larger parameters than comparable methods currently used. We exploit the achieved large parameter range to investigate the limits of entropy measures based on ordinal symbolics. Moreover, we discuss data simulations from this viewpoint.

Keywords: symbolic analysis; ordinal patterns; Permutation entropy; conditional entropy of ordinal patterns; Kolmogorov-Sinai entropy; algorithmic complexity

---

1. Introduction

Symbolic-based analysis techniques are efficient and robust research tools in the study of non-linear and possibly chaotic time-dependent systems. Initially, an experimental time series $x_0, x_1, \ldots, x_N$ with N in the natural numbers $\mathbb{N}$ is decoded into a sequence of symbols and, if needed, successive symbols are conflated into symbol words of length $m \in \mathbb{N}$. Subsequently, these symbol sequences or symbol word sequences are analyzed mainly by considering symbol (word) distributions or quantifiers based on the distributions, such as entropies. Anomalies in the symbol data or changes in the entropy can be used to detect temporal characteristics in the data. As a result of long-term data acquisition and high sample rates, the study of long-time series is becoming increasingly important in order to provide useful insights into long-term dynamics [1]. Efficient and robust algorithms are required that can analyze long time-series without decomposing the data into smaller series.

In this paper, we are especially interested in the relatively new ordinal symbolic approach which goes back to the innovative works of Bandt and Pompe [2] and Bandt et al. [3]. In the symbolization process, the ordinal approach regards the dependence of $d + 1$ equidistant data points resulting in ordinal patterns of order $d \in \mathbb{N}$. The ordinal approach is applied in many research areas to identify interesting temporal patterns that are hidden in the data. Application examples, techniques and further details are listed in the review papers of Kurths et al. [4], Daw et al. [5] and Zanin et al. [6]. In addition, see the contributions to the special topic "Recent Progress in Symbolic Dynamics and Permutation

Complexity. Ten Years of Permutation Entropy" of The European Physical Journal [7] and to the special issue "Symbolic Entropy Analysis and Its Applications" of Entropy [1]. The significant demand for literature on the ordinal idea is due to the fact that the method is easy to interpret and efficient to apply. However, large values of N, d and word lengths m lead to capacity or, on the contrary if N is reduced, undersampling problems. Here m is the number of successive ordinal patterns forming the words considered in the analysis.

This paper covers two main objectives: efficient algorithms for determining ordinal pattern (word) distributions and applicability of ordinal methods. The algorithms use overlapping information and binary representations of ordinal patterns and therefore realize an efficient determination of different entropy measures including and generalizing permutation entropies [2]. To the best of our knowledge, our algorithm not only allows larger parameters N, d and m than methods presently used but also compute a whole series of entropies based on ordinal words of different length and pattern order. A discussion of algorithmic complexity and runtime of the given algorithms in comparison to other ones is provided. Our algorithms are particularly useful for getting deeper insights into the general applicability of ordinal methods, which is the second objective of the paper. We discuss limits of estimating the complexity of finite time series and underlying systems on the base of ordinal patterns and words. There are different restrictions. First of all, if a system provides a large variety of ordinal patterns or ordinal pattern words with sufficiently large d or m, then a time series must be extremely long to represent the variety. This naturally bounds useful orders d and word lengths m (see e.g., [8,9] for further information). Even in the case that extremely long series and large d and m would be accessible, complexity estimation can be limited. Let us shortly explain reasons for that.

Central complexity measures for dynamical systems are related to the Kolmogorov-Sinai entropy being mathematically well-founded but being not easy from the computational and estimation viewpoint. There are different ways of approximating Kolmogorov-Sinai entropy on the base of ordinal pattern (word) distributions for sufficiently large d or m (see Bandt et al. [3] and Gutjahr and Keller [10] for ordinal pattern distributions and Keller et al. [11] for ordinal word distributions). Since there are systems with arbitrarily large, even infinite, Kolmogorov-Sinai entropy, there is no d or m which is large enough for reaching the Kolmogorov-Sinai entropies for all systems. Moreover, in case that data considered are directly an orbit of a time-discrete dynamical system (representing the system), calculation precision bounds the number of accessible ordinal patterns and words. The reason is that, independent from d and m, there are not more patterns and words than possible values of the system in the precision decided. The aspect of calculation precision is also interesting for data simulation including the problem of getting periodicities by rounding.

Many researchers use entropies based on ordinal pattern distributions as complexity measures themselves independent from the Kolmogorov-Sinai entropy, which is particularly justified by the problems mentioned and by the good performance of permutation entropy and similar measures [12,13]. For purely explorative data analysis as it is often given related to automatic learning, classification or comparison of data there is no a priori reason for avoiding large d and m.

The paper is organized as follows. We discuss the idea of ordinal patterns in Section 2, in particular, different pattern representations. Please note that the right choice of representation is substantial in developing our fast algorithms. Moreover, Section 2 provides the main entropy concept used in the paper. Section 3 is devoted to the efficient computation of ordinal patterns and ordinal pattern words from a time series. Here the special binary ordinal pattern representation mentioned above is described and used. Section 4 discusses different methods for establishing ordinal pattern and ordinal pattern word distributions. Furthermore, algorithmic complexity and runtime of our algorithms are investigated. Finally, Section 5 discusses the above specified limits of ordinal pattern-based entropies for complexity quantification in more detail. In this context, also aspects of data simulation are touched. In a certain sense, this paper is complementary to the recent paper [8] of Cuesta-Frau et. al. considering limits from the more practical viewpoint. Please consider the pseudocode descriptions of our algorithms given in the Appendix A and refer to MATLAB File Exchange [14] for realization in MATLAB 2018b.

2. Ordinal Pattern Representations and Empirical Permutation Entropy

Let $(\mathcal{X}, \preceq)$ be a totally ordered set. Mostly the set of real numbers $\mathbb{R}$ with the usual order $\leq$ is used in application. Nonetheless, our theory applies for the general case of ordinal data, for example on discrete-valued ratings or discrete rating scales as well. For our subsequent algorithmic analysis we demand that evaluation of the order relation $\preceq$ is not too complex and can be performed in constant time, i.e., independent from the compared elements.

We say that two vectors $x = (x_k)_{k=0}^d \in \mathcal{X}^{d+1}$ and $y = (y_k)_{k=0}^d \in \mathcal{X}^{d+1}$ share the same *ordinal pattern* of *order d* iff their components have the same order, i.e.,

$$x \sim_{\text{Ord}} y \Leftrightarrow \forall 0 \leq k, l \leq d : x_k \preceq x_l \Leftrightarrow y_k \preceq y_l .$$

The equivalence relation $\sim_{\text{Ord}}$ defines $(d+1)!$ equivalence classes on $\mathcal{X}^{d+1}$. Please note that the definition of $\sim_{\text{Ord}}$ and the following ideas can easily be extended to the case that x and y are from different ordered sets. This expands the usage of ordinal analysis to a broader range of applications, by other means incomparable properties can be compared by these methods.

We call a single-pair-comparison within x an *atomic* ordinal information. A pair (x_k, x_l) with $k < l$ is said to be *discordant* or an *inversion* if $x_k \succ x_l$ and *concordant* if $x_k \preceq x_l$. There are $d(d-1)/2$ of these elementary comparisons. Due to the transitive property of order relations, not all variations are possible to occur.

Please note that the set $\text{inv}(x) := \{(k,l) \in \{0,\ldots,d\} \mid k < l \wedge x_k \succ x_l\}$ of all inversions within x uniquely determines the regarding ordinal pattern.

2.1. Ordinal Representations

Each ordinal equivalence class is represented by an ordinal pattern. We describe the ordinal patterns formally as elements of a *pattern set* $\mathcal{P}_d := \mathcal{X}^{d+1}/\sim_{\text{Ord}}$. Concrete representations of the ordinal patterns have to be bijections from $\mathcal{P}_d$ to a suitable set. The choice of representation of the ordinal information within a pattern has an influence on various factors. These include computability, comparability and behavior under transformation as the pattern order d increases. The most common representations include the *permutation representation* and the *inversion vector representation*. They were considered from the beginning of Ordinal Pattern Analysis including Bandt [15] and Keller et al. [16]. Formally, they are given by the transformations

$$\text{P}: \mathcal{P}_d \to \mathcal{S}_{d+1} \qquad x \mapsto \pi = (\pi_k)_{k=0}^d, \qquad \pi_k = \#\{l \mid 0 \leq l \leq d \wedge (x_l \prec x_k \vee (x_l = x_k \wedge l < k))\},$$

$$\text{I}: \mathcal{P}_d \to \mathcal{I}_d \qquad x \mapsto i = (i_k)_{k=1}^d, \qquad i_k = \#\{l \mid 0 \leq l < k \wedge x_l \succ x_k\}$$

where $\mathcal{S}_{d+1}$ is the symmetric group on $\{0, 1, \ldots, d\}$, i.e., it contains all permutations of the numbers $0, 1, \ldots, d$ and where $\mathcal{I}_d = \times_{k=1}^d \{0, \ldots, k\}$. Please note that in the literature, the names of these representations vary; moreover some authors use the inverse permutation as the permutation representation instead.

Both transformations are based on order comparisons in different ways. The permutation representation assigns to each vector x the permutation which has the same order within its elements, i.e., for each element x_k of x and π_k of π the number of other elements in the respective vectors that are less or equal is the same. Therefore particular properties like the largest and smallest element can directly be determined from this representation. Moreover, the group structure on $\mathcal{S}_{d+1}$ with the permutation composition operation gives many possibilities to analyze the permutation vectors in algebraic terms (e.g., Amigó [17]). However, if a new element is appended to x the whole permutation vector has to be updated. This is a main disadvantage of the permutation representation that makes computation rather inefficient.

The inversion vector representation is chosen to avoid this problem. It assigns a vector i to each x where the k-th entry is the number of *inversions* of x_k with all previous elements in x. In other words, the underlying comparisons only refer to prior appended elements, therefore all preceding elements in i are not affected by the increase of d. This property allows a simple extension to patterns of infinite order d. Moreover, using the *factorial number system*, a special numeral system with mixed radix, the inversion vectors can easily be mapped to the numbers $0, \ldots, (d+1)! - 1$. The map is given by $\sum_{l=1}^{d} l! \cdot i_l$. The resulting number representation allows the most compact description of an ordinal pattern since it combines all ordinal information in a single integer.

For these representations as well as the representation presented in Section 3 one has to note the trade-off between accessibility of the atomic information and flexibility on the one hand and compactness of the representation on the other hand.

2.2. Ordinal Pattern Based Entropies

In the application of the ordinal approach, experimental data $x = (x_n)_{n=0}^{N} \in \mathcal{X}^{N+1}$ of length $N+1$ is decoded into a sequence of ordinal patterns and, if needed, successive patterns are conflated into pattern words of length $m \in \mathbb{N}$.

An important reason ordinal pattern-based measures are interesting for data analysis is that the ordinal pattern (word) sequences usually store much information about the systems dynamics if the data originates from an ergodic dynamical system. This is especially notable, if one focuses on quantifying the complexity of the underlying system. Complexity measures considered are mainly built up on the Shannon entropy of ordinal pattern (word) distributions. In addition, as already mentioned in the introduction, these measures are related to the Kolmogorov-Sinai entropy which is a central complexity measure for dynamical systems. In the following, we give precise definitions of these basic entropy concepts.

Let us regard $d \in \mathbb{N}$ as fixed and let p be some ordinal pattern in $\mathcal{P}_d$. To determine the absolute frequency h_p of how often p occurs in x, we identify the ordinal structures within each data segment $x^{(n,d)} := (x_k)_{k=n}^{n+d}, n = 0, \ldots, N-d$. The patterns associated with the segments are denoted by $p^{(n,d)}$. Then,

$$h_p := \# \left\{ n \in \{0, \ldots, N-d\} \mid p^{(n,d)} = p \right\}.$$

Let w be a word of $m \in \mathbb{N}$ successive patterns of order d and let $\mathcal{W}_{d,m}$ be the set of all possible ordinal words of length m and order d. We call the parameter pair (d, m) a *word configuration*. The absolute frequency h_w of some ordinal word $w \in \mathcal{W}_{d,m}$, is defined by

$$h_w = \# \left\{ n \in \{0, \ldots, N-d-m+1\} \mid w^{(n,d,m)} = w \right\}$$

where

$$w^{(n,d,m)} := \left(p^{(n+k,d)} \right)_{k=0}^{m-1}, n = 0, \ldots, N-d-m+1.$$

Please note that we write $p^{(n)}$ and $w^{(n,m)}$ when d is known from the context.

All entropy measures we consider in this paper, are based on quantities $H^{(d,m)}; d, m \in \mathbb{N}$. The $H^{(d,m)}$ are called *Empirical Shannon entropy of order d and word length m*, and defined by

$$H^{(d,m)} := - \sum_{w \in \mathcal{W}_{d,m}} \frac{h_w}{N-d-m+2} \ln \frac{h_w}{N-d-m+2}$$

$$= \ln(N-d-m+2) - \frac{1}{N-d-m+2} \sum_{w \in \mathcal{W}_{d,m}} h_w \ln h_w. \tag{1}$$

They quantify the complexity of ordinal patterns of order d conflated into pattern words of length m in a given time series. Please note that $\frac{h_w}{N-d-m+2}$ is the relative frequency of such words. To compute $H^{(d,m)}$; $d, m \in \mathbb{N}$ one has to determinate the pattern (word) distributions, which is our main subject of the next section.

3. A Pattern Representation for Efficient Computation

In this section, we describe how all desired ordinal patterns in x can be computed in an efficient and fast way. We will do so by taking advantage of the redundancy in terms of atomic ordinal information between the patterns. We introduce a pattern representation that allows us to do the following tasks in a simple and fast manner when $p^{(n,d)}$ is given:

- Computation of the successive pattern $p^{(n+1,d)}$,
- Computation of patterns $p^{(n,d-m)}$, $m = 1, \ldots, d-1$ of smaller order,
- Computation of the ordinal words $w^{(n,d-m,m+1)}$, $m = 1, \ldots, d-1$.

Furthermore, our representation is chosen so that pattern and word frequencies can be determined fast.

3.1. Computation of Successive Patterns

Example: Consider the first six data points of a time series as shown in Figure 1 and the ordinal order $d = 4$. Then, we have two ordinal patterns: the red colored pattern $p^{(0)}$ determined by the points $x^{(0)} = (x_k)_{k=0}^4$ and the blue colored pattern $p^{(1)}$ determined by $x^{(1)} = (x_k)_{k=1}^5$. Please note that both patterns are maximally overlapping and share the data points $x_1, \ldots, x_4$.

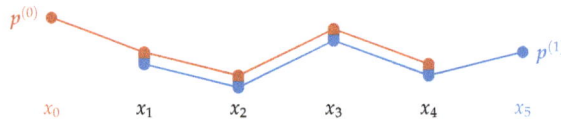

Figure 1. Successive ordinal patterns of order $d = 4$.

To take optimal advantage of this shared ordinal information, we subdivide the task of computing $p^{(1)}$ – when $p^{(0)}$ is given – into three parts:

Tasklist 1.

(i)   *remove atomic information regarding x_0,*
(ii)   *adapt the atomic information regarding x_1 to x_4,*
(iii) *add atomic information regarding x_5.*

Generally, these successive steps can be realized as follows: Consider the set $\text{inv}(x)$ of inversions within $x \in \mathcal{X}^{d+1}$ and the lower triangular matrix $M(x) \in \{0,1\}^{d \times d}$

$$M(x) := \begin{pmatrix} \mathbf{1}_{x_0 \succ x_1} & & 0 \\ \vdots & \ddots & \\ \mathbf{1}_{x_0 \succ x_d} & \cdots & \mathbf{1}_{x_{d-1} \succ x_d} \end{pmatrix},$$

where $\mathbf{1}$ is the indicator function. It takes the value of one if the statement in the subscript is true. Otherwise, the value is zero. For instance, the schematic illustrations in Figure 2 show the overlapping atomic information between successive matrix representations in general and for our previous example.

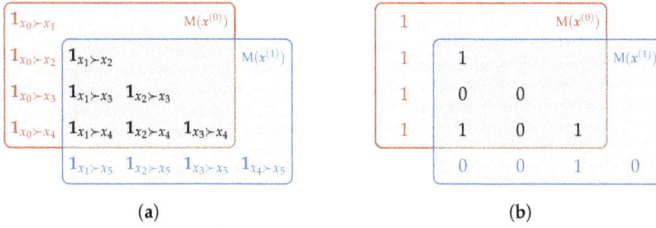

(a)   (b)

Figure 2. Schematic representation of the overlap between successive matrix representations $M(x^{(0)})$ and $M(x^{(1)})$ in general (**a**) and for our example (**b**) (compare to Figure 1).

We see that $M(x^{(n+1,d)})$ is obtained from $M(x^{(n,d)})$ in terms of Tasklist 1 by executing the following steps:

Tasklist 2.

(i)   *removing the first row and column of* $M(x^{(n,d)})$,

(ii)   *keeping the rest of* $M(x^{(n,d)})$,

(iii)   *appending a new row and column at the end with the atomic information* $1_{x_{n+k} \succ x_{n+d+1}}$ *for all* $k = 1, \ldots, d$.

While the matrix representation M is suitable for computing successive patterns, it lacks of compactness and is difficult to compare. We define a more compact representation which nevertheless directly contains all atomic information – the *binary vector representation*:

$$B(x) := M(x) \cdot (2^0 \quad \cdots \quad 2^d)^{\mathrm{T}} = \left(\sum_{0 \leq l < k} 1_{x_l \succ x_k} 2^l \right)_{k=1}^{d} \in \bigtimes_{k=1}^{d} \{0, 1, \ldots, 2^k - 1\}. \tag{2}$$

By this definition, we encode each row of M as a binary number where the least significant bit stands in the first column of M. Please note that the binary vector representation is closely related to inversion vectors by

$$I(x) = M(x) \cdot (1 \quad \cdots \quad 1)^{\mathrm{T}}.$$

By counting the ones in their binary representations, the elements of the binary vector can be transformed to inversion vector elements.

The key benefit of the binary system is that all operations on binary vectors can be performed as bit operations and can thereby be implemented in an efficient and fast way, especially in low-level programming languages. Here, we use *left* and *right bit shifts* in order to manipulate B and to use overlapping atomic information. By a right bit shift, the least significant bit gets dismissed. This corresponds to a division by 2 rounded down. Analogously, the left bit shift corresponds to a multiplication by 2 or, in other words, a concatenation of B with an additional 0 as the least significant bit. A bit shift on $n \in \mathbb{N}$ by i bits we denote by $n \gg i$ (resp. $n \ll i$) for a right (resp. left) bit shift. Denote that here $\ll$ is a mathematical operator and should not be confused with the "much smaller than" symbol. For example, consider the number 11 which is 1011 in binary representation. Then $1011_2 \gg 1 = 101_2$ which is 5 in decimal. Please note that bit shifts are monotonous functions.

The introduced representation simplifies Tasklist 2 to

Tasklist 3.

(i)   *perform a right bit shift on* $B(x^{(n,d)})$ *and remove the first entry,*

(ii)   *keep the rest of the vector,*

(iii)   *compute* $\sum_{k=1}^{d} 1_{x_{n+k} \succ x_{n+d+1}} 2^{k-1}$ *and append it to the end of the vector.*

The underlying relationship between consecutive binary vectors used in Tasklist 3(i) reads as follows: Let $b^{(n,d)} = (b_k^{(n,d)})_{k=1}^d = B(x^{(n,d)})$. Then

$$b_k^{(n,d)} = b_{k+1}^{(n-1,d)} \gg 1. \tag{3}$$

For the patterns of the example introduced in this section (see Figure 1), we have

$$B(x^{(0)}) = (1 \quad 3 \quad 1 \quad 11)^{\mathsf{T}} \text{ and}$$
$$B(x^{(1)}) = (3 \gg 1 \quad 1 \gg 1 \quad 11 \gg 1 \quad 4)^{\mathsf{T}} = (1 \quad 0 \quad 5 \quad 4)^{\mathsf{T}}.$$

3.2. Computation of Patterns of Smaller Order

As with inversion vectors, both, the matrix and binary vector representation, share the possibility to append new elements without changing the other elements. In particular, when the pattern $p^{(n,d)}$ is already computed in terms of M, the patterns of lower order $(p^{(n,d-k)})_{n=0}^{N-d}$ can be obtained by taking the $(d-k)$th leading principle submatrices of M as depicted in Figure 3a. In terms of B, the vector's last k elements have to be removed to obtain the pattern of order $d-k$.

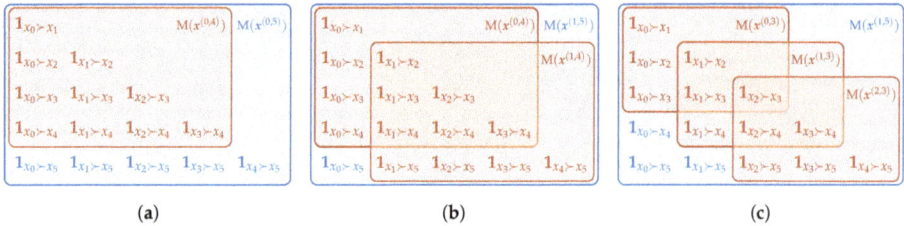

(a)   (b)   (c)

Figure 3. (a) Relationship between patterns of different order. (b) Difference between a word of length 2 and order 4 and the pattern of order 5. (c) Difference between a word of length 3 and order 3 and the pattern of order 5.

3.3. Computation of Ordinal Words

Ordinal words of length m combine the ordinal information of m successive patterns. Again, the scheme of atomic information shows the difference between one pattern of order d and two patterns of order $d-1$ which form a word of length 2. In the example seen in Figure 3b, the order between x_0 and x_5 is taken into account in $w^{(0,5,1)} = p^{(0,5)}$ but not in $w^{(0,4,2)} = (p^{(0,4)}, p^{(1,4)})$. Therefore, the corresponding bit $\mathbf{1}_{x_0 \succ x_5}$ in $w^{(0,5,1)}$ can be set to zero in pursuance of uniting all patterns that are the same except for $\mathbf{1}_{x_0 \succ x_5}$. The result corresponds with $w^{(0,4,2)}$. In the next step, $\mathbf{1}_{x_0 \succ x_4}$ and $\mathbf{1}_{x_1 \succ x_5}$ are set to zero to deduce $w^{(0,3,3)}$ (see Figure 3c). For general $k \in \{1, \ldots, d\}$ we have

$$w_k^{(n,d-m,1+m)} = \begin{cases} w_k^{(n,d,1)} & \text{for } k \leq d-m, \\ (w_k^{(n,d,1)} \gg k - (d-m)) \ll k - (d-m) & \text{for } k > d-m. \end{cases} \tag{4}$$

In direct consequence of the results in Section 3.2, for ordinal words we have

$$w_k^{(n,d,m+1)} = \begin{cases} w_k^{(n,d,1)} & \text{for } 1 \leq k \leq d, \\ \sum_{l=1}^d \mathbf{1}_{x_{k-l} \succ x_k} 2^{k-l} & \text{for } d+1 \leq k \leq d+m+1. \end{cases} \tag{5}$$

Since ordinal words can be treated as special ordinal patterns in our representation, we will often refer to both as patterns.

4. Implementation

In this part, we describe our algorithm that computes a whole series of Shannon entropies based on ordinal words of different word length and pattern order. Specifically, for a given *maximal order* $D \in \mathbb{N}$, the algorithm computes all Shannon entropies $H^{(d,m)}$ as defined in (1), where $d + m \leq D + 1$ with $1 \leq d, m \leq D$. When arranged in an array by pattern order d and word length m, this corresponds to all entries above, left, and on the main antidiagonal. These are the entropies regarding all ordinal words which contain information on at most D consecutive data points from x.

In Section 4.1, we will discuss different approaches and our choice. The following Section 4.2 is concerned with the algorithm's structure. Section 4.3 is dedicated to an analysis of the complexity of our algorithm. In the end we apply our algorithm on randomly generated artificial data and analyze its speed in Section 4.4.

4.1. Discussion of Methods

There are several possible approaches for computing the entropy of a given data vector, which differ in terms of time and memory consumption dependent on the maximal order D, data length N. For these parameters, we have usually N significantly larger than D but smaller than $D!$.

To compute the empirical entropy, we need the frequencies of the occurring patterns. Here, we will discuss different ways to compute them. We demand the methods to be adaptable to the transformations (4) and (5) in a sense that the data structure should not be recomputed completely after transforming the patterns.

A first method is to iterate through all patterns and increment a frequency counter when the respective pattern occurs. It has the advantage that there is no need to store all patterns at once; instead, each pattern can be computed separately. A naïve approach is to prepare an array, which length is the number of all possible patterns, and increment its corresponding entries when a certain pattern occurs. Since the number of possible patterns increases factorially in the pattern order D, this method is only practical for small orders. Overall, the array would be very sparse: on the one hand, the data length N is far smaller than $D!$ and thus the number of nonzero entries is relatively small. On the other hand, it was shown by Amigó [18] that the amount of forbidden patterns (i.e., patterns that cannot occur due to the inner structure of the underlying dynamics) also increases superexponentially, leading to further zero entries.

To avoid these zero entries, only the occurring patterns could be counted. This generally requires hashing for the key-value pairing that connects each pattern to its frequency. As an again simple and collision-free example, the hash can be chosen such that the patterns are indexed by their first occurrence. The disadvantage of this method is clearly that the indexing is not gained from information about the pattern itself; to achieve the hash value, it is necessary to search through all hashed patterns to find out whether the current pattern is new or did already occur. Other hash functions can be more effective, but also more complicated and not collision-free. However, it is not immediately clear how to construct them. In addition, the pattern transformations (4) and (5) would make it necessary to recompute all hashes in each step.

Since the problem is basically a searching problem, where the current pattern has to be found in the data structure for the frequencies, more advanced data structures such as search trees are promising approaches. Possible structures are B trees, AVL trees, where searching an element happens in $\mathcal{O}(D \cdot \log N)$ time. An extensive discussion can be found e.g., in standard text books such as the book by Cormen et al. [19]. However, search trees need an underlying total preorder on the set of ordinal patterns. Though all considered representations can be equipped with a total preorder, comparisons will typically have $\mathcal{O}(d)$ time complexity. This would lead to $\mathcal{O}(D \cdot N \cdot \log N)$ time complexity for each entropy, or $\mathcal{O}(D^3 \cdot N \cdot \log N)$ for all desired entropies.

For our computations, we have chosen a different approach where all patterns are given (or computed) beforehand with a total time complexity of $\mathcal{O}(D^2 \cdot N \cdot \log N)$. The choice is based on our idea to compute entropies for different pattern orders and word lengths by manipulating the representation of the patterns. We explain this idea in the following subsections in detail. Specifically by considering all patterns at once our method allows computing the entropies $(H^{(d,m)})_{m=1}^{D+1-d}$ in one run, which reduces the complexity magnitude of D by one. Though our methods result in the same asymptotic complexity for a single entropy, our experiments show that our approach is faster due to the possibility to vectorize the operations.

4.2. Structure of the Algorithm

We present the design of our algorithm in different ways. First we give a short overview of its structure with schematic illustrations. In the following four parts of the section, we explain key ideas of the algorithm in detail. A pseudocode implementation is provided in Algorithm A1 in the Appendix A. An implementation for MATLAB 2018b is provided by the authors on MATLAB File Exchange [14].

In Figure 4, the algorithm is depicted as a workflow diagram. The basic idea is to precompute all the words $w^{(\cdot,D,1)}$ and use them to compute all other word configurations by (4) and (5), i.e., $w^{(\cdot,d,m)}$ with $d \leq D$ and $m \leq D - d + 1$. For each word configuration we then compute the respective entropy. We start with $w^{(\cdot,D,1)} = p^{(\cdot,D)}$ since this is the word which contains all atomic information about the $D + 1$ considered points. As shown in the preceding section, other word configurations can be obtained by successively removing the atomic information through bit manipulations.

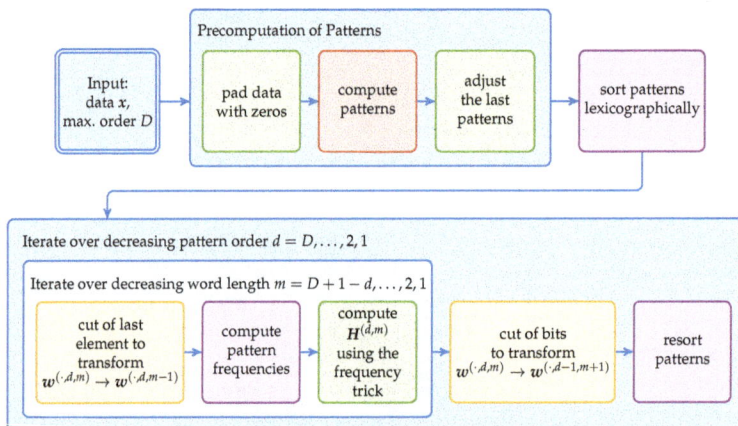

Figure 4. Schematic overview ofr our algorithm. The colors of the steps refer to different aspects of the algorithm discussed in the following subsections: The initial pattern computation (**red**), the processing of these patterns to other word configurations (**yellow**), the calculation of the pattern frequencies (**purple**) and the modifications to include all lower order patterns (**green**).

4.2.1. Initial Pattern Computation

The initial computation of the $p^{(\cdot,D)}$ is shown schematically in Figure 5. The corresponding pseudocode implementation can be found in Algorithm A2 in the Appendix A. It follows the idea of Tasklist 3, but operates in a different sequence. Here, at first the last entry $p_D^{(\cdot,D)}$ is computed for all patterns. The penultimate elements $p_{D-1}^{(\cdot,D)}$ are computed by performing a right bit shift on these last entries (using $k = D - 1$ in (3)). Sequentially all prior elements $p_{D-2}^{(\cdot,D)}, \ldots, p_2^{(\cdot,D)}, p_1^{(\cdot,D)}$ get computed in this way. In comparison to computing all patterns consecutively, the computation steps of this methods are highly parallelizable and vectorizable and thereby faster to perform.

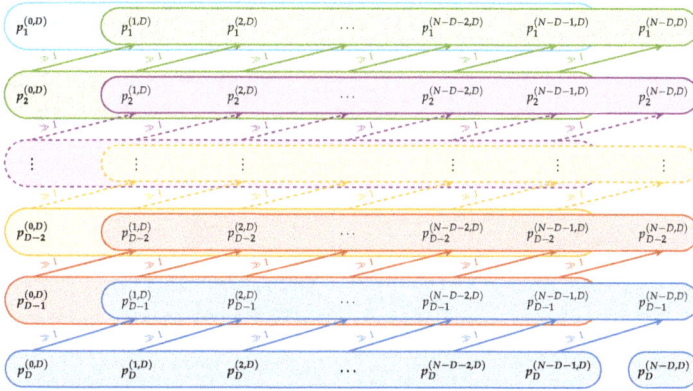

Figure 5. Computation scheme for the ordinal patterns, where each column corresponds to a binary vector. Each color shows one iteration step. The **bold** elements are computed directly while all other elements are derived from them by bit shifts. These are indicated by the arrows.

4.2.2. Obtaining other Word Configurations

The way our main algorithm processes through the different word configurations after computing the $p^{(\cdot, D)} = w^{(\cdot, D, 1)}$ is illustrated in Figure 6. In the main iteration using (4) the pattern order gets successively decreased while the word length gets increased. For a fixed pattern order d and word length m, all words with length smaller than m can be computed by simply cutting of last elements of the binary vectors as (5) indicates.

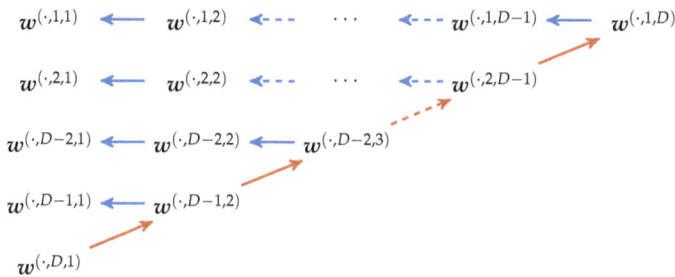

Figure 6. Computation of the different words in the algorithm. **Red** arrows denote applications of (4), while **blue** arrows mean usage of (5).

4.2.3. Computation of the Pattern Frequencies

With each list of words, we compute the word frequencies. In short, the frequencies are determined by sorting the patterns and by computing the length of blocks of the same pattern. Thus, as the first step, all words are sorted lexicographically. Next, we compute the elementwise differences

$$w^{(n,d,m)} - w^{(n-1,d,m)} = \left(w_1^{(n,d,m)} - w_1^{(n-1,d,m)} \quad w_2^{(n,d,m)} - w_2^{(n-1,d,m)} \quad \cdots \quad w_d^{(n,d,m)} - w_d^{(n-1,d,m)} \right)^{\mathrm{T}}$$

between successive patterns. Since the patterns are sorted, the difference vectors are zero vectors for identical patterns and have positive entries for different patterns. The signs of the cumulated differences within the patterns, given by

$$\left(\operatorname{sgn} \left(\sum_{l=1}^{k} \left(w_l^{(n,d,m)} - w_l^{(n-1,d,m)} \right) \right) \right)_{k=1}^{d}$$

indicate a pattern change for all word lengths $1, \dots, d$. The cumulation reflects the property that—due to the lexicographic order—changes in prior entries imply different patterns for all higher pattern lengths. As the positive signs indicate a change of patterns, the pattern frequencies for the words of length m can be determined by taking the difference between the indices of patterns that have positive signs in the m-th element. With the computed absolute frequencies h_p, the corresponding entropy can be computed with (1). Refer to Algorithm A3 in the Appendix A for a pseudocode implementation.

4.2.4. Inclusion of Missing Lower Order Patterns—The Frequency Trick

Not all ordinal words can be computed by transforming the $p^{(\cdot,D)}$ into other word configurations. This is due to the fact that applying (4) and (5) does not change the total number of patterns although the number of patterns contained in the data increases for other configurations. With a total data length $N + 1$, we get $N + 1 - D$ patterns $(p^{(n,D)})_{n=0}^{N-D}$ in the beginning. For arbitrary pattern order d and word length m, we get $N - d - m + 2$ words $(w^{(n,d,m)})_{n=0}^{N-d-m+1}$. Since $d + m - 1 \leq D$ holds by assumption, we have $N - d - m + 1 \geq N - D$. Thus, the last $D - d - m + 1$ patterns cannot be derived from the $p^{(\cdot,D)}$. We illustrate the missing patterns in Figure 7.

The procedure would become complicated if all the missing different sized patterns are computed and stored separately. Instead, we desire to compute all patterns by our methods. We achieve this by expanding the data vector with artificial data such that now the missing patterns of the enlarged data vector contain only artificial data and can be neglected. The word $w^{(N-1,1,1)}$ is the last occurring ordinal word in x of all configurations; it contains the ordinal information between the last points x_{N-1} and x_N. By going backwards through the computation process, it can be seen that $w^{(N-1,1,1)}$ is obtained from $w^{(N-1,D,1)}$. This pattern is determined by the points $x^{(N-1)} = (x_k)_{k=N-1}^{N+D}$ which are yet undefined for $N + 1 \leq k \leq N + D$. Therefore it suffices to pad these $N + D - (N + 1) = D - 1$ elements to x. With the padded data, we get N patterns in total, which is sufficient for each word configuration.

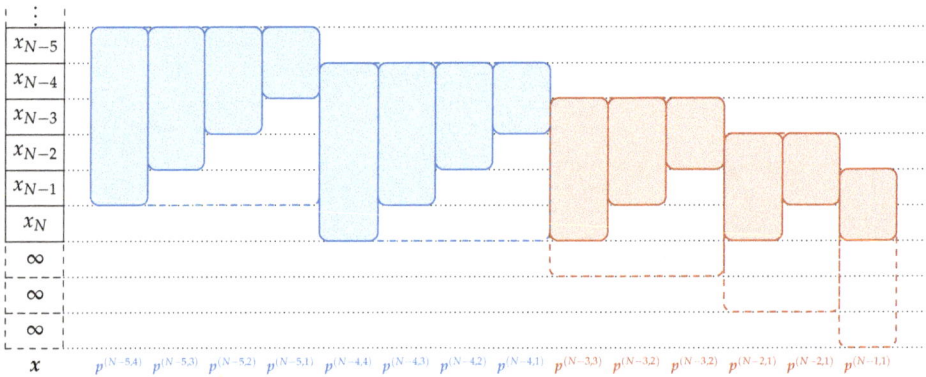

Figure 7. The patterns which are not covered by $p^{(\cdot,D)}$ for $D = 4$ and $m = 1$. The **blue** patterns are the two last patterns of $p^{(\cdot,D)}$ and the derived patterns of smaller order. The **red** patterns are the missing patterns. They can be derived from the dashed patterns constructed by padding $D - 1 = 3$ elements with value ∞ at the end of x.

For convenience, we define an element $\infty \notin \mathcal{X}$ to be the unique value such that $\infty \succ y$ for all $y \in \mathcal{X}$. By padding x with ∞, each atomic information about artificial elements as well as the regarding bit in the binary vector representation is always set to zero.

When the patterns are computed based on the padded data, each required word can be deduced from the N initially computed patterns. As stated above, we have to consider only the first $N - d - m + 2$ words $(w^{(n,d,m)})_{n=0}^{N-d-m+1}$ for the entropies. The last $d - m + 2$ words $(w^{(n,d,m)})_{n=N-d-m+2}^{N-1}$ partly contain ordinal information regarding the artificial data and have to be excluded from the entropy computations. Nonetheless, the latter are still needed for deducing words with other configurations and, thus, they cannot simply be deleted.

To prevent these words from distorting the word frequencies without removing them, we use a property of the entropy formula that we call the *frequency trick*. When the entropy is written in terms of absolute frequencies as in (1), patterns of absolute frequency 1 have no effect to the summation due to $\ln 1 = 0$. We take advantage of this by adjusting the binary vectors such that

$$h_{w^{(n,d,m)}} = 1 \text{ for } N - d - m + 2 \leq n \leq N - 1. \tag{6}$$

Thus, we can keep these patterns in our computing procedure. We can use the entropy Formula (1), although the number of pattern $\sum_{w \in W_{d,m}} h_w = N - 1$ is higher than the expected number of patterns $N - d - m + 2$.

We can make all words deduced from the initial patterns satisfy (6) only by adjusting the additional patterns $(p^{(n,D)})_{n=N-D+1}^{N-1}$. This can be done by replacing the artificial zero entries as follows:

$$p^{(N-D+1)} = \left(p_1^{(N-D+1)} \quad p_2^{(N-D+1)} \quad p_3^{(N-D+1)} \quad \cdots \quad p_{d-2}^{(N-D+1)} \quad p_{d-1}^{(N-D+1)} \quad 2^D \right),$$

$$p^{(N-D+2)} = \left(p_1^{(N-D+1)} \quad p_2^{(N-D+2)} \quad p_3^{(N-D+2)} \quad \cdots \quad p_{d-2}^{(N-D+2)} \quad 2^{D-1} \quad 2^D \right),$$

$$\vdots$$

$$p^{(N-2)} = \left(p_1^{(N-2)} \quad p_2^{(N-2)} \quad 2^3 \quad \cdots \quad 2^{D-2} \quad 2^{D-1} \quad 2^D \right),$$

$$p^{(N-1)} = \left(p_1^{(N-1)} \quad 2^2 \quad 2^3 \quad \cdots \quad 2^{D-2} \quad 2^{D-1} \quad 2^D \right).$$

The values are chosen in a way that they are larger than the greatest possible value for each entry. Indeed, in the binary vector representation the k-th entry is an integer based on k bits, which reaches its maximum with all bits being ones, which is $\sum_{0 \leq l < k} 2^l = 2^k - 1$. This choice makes it possible to identify any of the additional patterns independent from the data. Thus, each of the additional patterns is fully unique.

On top, each word that is deduced from one of the additional patterns is unique, as long as it contains ordinal information about artificial data points. Recall that the entries with artificial information has values larger than all entries with real information at the same index. Since bit shifts are monotonously increasing functions, applying (4) to each entry keeps the values of the artificial entries to be larger. Furthermore, by (5), only the last elements are removed. After applying both transformations, $p^{(N-D+1)}$ contains no more artificial ordinal information. This is consistent with the increase of the number of patterns when m decreases. The other patterns $p^{(N-D+2)}, \ldots, p^{(N-1)}$ stay unique. In summary, both transformations keep the uniqueness of the patterns as long as artificial information remains and (6) is satisfied.

4.3. Complexity Analysis

In the following, we analyze our algorithms both theoretically and practically. First we give an analysis in terms of Landau Big-$\mathcal{O}$ notation dependent on the parameters N and D. It is summarized in Figure 8. Take into account that in general it is not trivial to extend Big-$\mathcal{O}$ notation to the multidimensional case (c. f. [20]). Nonetheless, since $D \ll N$ holds, the asymptotics of our algorithm is mainly determined by N. Therefore, the one dimensional definitions of Big-$\mathcal{O}$ notation apply. Further note that in terms of complexity, the base of considered logarithms has no influence. Therefore we denote the logarithm with the generic log in this section.

We analyze the algorithm on the individual parts as divided in Figure 4. In terms of time complexity, the initial computation of the patterns discussed in Section 4.2.1 needs $N - D + 1$ comparisons for the last pattern row. For each of the $D - 2$ following rows, one additional comparison and $N - D - 1$ bit shifts are performed, which leads to $DN - D^2 + 2D - 2N + 2$ bit shifts and $N - 1$

comparisons. Assuming both operations can be performed in constant time, we get a time complexity of $\mathcal{O}(D^2 + D \cdot N)$. Since usually N is significantly larger that D, it can be simplified to $\mathcal{O}(D \cdot N)$. In comparison, the naïve approach of finding all ordinal patterns in the binary vector representation by performing all $\binom{D}{2}$ comparisons for each pattern has a worse total time complexity of $\mathcal{O}(D^2 \cdot N)$.

For computing the other representations presented in Section 2 patternwise, there exist $\mathcal{O}(D \log D \cdot N)$ algorithms; the permutation representation can be obtained by sorting the data, for the inversion vector representation Knuth [21] gave an $\mathcal{O}(D \log D)$ conversion from permutations. It is known that $D \log D$ is a lower boundary for sorting algorithms based on comparisons [19]. Thus in terms of complexity, these algorithms are optimal for these representations when computing each pattern by itself. Adapting information from previous patterns allowed us to achieve the improvement in terms of linear time complexity in D. In similar ways, there are ways to obtain linear time complexity for permutations and inversion vectors. The latter was shown by Keller et al. [16], their idea can be adapted for the former straightforward.

Precomputation of Patterns
| Padding | D | $\Big\}D \cdot N$
| Pattern Computation | $D \cdot N$
| Pattern Adjustments | D^2
Pattern Sorting | | $D \cdot N \cdot \log N$
Iteration over d
| Iteration over m
| | Transform $w^{(\cdot,d,m)} \to w^{(\cdot,d,m-1)}$ | $-$ | $\Big\}D \cdot N$ | $\Big\}D^2 \cdot N \log N$
| | Compute Frequencies | N
| | Compute Entropies | N
| Transform $w^{(\cdot,d,m)} \to w^{(\cdot,d-1,m+1)}$ | $D \cdot N$
| Resort Patterns | $D \cdot N \log N$

Figure 8. Complexity Analysis of our Algorithm, where D describes the maximal pattern order and N describes the length of the data vector.

The padding of the artificial data can be realized in $\mathcal{O}(D)$ insertions which amortized needs $\mathcal{O}(D)$ time. the data is stored in a dynamic array. The adjustment of the last patterns concerns $\frac{1}{2}D(D-1) \in \mathcal{O}(D^2)$ elements. In total, the precomputation of the patterns happens in $\mathcal{O}(D \cdot N)$ time.

In the next step, the patterns get sorted. For this, typical sorting algorithms can be used. Pattern comparisons need D comparisons of their elements in the worst case. Hence the sorting needs $\mathcal{O}(D \cdot N \log N)$. Please note that in practice, there are less comparisons needed. The probabilities of the first k elements being equal is quickly decreasing in k; in particular, it is given by $\sum_{w \in \mathcal{W}_{d,m}} \Pr(w)^2$, where $\Pr$ denotes the (usually) unknown probability of the word w.

The inner loop takes $\mathcal{O}(N)$ time: The transformation takes no additional time since it can be reached by simply ignoring the respective last elements. To attain the changes between the sorted patterns and thereby the frequencies, an iteration through all patterns is needed; for the entropy, all the $\mathcal{O}(N)$ pattern frequencies must be processed. All iterations need $\mathcal{O}(D \cdot N)$ time altogether. Each of the subsequent transformations of the patterns consists of D bit shifts. Together with the resorting of the transformed patterns, all $\mathcal{O}(D)$ iterations take $\mathcal{O}(D^2 \cdot N \log N)$ time in total.

In summary, our algorithm has a time complexity of $\mathcal{O}(D^2 \cdot N \log N)$ for $\frac{1}{2}D(D+1)$ entropies. In principle, for a single entropy each of the loops has to be performed only once, leading to the complexity of $\mathcal{O}(D \cdot N \log N)$.

4.4. Runtime Analysis

In this section, we test our algorithm on artificial data in order to evaluate its performance and compare it to other available programs. The artificial data is generated by a simple random number generator which gives uniformly distributed random numbers on $[0, 1]$. The data is chosen to be uniformly distributed because under this assumption also the ordinal patterns are uniformly distributed. This corresponds to the Shannon entropy taking its maximum. In this sense, the case of uniformly distributed data can be considered as the worst case scenario from the computational viewpoint. In contemplation of investigating the independence of the results from the distribution, we additionally performed our algorithms on normally distributed data. Our computations were performed on MATLAB 2018b on a Windows 10 computer equipped with an Intel Core i7-3770 CPU and 16 GB RAM. We repeated the computations fifty times in order to exclude external distortions and took the mean value of the results.

4.4.1. Computational Time for Pattern and Entropy Computation

First, we restrict our analysis to the pattern computation. We compare our binary vector approach with implementations which determines the inversion and permutation vector representations. For all three representations, we compare the naïve, patternwise approach with the successive methods introduced in Section 3.

As shown in the prior section, the computational time depends on the data length N and the maximal pattern order D. In our implementation in MATLAB, both parameters are limited due to the following reasons. In a binary vector of order D, its elements are integers with up to D bits. Since arithmetic computations are needed when the bit shifts are performed, D is physically limited to 64 when using unsigned long integers. In practice, it is even limited to 51 since internally, all arithmetic calculations are performed on floating-point numbers in MATLAB. From the 64 bits only 52 bits are reserved for the mantissa from which one bit is needed for the frequency trick. Recall that the maximal possible precision in Matlab is given by approximately $2.2204 \cdot 10^{-16}$.

The data length is in our implementation limited by the machine's memory. In our case, a 16 GB RAM leads to approximately $4 \cdot 10^9$ array elements. Therefore, to store all N patterns of order D, $N \approx 10^8$ data points are possible to process on our machine.

Clearly, for MATLAB or other programming languages there are several possibilities for memory management and high-precision integer arithmetics (c. f. the vpi toolbox [22]), which are offered either by the maintainer or by third-party programs. Anyhow, we restricted ourselves to basic methods of MATLAB.

We tested the runtime of our algorithm for varying N and D on randomly generated data vectors by usage of MATLABs `tic` and `toc` commands, which measure the time between both function calls. The results can be seen in Figure 9a–d. We compare uniformly distributed Figure 9a,b and normal distributed Figure 9c,d data. Both distributions lead to similar results. For both, the asymptotic tendency found in the previous section is observable in the graphs: All implementations have linear tendency in N. The successive implementations are also linear in D, whereas a superlinear behavior can be examined on the naïve approaches. For each representation, the successive implementation is significantly higher than its naïve counterpart. In addition, the successive binary vector approach is the fastest in total. On average, it is 5.38 times faster than the successive algorithm for the inversion vectors and 5.89 times faster than the respective permutation vector algorithm.

In addition to the runtime tests for the patterns, we tested the time consumption of the whole program. Again, the theoretical considerations of Section 4.3 are confirmed. For increasing data length N, the runtime follows an $N \cdot \log N$ curve. We assume that the deviation for small N is due to the strictly linear time complexity of pattern computation. The proportion of computing time for the patterns tends to 10%. For varying order D, the quadratic regression we did on our results for the parameter D gives an almost perfect fit (compare Figure 9f). With increasing maximal pattern order D, the proportion of computing time for the pattern calculation relative to the total time decreases considerably. Most of the computation time is spent on the frequency computation and the sorting of the patterns. Therefore, this part seems to be a promising starting point for future improvements.

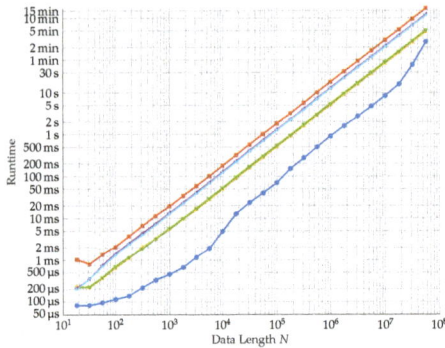

(a) Runtime of pattern computation for fixed $D = 10$ and uniformly distributed random data with changing length N.

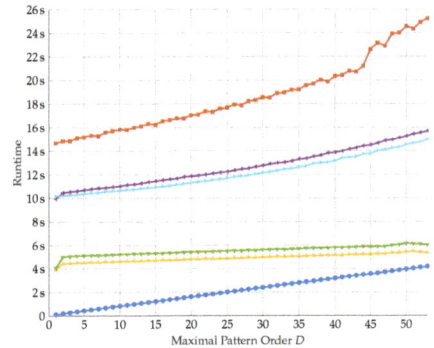

(b) Runtime of pattern computation for changing D and uniformly distributed random data with fixed length $N = 10^6$.

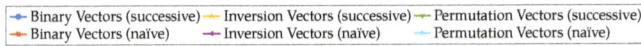

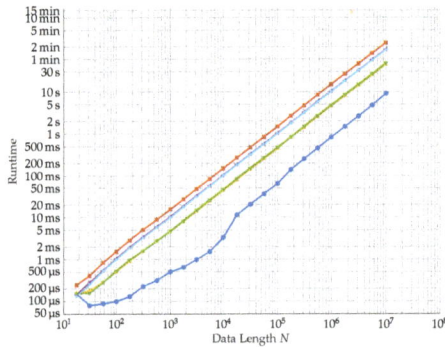

Legend: — Binary Vectors (successive) — Inversion Vectors (successive) — Permutation Vectors (successive) — Binary Vectors (naïve) — Inversion Vectors (naïve) — Permutation Vectors (naïve)

(c) Runtime of pattern computation for fixed $D = 10$ and normal distributed random data with changing data length N.

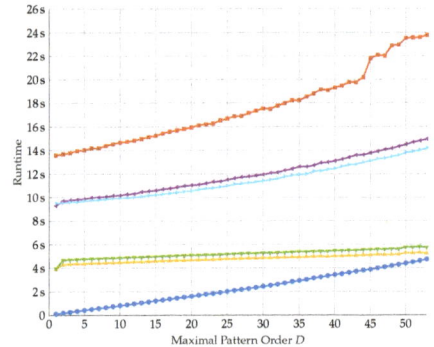

(d) Runtime of pattern computation for changing D and normal distributed random data with fixed data length $N = 10^6$.

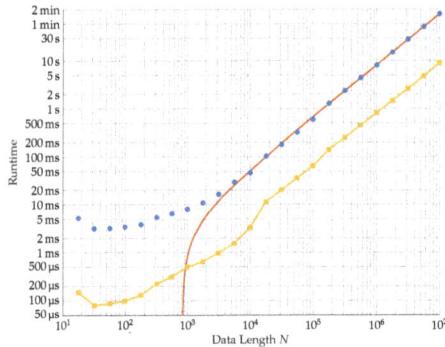

(e) Runtime of entropy computation for fixed $D = 10$ and uniformly distributed random data with changing data length N (**blue** dots). For comparison, the **yellow** curve marks the time for the pattern computation. The **red** line is a nonlinear fit with the fitting model $a \cdot N \ln N + b$ and $a = -3.3 \cdot 10^{-3}$ and $b = 5.8615 \cdot 10^{-7}$; $MSC = 3.925 \cdot 10^{-3}$.

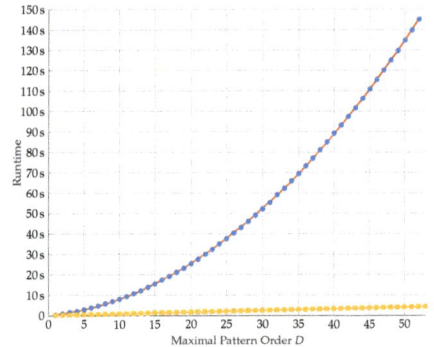

(f) Runtime of entropy computation for changing D and uniformly distributed random data with fixed data length $N = 10^6$ (**blue** dots). For comparison, the **yellow** curve marks the time for the pattern computation. The **red** line is a quadratic fit with the fitting model $a \cdot D^2 + b \cdot D + c$ and $a = 0.04789$, $b = 0.2983 \cdot 10^{-7}$ and $c = 0.1667$; $SSE = 1.4487$, $R^2 = 1$.

Figure 9. Runtime analysis of the entropy and pattern computation dependent on the data length N and the maximal pattern order D. The resulting runtimes are averaged over 50 trials.

4.4.2. Comparison with Other Implementations

We compare our algorithm to the implementations by V. Unakafova (see *PE.m*, [23], detailed explanations in the accompanying literature [24]), G. Ouyang (*pec.m*, [25]) and A. Müller (*petropy.m* [26], description in [9]) in the context in which the programs are comparable. None of the other implementations can compute entropies for ordinal words. The program *pec.m* computes a single entropy for a given pattern order and supports time delays. *petropy.m* again gives only one entropy, but allows multiple time delays. In addition it offers several ways to treat equal values in the data. The implementation in *PE.m* computes the permutation entropy for sliding windows over the data set, giving the entropy for each window.

Therefore we restrict our comparisons to the case of simple ordinal patterns ($m = 1$). Though our algorithms can be extended easily to support time delays, we will not use time delays in any implementation. At last the sliding window in *PE.m* was chosen such that the number of windows equals one.

For further comparability, we divide the runtime for our algorithm by the number of computed entropies $\frac{1}{2}D \cdot (D + 1)$ to get an approximation for the mean time needed for a single entropy. In addition, we tested a modified version of our algorithm that computes the entropy for a single word configuration.

Figure 10 shows the results of our analysis. Again, we analyzed the runtime in dependence of the data length N on the left and pattern order D on the right.

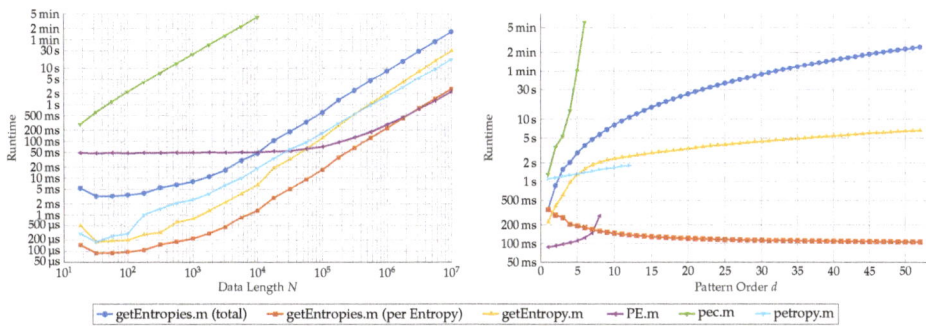

(**a**) Runtime comparison of different implementations for permutation entropy with fixed $D = 8$ and uniformly distributed random data with changing data length N. Measurements of *pec.m* were only performed up to $N = 10^4$ due to the long runtimes.

(**b**) Runtime comparison of different implementations for permutation entropy with changing D and uniformly distributed random data with fixed data length $N = 10^6$. Again, the runtime of *pec.m* was only measured for $D < 8$ because of its bad performance. *PE.m* is only implemented for $D < 9$; *petropy.m* works only if $D \cdot \log_{10} D > 15$ which means $D < 14$.

Figure 10. Comparision of the performance of different MATLAB scripts for permutation entropy computation. The **blue** curve is the total runtime of our approach while the **red** curve gives the runtime per entropy. The modified version for a single entropy is marked **yellow**. Both are compared with the approaches from V. Unakafova (**purple**), G. Ouyang (**green**) and A. Müller (**cyan**).

In *pec.m*, the first method from our discussion in Section 4.1 was chosen. This led to very long computing times even for small N and D, making it by far the slowest approach (the green line in Figure 10).

The performance of *PE.m* (purple line) varies. It is clearly the fastest of the compared approaches for small pattern orders. Though, for $d = 8$, bigger data sets are needed to compete with the theoretical average time per entropy of our method. Responsible for the high computation speed is mainly the usage of lookup tables to determine succeeding patterns. While this method clearly reduces computation time, these tables have to be available and increase superexponentially in size, which limits its usage to pattern orders of size 8 or smaller. With increasing size of the lookup table, its speed advantages decrease.

The implementation of *petropy.m* (cyan line) has a similar performance to our modified one-entropy-programm, mainly because the program uses a similar approach to ours. It uses the same method of frequency determination. However, it uses the factorial number representation for the ordinal patterns. This choice leads to the upper limit of the pattern order since these numbers can be as high as $D!$. With an maximal possible integer value of 2^{53} in MATLAB, this results in $D = 13$ as the largest possible pattern order, a slightly higher range for D than *PE.m*.

In direct comparison, our approach that computes all $\frac{D\cdot(D+1)}{2}$ entropies takes (apart of the by far slowest *pec.m*) the most time for computing. On the other hand, the average time per entropy is on wide parameter ranges the fastest. Only for small d, the approach of V. Unakova is faster. The modified version competes well with the other approaches. Nevertheless it is slower than the average computing time per entropy since the structures used for optimizing the computation of multiple entropies cannot be exploited.

Clearly, the main advantage of our algorithm lies in the extended parameter range for the pattern order d. If over the top the behavior of the entropy in dependence of d is from particular interest, our approach saves much time compared to computing alls the entropies serially.

5. Limits of Ordinal Pattern Based Entropies

As already mentioned in the introduction, most complexity measures for dynamical systems are strongly linked to the Kolmogorov-Sinai entropy (abbrev KSE). In our following discussion of ordinal pattern-based entropies in that context, we forgo to give detailed definitions of this concept. Instead, we refer for details and more background of the following to [13]. Furthermore we restrict the following considerations to the most simple case of a one-dimensional dynamical system. Please note that with the appropriate generalizations, stochastic processes could be included in the discussion withal.

By a (measurable) *dynamical system* $(\mathcal{X}, \mathcal{A}, T, \mu)$ we understand a probability space $(\mathcal{X}, \mathcal{A}, \mu)$ equipped with a map $T : \mathcal{X} \hookleftarrow$ being measurable with respect to $\mathcal{A}$, where μ is invariant with respect to T. The latter means that $\mu(T^{-1}(A)) = \mu(A)$ for all events $A \in \mathcal{A}$ and describes stationarity of the system. X is considered as the *state space* and T provides the dynamics. A dynamical system as given produces a time series $x = (x_n)_{n=0}^N$, when $x_0 \in \mathcal{X}$ is given and x_n is the n-th iterate of x_0 with respect to T for $n = 1, 2, \ldots, N$. If the probability measure is ergodic, as we assume in the following, statistical properties of the given system can be assessed from such time series for sufficiently large N. In particular, probabilities of patterns can be estimated by their relative frequencies within the time series. With this setting there are several ways for estimating the Kolmogorov-Sinai entropy from ordinal pattern-based entropies given a time series from the system.

In theory, the following quantifiers had shown to be good estimators of the Kolmogorov-Sinai entropy for sufficiently large N, d and m:

$$\frac{1}{d}H^{(d,1)} \text{ if } T \text{ is an interval map with countably many monotone pieces,} \tag{7}$$

$$\frac{1}{m}H^{(d,m)} \text{ and } H^{(d,m+1)} - H^{(d,m)} \text{ generally.} \tag{8}$$

Please note that $\frac{1}{d}H^{(d,1)}$ is called the *empirical Permutation entropy of order d* (abbrev. ePE) and $H^{(d,2)} - H^{(d,1)}$ the *Empirical Conditional entropy of Ordinal Patterns of order d* (abbrev. eCE). Both are estimators of the respective entropies of the dynamical system which base on the time series x and whose precise definitions are specified in the given literature.

For (7) the statement follows directly from a recent result of Gutjahr and Keller [10] generalizing the celebrated result of Bandt et al. [3] that for piecewise monotonous interval maps the KSE and the Permutation entropy are coinciding. As reported in [12,13], the latter entropy seems to be a good estimator of the KSE for sufficiently large d with some plausible reasoning, but this point is not completely understood.

The problem in the statement above is what sufficiently large means. First of all, in the case of existence of many different ordinal patterns in a system, N must be extremely large in order to realize them all in a time series (for practical recommendations relating N and d, see e.g., [8,9]). Even if arbitrarily long time series would be possible, there would be serious problems to reach the KSE. Let us demonstrate this for the ePE (compare (7)).

For $d \in \mathbb{N}$ it holds

$$\tfrac{1}{d} H^{(d,1)} \leq \frac{\ln((d+1)!)}{d} = \frac{\sum_{k=1}^{d+1} \ln k}{d} \tag{9}$$

since maximal Shannon entropy is given for the uniform distribution. The right side of this formula is very slowly increasing. For example, for $d = 100$ it is less than 4, saying that for maps with KSE larger than 4 one needs ordinal patterns of order larger than 140 to be able to get a sufficiently good estimation. The larger the KSE is, the larger the d for its reliable estimation theoretically must be, and the KSE can be arbitrarily large. One reason for this is that if a map T has some KSE c, then its n-th iterate $T^{\circ n}$ has KSE $n \cdot c$ (see e.g., [27]).

Formula (9) is also interesting from the simulation viewpoint. Given an (one-dimensional) dynamical system and some precision, then usually simulations stick to the precision: Given a value in its precision, the further iterates in their precision are determined. This shows that the precision bounds the number of possible ordinal patterns independent of d.

For example, consider the state space $[0,1]$ and the usage of double-precision IEEE 754 floating-point numbers, which is the standard data type for numerical data in the most programming languages. Recall that these numbers have the form $s \cdot m \cdot 2^e$, where s, m and e are coded within 64 bits of information. One bit is reserved for the sign s and 11 bits are used for the exponent e. The remaining 52 bits are left for the fraction value (mantissa) m. The exponent generally ranges from -1022 to 1023, for our particular state space, only negative exponents are needed. For each choice for the exponent, there are 2^{52} possible values of the mantissa. Therefore, there are in total $1022 \cdot 2^{52} \approx 4.6 \cdot 10^{18}$ distinct numbers possible in a simulation, and more distinct ordinal patterns cannot be found. Since already for $d = 20$ there are $21! \approx 5.1 \cdot 10^{19} > 4.6 \cdot 10^{18}$ possible ordinal patterns, simulations do not provide an ePE larger than $\frac{\ln 21!}{d} \leq \frac{\ln 21!}{20} \leq 2.269$. Because there are one-dimensional maps with arbitrary KSE, this limits the estimation of the KSE by the simulation.

The kind of simulation described, which we refer to as the naïve simulation, does not reproduce the real-world situation. The measuring process itself goes hand in hand with precision loss and therefore measuring errors occur. By iterating through these erroneous values, the error cumulates. In consequence due to the chaotic behavior, the values obtained after a few iterations have nothing left in common with the real values. For example, when identifying a real-world system with a dynamical model, the simulation can in consequence produce periodic sequences of values, although the sequence of measured values does not contain periodicities. This can reduce the number of ordinal patterns and so corresponding entropies.

In real-world systems, the measuring process and its belonging errors have no influence to the system. There—under assumption of other sources of noise—the system generates its time-dependent values of arbitrary accuracy exactly. Only in the moment where we want to know a certain iterate, we measure it with the given limited precision. Therefore, we desire to have a simulation where the iteration itself can be performed exactly for starting values of unlimited precision.

Here for demonstration purposes we are especially interested in the system $([0,1], \mathcal{B}, L, \mu)$, where $\mathcal{B}$ are the Borel sets on $[0,1]$, L is the logistic map defined by $L(x) = 4x(1-x)$ for $x \in [0,1]$ and μ is the probability measure on $\mathcal{B}$ with a density p on $[0,1]$ given by $p(x) = \frac{1}{\pi \sqrt{x(1-x)}}$ for $x \in]0,1[$.

To mimic the real situation, we generate a sequence $x'_0, x'_1, x'_2, \dots$ by using other systems that are topologically (semi-)conjugated to the logistic map. These systems show to be suitable to our simulation requirements.

In the first step, we take advantage of the well-known fact that the system $([0,1], \mathcal{B}, L, \mu)$ is semi-conjugated to the angle doubling map on $[0, 2\pi]$ which is given by $A(\beta) = 2\beta$ mod 2π equipped with the Lebesgue measure (compare [28] for this and the following statements). The semi-conjugacy is given by the map ψ with

$$\psi(\beta) = \sin^2(\beta)$$

for $\beta \in [0, 2\pi]$. This means that

$$L(\psi(\beta)) = \psi(A(\beta)) \tag{10}$$

for all $\beta \in [0, 2\pi]$, i.e., applying L on $[0, 1]$ corresponds to doubling the arc length of a segment on the unit cycle and thereby doubling the angle. Furthermore, μ is the image of λ, i.e., $\mu(B) = \lambda(\phi^{-1}(B))$ for all $B \in \mathcal{B}$. The relationship is illustrated in Figure 11.

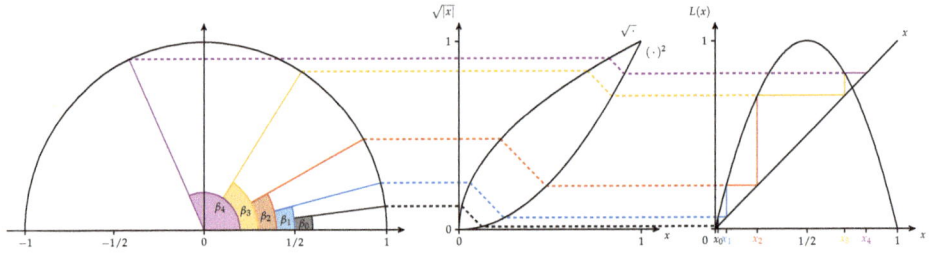

Figure 11. Semi-Conjugacy between the orbits of the logistic map L and the angle-doubling function.

As the second step, we use another conjugacy, namely that the angle doubling map is conjugated to the shift function on all infinite 0-1-sequences, i.e., $\{0, 1\}^\infty$ equipped with the $(\frac{1}{2}, \frac{1}{2})$-Bernoulli measure ν. The latter assigns to each cylinder set of size n of $\{0, 1\}^\infty$ the measure 2^{-n}. The shift is defined as

$$S : \{0, 1\}^\infty \hookleftarrow , (b_1, b_2, b_3, \dots) \mapsto (b_2, b_3, b_4, \dots).$$

The conjugation between a 0-1-sequence $(b_j)_{j\in\mathbb{N}}$ and an angle $\beta \in [0, 2\pi]$ is given by the binary expansion

$$\beta = \phi((b_j)_{j\in\mathbb{N}}) = 2\pi \sum_{j=1}^{\infty} b_j 2^{-j}.$$

Please note that ϕ is only almost everywhere bijective, for example $\phi((1, 0, 0, \dots)) = \phi((0, 1, 1, \dots)) = \pi$. Analogously to (10), it follows that

$$A(\phi((b_j)_{j\in\mathbb{N}})) = \phi(S((b_j)_{j\in\mathbb{N}})) \tag{11}$$

for all $(b_j)_{j\in\mathbb{N}} \in \{0, 1\}^\infty$. Concluding (10) and (11), the following diagram commutes:

$$
\begin{array}{ccc}
(\{0,1\}^\infty, \nu) \xrightarrow{\phi} & ([0,1], \lambda) \xrightarrow{\psi} & ([0,1], \mu) \\
\downarrow{\scriptstyle S} & \downarrow{\scriptstyle A} & \downarrow{\scriptstyle L} \\
(\{0,1\}^\infty, \nu) \xrightarrow{\phi} & ([0,1], \lambda) \xrightarrow{\psi} & ([0,1], \mu)
\end{array}
$$

We can generate an orbit $x_0, x_1, x_2, \dots$ for L as follows: Take a random sequence $b_1, b_2, b_3, \dots$, of elements in $\{0, 1\}$ meaning that each b_n is taken from the symbols $\{0, 1\}$ with equal probability and in an independent way and let

$$x_n = \psi(\phi(S^{\circ n}((b_j)_{j\in\mathbb{N}}))) = \sin^2\left(2\pi \sum_{j=1}^{\infty} b_{j+n} 2^{-j}\right).$$

Let $p \in \mathbb{N}$ be the number of binary digits defined by a given precision. The approximation up to precision p given by

$$x'_n = \sin^2\left(2\pi \sum_{j=1}^{p} b_{j+n} 2^{-j}\right).$$

provides a sequence $x'_0, x'_1, x'_2, \ldots$ as desired. The pleasant point is that in the simulation we can start from $b_1, b_2, \ldots, b_p$ and than can append b_{p+1} by random choice and delete b_1, then add b_{p+2} and delete b_2, then add b_{p+3} and delete b_3, etc. We refer to this simulation (see Algorithm 1) as the advanced one.

Algorithm 1: LOGISTIC-MAP-SIMULATION

    Input: number of iterations n, precision p.
    Output: array L with simulated values of the logistic map.
    begin
        choose start value $\alpha \sim U(0,1)$ randomly ;   // $\beta = 2\pi \cdot \alpha$
        $\alpha_p \longleftarrow \lfloor 2^p \cdot \alpha \rfloor \cdot 2^{-p}$;   // round to precision
        $L[0] \longleftarrow \sin^2(2\pi \cdot \alpha_p)$;
        for $k = 1, \ldots, n$ **do**
            $b \sim B(1, \frac{1}{2})$ chosen randomly ;
            $\alpha_p \longleftarrow (2\alpha_p \bmod 1) + b \cdot 2^{-p}$;   // add new precision digit
            $L[k] \longleftarrow \sin^2(2\pi \cdot \alpha_p)$;
        end
    end

We have determined the ePE and the eCE for time series related to the logistic map L and its iterates $L^{\circ 2} = L \circ L, L^{\circ 3} = L \circ L \circ L$ in dependence on the order d. The results based on time series of length 10^7 are presented by Figure 12. The time series behind the pictures on the left side ((a) and (c)) was obtained by the naïve simulation and on the right side ((b) and (d)) by the advanced simulation, both under the double precision described above. In (a) and (b), we have added **blue** colored curves showing the upper bound of the ePE (9) for d. Please note that the results were obtained by our algorithm from Section 4. It allowed to compute all values at once for each curve.

First of all, for L and $L^{\circ 2}$ (**red** and **yellow** colored curves) with KSE $\ln 2 \approx 0.693147$ and $2 \ln 2 \approx 1.38629$, respectively, the values of eCE restore the KSE much better than the ePE. For a very good performance of the eCE, consider d in the interval between 6 and 16 for L and between 6 and 9 for $L^{\circ 2}$. For d left of the intervals, the ordinal patterns seem to be too short to capture the full complexity, for d right of the intervals the time series seem to be to short relative to d to capture the ordinal pattern complexity. As already reported in [12,13] for L, convergence of the ePE seems to be too slow to get good estimations for moderately long time series.

It is not excluded that for L or $L^{\circ 2}$ the naïve algorithm sometimes runs in a loop or, more general, reduces variety of ordinal patterns. For $L^{\circ 3}$, we see such behavior in Figure 12 (cf. the **purple** curves), namely larger empirical entropies are reached for the advanced simulation than for the naïve one. It is remarkable that the eCE for $d = 8$ is near to the KSE $3 \ln 2 \approx 2.07944$, it seems however that the set of too small and the set of too large d as described above are overlapping, such that the KSE is not reached completely.

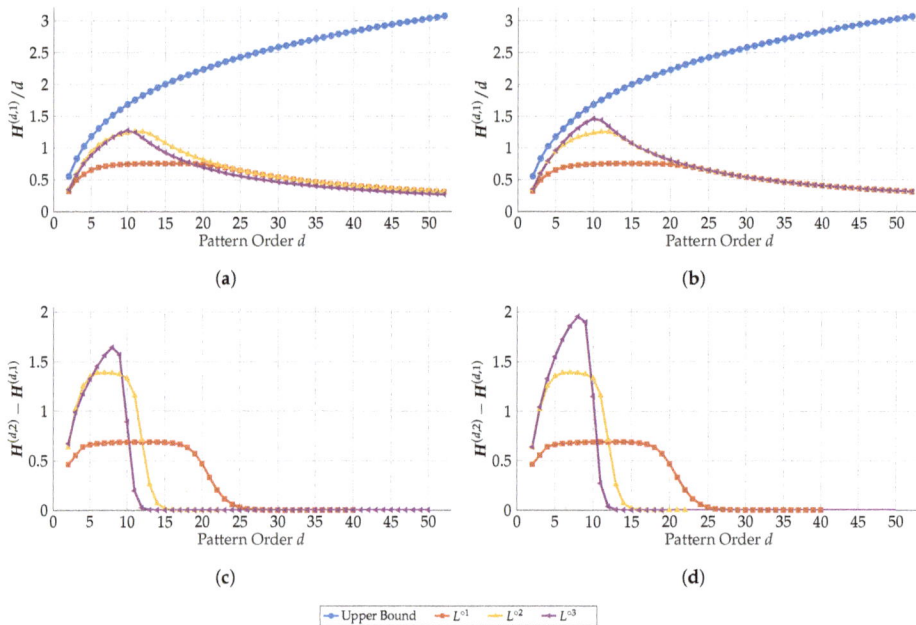

Figure 12. ePE (**a,b**) and empirical conditional entropy of ordinal patterns (**c,d**) for the logistic map $L = L^{\circ 1}$ and its iterates $L^{\circ 2}, L^{\circ 3}$ in dependence on the pattern order d: Left side pictures (**a,c**) show results based on the naïve simulation and right side pictures (**b,d**) on the advanced simulation, each based on $N = 10^7$ iterates. For comparison, in (**a,b**) the upper bound of the ePE (9) for d was added (**blue** curves).

Our exemplary considerations underline that some deeper thinking about the (mathematical) structure of systems and measurements, possibly also on the base of symbolic dynamics, is important for simulating data and beyond. To develop powerful and reliable tools for data analysis, moreover, a better understanding of ordinal pattern-based entropies for complexity quantification and their relationship to other complexity measures remains a permanent challenge.

Author Contributions: A.B.P. wrote the main parts of the paper and the complete software, and prepared all figures. I.S. substantially contributed to Sections 1 and 2.2 and K.K. to Sections 2.2 and 5. Moreover, K.K. supervised writing the paper. All authors have read and approved the final manuscript.

Funding: This research received no external funding.

Acknowledgments: The first author thanks Max Bannach for fruitful discussions about complexity and data structures as well as for insights from a theoretical computer scientists point of view.

Conflicts of Interest: The authors declare no conflict of interest.

Appendix A. Algorithms

Algorithm A1: GET-ENTROPIES

Input: Data array $x \in \mathcal{X}^{N+1}$, pattern order D.

Output: Array H of Shannon entropies.

begin

    `;/* Infinity padding for lower order patterns */`

    $x[N+1:N+D-1] \longleftarrow \infty$;

    $p \longleftarrow$ COMPUTE-BINARY-PATTERNS$[x, D]$;

    `;/* adjustments for the frequency trick */`

    for $k = 1, \ldots, D-2$ **do**

        for $l = D+1-k, \ldots, D$ **do**

            $p[N-D+k, l] \longleftarrow 2^l$;

        end

    end

    sort the rows of p in lexicografical order ;

    for $d = D, D-1, \ldots, 1$ **do**   `// iterate over pattern orders`

        for $m = D+1-d, D-d, \ldots, 1$ **do**   `// iterate over word lengths`

            $h \longleftarrow$ GET-PATTERN-FREQUENCIES$[p, d, m]$;

            $H[d, m] \longleftarrow \ln(N-d-m+2) - \frac{1}{N-d-m+2} h^{\mathrm{T}} \ln[h]$;

        end

        `;/* Transform the patterns to lower order words by Equation (4) */`

        for $n = 0, \ldots, N-2$ **do**

            for $k = 1, \ldots, D-d+1$ **do**

                $p[n, d+k-1] \longleftarrow (p[n, d+k-1] \gg k) \ll k$

            end

        end

        resort the rows of p in lexicografical order ;

    end

end

Algorithm A2: COMPUTE-BINARY-PATTERNS

Input: Data array $x \in \mathcal{X}^{N+1}$, pattern order D.

Output: Array of patterns $p \in \mathbb{N}^{N+1-D \times D}$.

begin

    $N \longleftarrow length[x] - 1$;

    for $n = 0, \ldots, N-D$ **do**

        $p[n, D] = \sum_{l=0}^{D-1} \mathbf{1}_{x_{n+l} \succ x_{n+D}} 2^l$;

    end

    for $k = D-1, \ldots, 0$ **do**

        $p[0, k] \longleftarrow \sum_{l=0}^{k-1} \mathbf{1}_{x_l \succ x_k} 2^l$;

        $p[1 : N-D, k] \longleftarrow p[0 : N-D-1, k+1] \gg 1$;

    end

end

Algorithm A3: GET-PATTERN-FREQUENCIES

Input: Array of patterns $p \in \mathbb{N}^{N+1-D \times D}$, pattern order d, word length m.

Output: Array of frequencies h.

begin

    $i \longleftarrow 0$;

    for $n = 0, \ldots, N - D$ **do**

       if $\|p[n+1, 0 : D-1] - p[n, 0 : D-1]\| > 0$ **then**

          $i \longleftarrow i + 1$;

       end

       $h[i] \longleftarrow h[i] + 1$;

    end

end

References

1. Alcaraz Martínez, R. Symbolic Entropy Analysis and Its Applications. *Entropy* **2018**, *20*, 568. [CrossRef]
2. Bandt, C.; Pompe, B. Permutation entropy—A natural complexity measure for time series. *Phys. Rev. E* **2002**, *88*, 174102. [CrossRef]
3. Bandt, C.; Keller, G.; Pompe, B. Entropy of interval maps via permutations. *Nonlinearity* **2002**, *15*, 1595–1602. [CrossRef]
4. Kurths, J.; Schwarz, U.; Witt, A.; Krampe, R.T.; Abel, M. Measures of complexity in signal analysis. *AIP Conf. Proc.* **1996**, *375*, 33–54.
5. Daw, C.S.; Finney, C.E.A.; Tracy, E.R. A review of symbolic analysis of experimental data. *Rev. Sci. Instrum.* **2003**, *74*, 915–930. [CrossRef]
6. Zanin, M.; Zunino, L.; Rosso, O.A.; Papo, D. Permutation entropy and its main biomedical and econophysics applications: A review. *Entropy* **2012**, *14*, 1553–1577. [CrossRef]
7. Amigó, J.M.; Keller, K.; Kurths, J. Recent progress in symbolic dynamics and permutation complexity. Ten years of permutation entropy. *Eur. Phys. J. Spec. Top.* **2013**, *222*, 241–598. [CrossRef]
8. Cuesta-Frau, D.; Murillo-Escobar, J.P.; Orrego, D.A.; Delgado-Trejos, E. Embedded Dimension and Time Series Length. Practical Influence on Permutation Entropy and Its Applications. *Entropy* **2019**, *21*, 385. [CrossRef]
9. Riedl, M.; Müller, A.; Wessel, N. Practical considerations of permutation entropy. *Eur. Phys. J. Spec. Top.* **2013**, *222*, 249–262. [CrossRef]
10. Gutjahr, T.; Keller, K. Equality of Kolmogorov-Sinai and permutation entropy for one-dimensional maps consisting of countably many monotone parts. *Discret. Contin. Dyn. Syst. A* **2019**, *39*, 4207–4224. [CrossRef]
11. Keller, K.; Maksymenko, S.; Stolz, I. Entropy determination based on the ordinal structure of a dynamical system. *Discrete Contin. Dyn. Syst. B* **2015**, *20*, 3507–3524. [CrossRef]
12. Unakafov, A.; Keller, K. Conditional entropy of ordinal patterns. *Phys. D* **2014**, *269*, 94–102. [CrossRef]
13. Keller, K.; Mangold, T.; Stolz, I.; Werner, J. Permutation Entropy: New Ideas and Challenges. *Entropy* **2017**, *19*, 134. [CrossRef]
14. Piek, A.B. Fast Ordinal Pattern and Permutation Entropy Computation. MATLAB Central File Exchange. Available online: https://www.mathworks.com/matlabcentral/fileexchange/71305-fast-ordinal-pattern-and-permutation-entropy-computation (accessed on 15 May 2019).
15. Bandt, C. Ordinal time series analysis. *Ecol. Model.* **2005**, *182*, 229–238. [CrossRef]
16. Keller, K.; Sinn, M.; Emonds, J. Time Series from the Ordinal Viewpoint. *Stochast. Dyn.* **2007**, *7*, 247–272. [CrossRef]
17. Amigó, J.M.; Monetti, R.; Aschenbrenner, T.; Bunk, W. Transcripts: An algebraic approach to coupled time series. *Chaos* **2012**, *22*, 013105. [CrossRef] [PubMed]
18. Amigó, J.M. *Permutation Complexity in Dynamical Systems. Ordinal Patterns, Permutation Entropy and all that*; Springer Series in Synergetics; Springer: Dordrecht, The Netherlands, 2010; ISBN 978-3-642-04083-2.
19. Cormen, T.H.; Leiserson, C.E.; Rivest, R.L.; Stein, C. *Introduction to Algorithms*, 3rd ed.; MIT Press: Cambridge, MA, USA, 2009; pp. 191–194, 304, 333, 484–504; ISBN 978-0-262-03384-8.

20. Howell, R.R. *On Asymptotic Notation with Multiple Variables*; Technical Report; Dept. of Computing and Information Sciences, Kansas State University: Manhattan, KS, USA, 2008. Available online: http://people. cs.ksu.edu/~rhowell/asymptotic.pdf (accessed on 13 March 2019).

21. Knuth, D.E. *The Art of Computer Programming Volume 3: Sorting and Searching*; Addison Wesley Longman Publishing Co., Inc.: Redwood City, CA, USA, 1998; ISBN 0-201-89685-0.

22. D'Errico, J. Variable Precision Integer Arithmetic. MATLAB Central File Exchange. Available online: https://www.mathworks.com/matlabcentral/fileexchange/22725-variable-precision-integer-arithmetic (accessed on 15 May 2019).

23. Unakafova, V. Fast Permutation Entropy. MATLAB Central File Exchange. Available online: https:// www.mathworks.com/matlabcentral/fileexchange/44161-permutation-entropy--fast-algorithm (accessed on 15 May 2019).

24. Unakafova, V.; Keller, K. Efficiently Measuring Complexity on the Basis of Real-World Data. *Entropy* **2013**, *15*, 4392–4415. [CrossRef]

25. Ouyang, G. Permutation Entropy. MATLAB Central File Exchange. Available online: https://www. mathworks.com/matlabcentral/fileexchange/37289-permutation-entropy (accessed on 15 May 2019).

26. Müller, A. PETROPY—Permutation Entropy. MATLAB Central File Exchange. Available online: http://tocsy.pik-potsdam.de/petropy.php (accessed on 15 May 2019).

27. Walters, P. *An Introduction to Ergodic Theory*; Springer: New York, NY, USA, 2000; pp. 91–92, ISBN 978-0-387-95152-2.

28. Choe, G.H. *Computational Ergodic Theory*; Springer: Berlin/Heidelberg, Germany, 2005; pp. 62–64, ISBN 978-3-540-23121-9.

entropy

MDPI

Article

Identification of Auditory Object-Specific Attention from Single-Trial Electroencephalogram Signals via Entropy Measures and Machine Learning

Yun Lu, Mingjiang Wang *, Qiquan Zhang and Yufei Han

Key Laboratory of Shenzhen Internet of Things Terminal Technology, Harbin Institute of Technology Shenzhen Graduate School, Shenzhen 518055, China; luyun@stu.hit.edu.cn (Y.L.); zhangqiquan_hit@163.com (Q.Z.); hyf122168@163.com (Y.H.)
* Correspondence: mjwang@hit.edu.cn; Tel.: +86-755-260-33791

Received: 17 April 2018; Accepted: 16 May 2018; Published: 21 May 2018

Abstract: Existing research has revealed that auditory attention can be tracked from ongoing electroencephalography (EEG) signals. The aim of this novel study was to investigate the identification of peoples' attention to a specific auditory object from single-trial EEG signals via entropy measures and machine learning. Approximate entropy (ApEn), sample entropy (SampEn), composite multiscale entropy (CmpMSE) and fuzzy entropy (FuzzyEn) were used to extract the informative features of EEG signals under three kinds of auditory object-specific attention (Rest, Auditory Object1 Attention (AOA1) and Auditory Object2 Attention (AOA2)). The linear discriminant analysis and support vector machine (SVM), were used to construct two auditory attention classifiers. The statistical results of entropy measures indicated that there were significant differences in the values of ApEn, SampEn, CmpMSE and FuzzyEn between Rest, AOA1 and AOA2. For the SVM-based auditory attention classifier, the auditory object-specific attention of Rest, AOA1 and AOA2 could be identified from EEG signals using ApEn, SampEn, CmpMSE and FuzzyEn as features and the identification rates were significantly different from chance level. The optimal identification was achieved by the SVM-based auditory attention classifier using CmpMSE with the scale factor $\tau = 10$. This study demonstrated a novel solution to identify the auditory object-specific attention from single-trial EEG signals without the need to access the auditory stimulus.

Keywords: auditory attention; entropy measure; linear discriminant analysis (LDA); support vector machine (SVM); auditory attention classifier; electroencephalography (EEG)

1. Introduction

Existing relevant research has revealed that auditory objects [1], as neural representational units encoded in the human auditory cortex [2], are involved with high-level cognitive processing in the cerebral cortex, such as top-down attentional modulation [3]. Top-down attention is a selection process that focuses cortical processing resources on the most relevant sensory information in order to enhance information processing. There are many reports that auditory attention can be detected from brain signals, such as invasive electrocorticography [4], non-invasive magnetoencephalography (MEG) [5,6] and electroencephalography (EEG) [7,8]. These findings provide hard evidence that peoples' attention to a specific auditory object, which is referred to as the auditory object-specific attention, can be identified from brain signals. As the reflection of electrical activity in the cerebral cortex, EEG signals contain a wealth of information which is closely relating to advanced nervous activities in human brain such as learning, memory and attention [9]. Owing to the advantages of relatively low cost, easy to access and high temporal resolution, EEG signals are of much more practical value for the study of auditory object-specific attention [10].

In recent years, many attempts have been made to identify auditory object-specific attention from EEG signals. According to a review of the available literature, there are mainly three approaches to achieving the identification of auditory object-specific attention from ongoing EEG signals:

- The first approach is the use of system identification, which is mainly to build a linear forward map from the auditory stimuli of specific acoustic features on EEG signals. This is a direct method to estimate EEG signals [11]. The auditory object-specific attention can be inferred from the estimated EEG signals [5].
- The second approach is the use of stimulus reconstruction, which is mainly to reconstruct specific acoustic features (temporal envelopes) of auditory stimuli from the ongoing EEG signals [12]. Recently, this classical method has been extensively used to study the processing of speech perception in EEG signals. The auditory object-specific attention can be identified based on the reconstructed the acoustic features [7].
- The third approach is to extract the informative features of EEG signals and/or auditory stimuli and then exploit machine learning algorithms to train a classifier for the detection of auditory attention [10,13]. The informative features can be the cross-correlation between EEG signals and an auditory stimulus' envelope, the power in EEG signal bands, the measure of auditory event-related potentials [14] and so on. Machine learning algorithms, such as linear discriminant analysis (LDA) [15], regularized discriminant analysis (RDA) [13], support vector machine (SVM) [16,17], neural networks [18] and so on, have been reported in the published studies.

According to the available research, the first and second approaches must exploit EEG signals and acoustic features of auditory stimulus to achieve the identification of auditory object-specific attention. The auditory attention identification from EEG signals usually requires that the acoustic features of auditory stimuli be known in advance, for instance, the speech envelopes in the study of Horton [15]. Because the identification of auditory attention from EEG signals were based on that the cortical oscillations phase locked to the envelope of the auditory stimuli [19] and the temporal envelope of the auditory stimuli could be reconstructed from individual neural representations [2]. However, machine learning techniques recently have made remarkable progress and can achieve unprecedented accuracy for classification tasks [20]. With the help of machine learning, the third approach has the potential to achieve the detection of auditory object-specific attention by exploiting the enough useful information from EEG signals alone.

Using entropy measures to extract the informative features of EEG signals for brain-state monitoring and brain function assessment are becoming the hot research topics. EEG signals are commonly accepted to be non-stationary, nonlinear and multicomponent in nature. As typical nonlinear analysis methods, entropy measures in EEG signals may be much more appropriate to capture the imperceptible changes in different physiological and cognitive states of human brain. For example, Mu et al. studied the detection of driving fatigue using four entropy measures, i.e., spectrum entropy, approximate entropy (ApEn), sample entropy (SampEn) and fuzzy entropy (FuzzyEn), to extract features of EEG signals and reached an average recognition accuracy of 98.75% [21]. Hosseini et al. proposed the use of ApEn and wavelet entropy in EEG signals for emotion state analysis and the research found that ApEn and wavelet entropy were capable of discriminating emotional states [22]. Shourie et al. adopted ApEn to investigate the differences between EEG signals of artists and non-artists during the visual perception and mental imagery of some paintings and at resting condition. The research found that ApEn was significantly higher for artists during the visual perception and the mental imagery when compared with nonartists [23,24]. Alaraj and Fukami exploited ApEn to quantitatively evaluate the wakefulness state and the results showed that ApEn outperformed other conventional methods with respect to the classification of awake and drowsy subjects [25]. To date, applying entropy measures to the quantification of EEG signals has been proved to be a powerful tool to identify mental tasks and reveal cerebral states.

In continuation of the aforementioned studies, we move one step ahead in this study and explore the feasibility of using entropy measures on EEG signals to identify auditory object-specific attention, without the need for the acoustic features of auditory stimulus. Using four well-established entropy measures, i.e., ApEn, SampEn, composite multiscale entropy (CmpMSE) and FuzzyEn, to extract the informative features of EEG signals, we investigate the changes of these entropy measures in EEG signals under different auditory object-specific attention. Then, we use machine learning to train an auditory attention classifier for the identification of auditory object-specific attention. Based on preliminary experiment research, we demonstrate a novel solution to the identification of auditory object-specific attention from the ongoing EEG signals by the use of auditory attention classifier. The study of identification of auditory object-specific attention not only has great research value on monitoring the cognitive and physiological states of human brain, but also has great potential of the realization of assistive hearing technology with neural feedback.

2. Methods

2.1. Subjects

Thirteen subjects (aged 21 to 28 years, four females) participated in this study. All subjects were normal-hearing and right-handed college students and none had a history of neurological illness, which were confirmed by questionnaires. The experiment procedures were approved by the ethics committee of Harbin Institute of Technology Shenzhen Graduate School and all experiments were performed in accordance with relevant guidelines and regulations. Informed consent forms were signed by the subjects before the experiments were performed.

2.2. Experimental Design

In the study two audio signals were used as the auditory stimuli and the durations of both the audio signals were 60 s. While the subjects focused their attention on listening to the specific auditory stimulus, the corresponding specific auditory object was emerging in their auditory cortex. The audio signal A was the roaring sound of tiger corresponding to the Auditory Object1; the audio signal B was a segment of a stand-up comedy corresponding to the Auditory Object2. The audio signals were binaurally played with ER4 (Etymotic Research Inc., Elk Grove Village, IL, USA) in-ear earphones. The subjects were instructed to keep their attention on the auditory stimuli with their eyes closed while their scalp EEG signals were being recorded.

In this study 8-channel EEG signals were recorded using the ENOBIO 8 system (Neuroelectrics, Barcelona, Spain) with dry electrodes. The EEG signals were sampled at 500 Hz with band pass filter 0.540 Hz from eight sites on the scalp. According to the international standard 10–20 system, the eight electrode sites were selected to be T7 and T8 in the temporal region, P7 and P8 in the posterior temporal region, P3 and P4 in the parietal region, Cz in the central region and Fz in the frontal region, respectively. In order to minimize the possibility of the movements of subject's body as much as possible during experiment, the subjects were asked to sit in a comfortable chair.

The experiments were conducted in a soundproof room. Each subject was required to undergo three EEG measurement protocols in a random order and there were totally 39 EEG measurements in this study, each of 60 s in length. The three 60-s EEG measurement protocols corresponded to three kinds of auditory object-specific attention. The first EEG measurement protocol, during which the subject was instructed to keep calm and his brain was in a resting state without any audio signal playing, corresponded to Rest. The second EEG measurement protocol, during which the subject was instructed to keep his attention on the auditory stimulus with the audio signal A playing, corresponded to Auditory Object1 Attention (AOA1). The third EEG measurement protocol, during which the subject was instructed to keep his attention on the auditory stimulus with the audio signal B playing, corresponded to Auditory Object2 Attention (AOA2). For each subject the three EEG measurement protocols were randomly performed to reduce EEG signals to be contaminated by a fixed order of auditory tasks or the dominance of ears.

2.3. Entropy Measures in EEG Signals

ApEn, proposed by Pincus [26], is considered as a complexity measure of time series. ApEn has the ability of measures of predictability based on evaluating the irregularity of time series. The more similar patterns the time series has, the less irregular the time series are and the more likely the time series are to be predictable. The computation method of ApEn is firstly involved in the phase space reconstruction, in which the time series are embedded into phase spaces of dimension m and $m + 1$, respectively; and then calculates the percentages of similar vectors in phase spaces with acceptable matches. SampEn, proposed by Richman et al. [27], is a modified version of ApEn. SampEn has better performance over ApEn in the consistency and dependence on data length. CmpMSE was proposed by Wu [28]. For a predefined scale factor, a set of k-th coarse-grained time series based on the composite averaging method are reconstructed from original time series. The sample entropies of all coarse-grained time series are calculated, and then CmpMSE is defined as the mean of all the sample entropies. FuzzyEn used the fuzzy membership function to obtain a fuzzy measurement of two vectors' similarity [29]. The family of exponential function was usually used as the fuzzy function and it was continuous and convex so that the similarity does not change abruptly [30].

Before the calculation of ApEn, SampEn, CmpMSE and FuzzyEn, EEG signals were preprocessed by linear detrending, which the polynomial curve fitting were used as the trend terms of EEG signals and then subtracted. Then, a 9-level wavelet decomposition was performed, using Daubechies (db4) wavelets as the wavelet function which was suitable for detecting changes of the EEG signals [31]. Two of the highest detail coefficients D2 (62.5–125 Hz) and D1 (125–250 Hz) [31] and the approximation coefficients A9 (0–0.49 Hz) were considered as noises. The denoised EEG signals were recovered by the detail coefficients from D3 and D9 and the effective frequency band of the denoised EEG signals was considered to be 0.5–62.5 Hz. The recording time of EEG signals was 60 s, which corresponded to the durations of both the audio stimuli. Because of the sampling rate was 500, the data length used for the calculation of entropy measures in EEG signals was $N = 30,000$. For the calculation of ApEn, SampEn and FuzzyEn, the embedding dimension $m = 2$ and the tolerance $r = 0.15 \times$ STD, where STD was the standard deviation of EEG signals. For the calculation of CmpMSE, the scale factor was selected to $\tau = 30$ and the parameter m and r were the same as ApEn and SampEn. The assignment of the parameters for the calculation of ApEn, SampEn, CmpMSE and FuzzyEn, are shown in Table 1.

Table 1. The assignment of the parameters for the calculation of approximate entropy (ApEn), sample entropy (SampEn), composite multiscale entropy (CmpMSE) and fuzzy entropy (FuzzyEn). STD is denoted as the standard deviation of electroencephalography (EEG) signals.

Entropy Measures	Data Length N	Time Delay t	Embedding Dimension m	Tolerance r	Scale Factor τ	Reference
ApEn	30,000	1	2	$0.15 \times$ STD	–	[27]
SampEn	30,000	1	2	$0.15 \times$ STD	–	[27]
CmpMSE	30,000	1	2	$0.15 \times$ STD	30	[28]
FuzzyEn	30,000	1	2	$0.15 \times$ STD	–	[30]

2.4. Statistical Analysis Methods

The statistical analysis methods included the multiple-sample tests for equal variances, Shapiro-Wilk W test for normal distribution, parametric or non-parametric analysis of variance and multiple comparisons, which were used to evaluate ApEn, SampEn, CmpMSE and FuzzyEn in EEG signals under auditory object-specific attention of Rest, AOA1 and AOA2. The multiple-sample tests for equal variances were to use the Bartlett null hypothesis test that the values of the entropy measure in EEG signals under examination had the same variance. The Shapiro-Wilk W test was used to examine normality of the values of ApEn, SampEn, CmpMSE and FuzzyEn under the null hypothesis that the values of the entropy measures under examined obeyed normal distributions. If the values of the entropy measures in EEG signals under auditory object-specific attention of Rest,

AOA1 and AOA2 had equal variances and obeyed normal distributions, the parametric analysis of variances and multiple comparisons were performed to determine whether the values of the entropy measures were significantly different from each other for Rest, AOA1 and AOA2 of auditory object-specific attention; otherwise, the Kruskal-Wallis test (an extension of the Wilcoxon rank sum test) and multiple comparisons were used. Besides, Bonferroni correction was used to counteract the problem of multiple comparisons.

2.5. *Auditory Attention Classifier Based on Entropy Measures and Machine Learning*

Machine learning, which learns from data to make data-driven predictions or decisions, is now widely used for the analysis of EEG signals [32]. Machine learning is the method that is used to train a model and develop related algorithms with the input features and lend itself to prediction. In this study three kinds of auditory object-specific attention were investigated, i.e., Rest, AOA1 and AOA2. To demonstrate auditory attention classifier, we used ApEn, SampEn, CmpMSE and FuzzyEn to extract the information of EEG signals as features and then exploited LDA and SVM to construct the auditory attention classifiers. As a classical machine learning method, LDA is a statistical classifier which achieves a linear decision boundary based on the within and between class scatter matrices [33]. LDA performs the discrimination of different classes by maximizing the between class scatter and minimizing the within class scatter. LDA has been commonly used for EEG signals classification, which allows for fast and massive processing of data samples [34,35]. Like LDA, SVM also is one of the most machine learning methods, which can be not only used for the linear classification but also for the non-linear classification using a specific kernel function.

The thirteen subjects' EEG signals were recorded and each subject underwent three EEG measurement protocols. Therefore, a total of 39 samples were available for the auditory attention classifier. To facilitate the training and testing of the auditory attention classifier, LDA and SVM are employed for supervised learning. After training, the LDA-based and SVM-based auditory attention classifiers were capable of identifying the auditory object-specific attention of Rest, AOA1 and AOA2. In order to assess the identification of auditory object-specific attention, we used the leave-one-out cross-validation (LOOCV) approach to examine the identification accuracy of auditory object-specific attention of Rest, AOA1 and AOA2.

3. Results

3.1. *Statistical Results of Entropy Measures in EEG Signals*

According to the statistical results of multiple-sample tests for equal variances, the values of ApEn, SampEn, CmpMSE and FuzzyEn all obeyed equal variances (all $p > 0.05$) with respect to Rest, AOA1 and AOA2 of auditory object-specific attention. According to the statistical results of Shapiro-Wilk W test, for ApEn and CmpMSE in EEG signals under auditory object-specific attention of Rest, AOA1 and AOA2, the vast majority of cases of ApEn and CmpMSE obeyed normal distributions ($p > 0.05$), except in the case of ApEn in EEG signals with P8 channel ($p = 0.012$) under the auditory object-specific attention of AOA2, and except in the case of CmpMSE in EEG signals with Fz channel ($p = 0.031$) under auditory object-specific attention of AOA1. For SampEn in EEG signals under auditory object-specific attention of Rest, AOA1 and AOA2, the majority of cases of SampEn obeyed normal distributions, except in the case of SampEn in EEG signals with Fz channel ($p = 0.038$) under auditory object-specific attention of Rest and with Cz, Fz and P3 channels ($p = 0.036, 0.013$ and 0.037, respectively) under auditory object-specific attention of AOA2. For FuzzyEn in EEG signals under auditory object-specific attention of Rest, AOA1 and AOA2, the vast majority of cases of FuzzyEn obeyed normal distributions, except in the case of FuzzyEn in EEG signals with T7, T8 and P8 channels ($p = 0.013, 0.005$ and 0.0003, respectively) under auditory object-specific attention of AOA2.

Thus, according to the normality and non-normality of these entropy measures, the parametric and non-parametric analysis of variance were carried out respectively to test the significance of difference degree of these entropy measures among Rest, AOA1 and AOA2. For ApEn of P8 channel, the values of ApEn in EEG signals did not obey normal distribution, so the Kruskal-Wallis test was carried out; but for ApEn of the other channels, the values ApEn in EEG signals obeyed normal distributions, so the one-way analysis of variance was carried out. Like ApEn, for SampEn, CmpMSE and FuzzyEn the same statistical methods were applied to test the significance of difference degree of these entropy measures among Rest, AOA1 and AOA2.

Figure 1 shows the statistical results of the values of ApEn, SampEn, CmpMSE and FuzzyEn in EEG signals under auditory object-specific attention of Rest, AOA1 and AOA2. The values of these entropy measures are given as means ± standard errors. The *p* value denotes the levels of significance of the difference of group means and the small *p* value of 0.05 indicates that the values of the entropy measures among Rest, AOA1 and AOA2 significantly differ ($p < 0.05$). For ApEn and CmpMSE, the all *p* values of eight channels are less than 0.05 and indicate that there are significant differences ($p < 0.05$) of ApEn and CmpMSE in EEG signals under auditory object-specific attention of Rest, AOA1 and AOA2. For SampEn the *p* values of T7, P7, T8, P8 and Fz channels are less than 0.05 and the other channels' *p* values are greater than 0.05. For FuzzyEn the *p* values of most channels are less than 0.05, except for P7 channel.

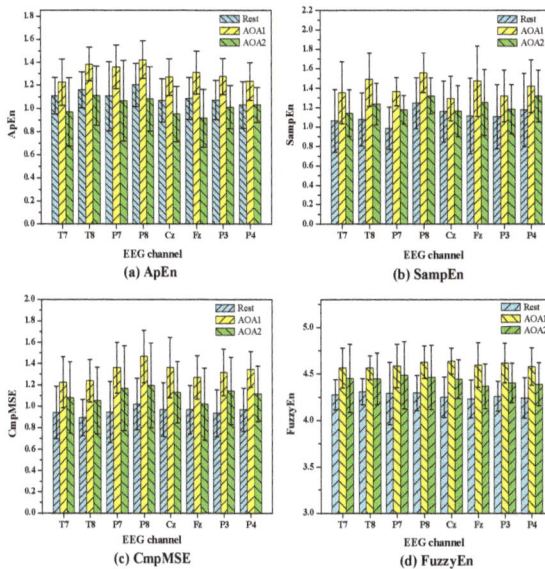

Figure 1. The statistical results of the values of the entropy measures in EEG signals under auditory object-specific attention of Rest, Auditory Object1 Attention (AOA1) and Auditory Object2 Attention (AOA2). (**a**) ApEn; the *p* values of eight channels are 0.021, 0.002, 0.028, 0.001, 0.001, 0.001, 0.001 and 0.006, corresponding to T7, T8, P7, P8, Cz, Fz, P3 and P4, respectively; (**b**) SampEn; the *p* values of eight channels are 0.049, 0.01, 0.001, 0.002, 0.213, 0.017, 0.105 and 0.142, respectively; (**c**) CmpMSE; the *p* values of eight channels are 0.045, 0.003, 0.007, 0.002, 0.003, 0.009, 0.002 and 0.001, respectively; (**d**) FuzzyEn; the *p* values of eight channels are 0.021, 0.005, 0.051, 0.004, 0.0001, 0.0005, 0.0001 and 0.0009, respectively. The values of these entropy measures are given as means ± standard errors. The small *p* value of 0.05 indicates that the values of the entropy measures among Rest, AOA1 and AOA2 significantly differ.

To further investigate which pairs of means were significantly different for Rest, AOA1 and AOA2, the multiple comparisons test was carried out. Because there were three kinds of auditory object-specific attention, for multiple comparisons test each entropy measure was testing three ($3 \times 2/2$)

independent hypotheses. Therefore, here a p value of <0.015 (0.05/3) was considered statistically significant using the Bonferroni criterion. Table 2 shows the p values of multiple comparisons between the auditory object-specific attention of Rest, AOA1 and AOA2. In Table 2, it is clearly observed that there are significant differences in ApEn of T8, P8, Cz, Fz, P3 and P4 channels between AOA1 and AOA2, in ApEn of P3 and P4 channels between Rest and AOA1; in SampMSE of T8, P7, P8 and Fz channels between Rest and AOA1; in CmpMSE of T8, P7, P8, Cz, Fz, P3 and P4 channels between Rest and AOA1; in FuzzyEn of T7, T8, P8, Cz, Fz, P3 and P4 channels between Rest and AOA1, in FuzzyEn of Cz, Fz, P3 and P4 channels between AOA1 and AOA2. It must be noted that, for SampEn of Cz, P3 and P4 channels and FuzzyEn of P7 channel, there was no need to perform the multiple comparisons because there were no significant differences in SampEn of Cz, P3 and P4 channels and FuzzyEn of P7 channel among Rest, AOA1 and AOA2 for the parametric or non-parametric analysis of variance.

As shown in Figure 1, it is clearly observed that there are obvious differences in the mean values of these entropy measures under auditory object-specific attention of Rest, AOA1 and AOA2. For example, the mean values of ApEn, SampEn and CmpMSE under auditory object-specific attention of AOA1 are obviously higher than those under auditory object-specific attention of Rest and AOA2. In the viewpoint of mathematics, Figure 1 has a certain intrinsic correlation with Table 2. For instance, for ApEn of P4 channel in Table 2, the p values of between Rest and AOA1 and between AOA1 and AOA2 are less than 0.015, which indicate that the differences of the values of ApEn between Rest and AOA1 and between AOA1 and AOA2 are significant. At the same time, in Figure 1 the significant differences of the mean values of ApEn of P4 channel between Rest and AOA1 and between AOA1 and AOA2 are observed. Therefore, the size of the p values as shown in Table 2, to a certain extent, can indicate the discriminating power of the entropy measures. The smaller the p values, the stronger the discriminating power of the entropy measures may be. As shown in Table 2, the p values of CmpMSE, on the whole, are less than those of ApEn and SampEn.

Table 2. The p values of multiple comparisons between the auditory object-specific attention of Rest, AOA1 and AOA2 for ApEn, SampEn and CmpMSE.

		T7	T8	P7	P8	Cz	Fz	P3	P4
	Rest vs. AOA1	0.382	0.015	0.078	0.041	0.035	0.024	0.011	0.014
ApEn	Rest vs. AOA2	0.260	0.793	0.936	0.430	0.296	0.109	0.650	0.999
	AOA1 vs. AOA2	0.015	0.003	0.036	0.001	0.001	0.001	0.001	0.013
	Rest vs. AOA1	0.046	0.001	0.001	0.002	–	0.013	–	–
SampEn	Rest vs. AOA2	0.777	0.256	0.021	0.660	–	0.430	–	–
	AOA1 vs. AOA2	0.183	0.042	0.027	0.023	–	0.246	–	–
	Rest vs. AOA1	0.034	0.002	0.005	0.002	0.002	0.010	0.001	0.001
CmpMSE	Rest vs. AOA2	0.413	0.216	0.182	0.318	0.296	0.800	0.107	0.191
	AOA1 vs. AOA2	0.393	0.122	0.268	0.071	0.093	0.058	0.204	0.025
	Rest vs. AOA1	0.001	0.001	–	0.001	0.001	0.001	0.001	0.001
FuzzyEn	Rest vs. AOA2	0.051	0.050	–	0.047	0.038	0.045	0.021	0.016
	AOA1 vs. AOA2	0.249	0.138	–	0.138	0.001	0.004	0.002	0.008

3.2. Individual-Level Analysis of Entropy Measures in EEG Signals

To demonstrate the individual-level identification of Rest, AOA1 and AOA2, we first carried out the individual-level analysis of ApEn, SampEn, CmpMSE and FuzzyEn in EEG signals under auditory object-specific attention of Rest, AOA1 and AOA2. Figure 2 presents the values of four subjects' ApEn, SampEn, CmpMSE and FuzzyEn in EEG signals of P3 channel under auditory object-specific attention of Rest, AOA1 and AOA2 and the EEG signals were selected from four representative subjects.

As shown in Figure 2, it is clearly observed that the values of ApEn, SampEn, CmpMSE and FuzzyEn show obvious differences with respect to Rest, AOA1 and AOA2 of auditory object-specific attention. For the subjects A, B, C and D, the values of ApEn, SampEn, CmpMSE and FuzzyEn on auditory object-specific attention of AOA1 are greater than those of the entropy measures on auditory object-specific attention of Rest and AOA2.

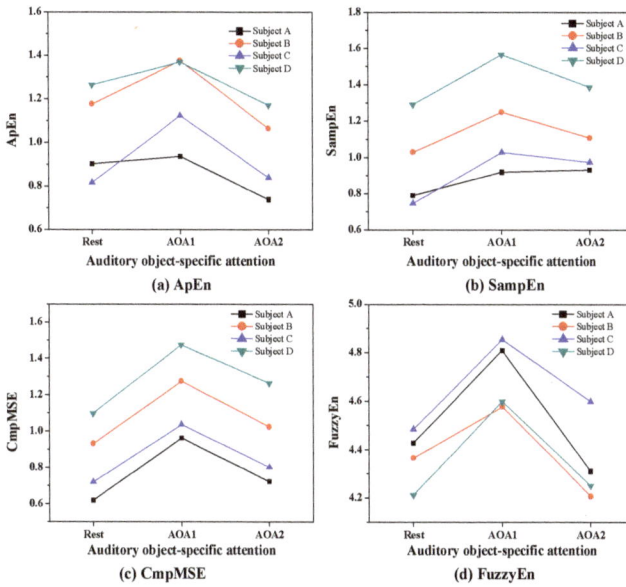

Figure 2. Individual-level analysis of (**a**) ApEn, (**b**) SampEn, (**c**) CmpMSE and (**d**) FuzzyEn in EEG signals of P3 channel under auditory object-specific attention of Rest, AOA1 and AOA2.

The values of SampEn and CmpMSE on auditory object-specific attention of AOA2 are greater than those of the entropy measures on auditory object-specific attention of Rest, and yet the values of ApEn on auditory object-specific attention of AOA2 are lower than that of ApEn on auditory object-specific attention of Rest.

Through this individual-level analysis of entropy measures on auditory object-specific attention of Rest, AOA1 and AOA2, it was clear that ApEn, SampEn, CmpMSE and FuzzyEn in EEG signals could be used as informative indicators to determine the auditory object-specific attention.

3.3. Identification of Auditory Object-Specific Attention by Auditory Attention Classifier

In order to demonstrate the discriminating power with respect to Rest, AOA1 and AOA2 of auditory object-specific attention, the identification of auditory object-specific attention was investigated by two auditory attention classifiers, one used LDA to construct the auditory attention classifier and the other used SVM to construct the auditory attention classifier. The LDA-based auditory attention classifier is designed using a multiclass classification method with open source code [36]. The SVM-based auditory attention classifier is designed using the LIBSVM toolbox [37]. To statistically evaluate whether the identification rates were significantly different from the chance level (33.3%), the chi-squared test was used, with the null hypothesis that the identification rates was dependent of the chance level.

Table 3 shows the identification rates of Rest, AOA1 and AOA2 of auditory object-specific attention by the LDA-based and SVM-based auditory attention classifiers using ApEn, SampEn, CmpMSE and FuzzyEn in EEG signals of eight channels as features. For the LDA-based auditory attention classifier, the average identification rates of the auditory object-specific attention are 48.7%, 46.2%, 43.6% and 46.2%, corresponding to ApEn, SampEn, CmpMSE and FuzzyEn, respectively.

For ApEn, SampEn, CmpMSE and FuzzyEn the p values of chi-squared test are 0.146, 0.076, 0.079 and 0.076, respectively. For the SVM-based auditory attention classifier, the average identification rates of the auditory object-specific attention are 56.4%, 56.4%, 53.8% and 58.9%, corresponding to ApEn, SampEn, CmpMSE and FuzzyEn, respectively. The corresponding p values of chi-squared test are 0.026, 0.026, 0.017 and 0.013, respectively. For the SVM-based auditory attention classifier using ApEn, SampEn, CmpMSE and FuzzyEn as features, the identification rates are significantly different ($p < 0.05$) from the chance level.

Table 3. The identification rates of Rest, AOA1 and AOA2 of auditory object-specific attention by the linear discriminant analysis-based (LDAb) and support vector machine-based (SVMb) auditory attention classifiers (AAC) using ApEn, SampEn, CmpMSE and FuzzyEn in EEG signals as features. The p values indicate the level of significance between the identification rates and chance level (33.3%). LDAb AAC, LDA-based auditory attention classifier; SVMb AAC, SVM-based auditory attention classifier; Average IdR, average identification rate.

	Identification Rate	ApEn	SampEn	CmpMSE	FuzzyEn
	Rest (%)	38.5	38.5	30.8	53.9
	AOA1 (%)	61.5	69.2	69.2	61.5
LDAb AAC	AOA2 (%)	46.2	30.8	30.8	23.1
	Average IdR (%)	48.7	46.2	43.6	46.2
	p value	0.146	0.076	0.079	0.076
	Rest (%)	53.8	46.2	69.2	69.2
	AOA1 (%)	69.2	69.2	61.5	61.5
SVMb AAC	AOA2 (%)	46.2	53.8	30.8	46.2
	Average IdR (%)	56.4	56.4	53.8	58.9
	p value	0.026	0.026	0.017	0.013

As shown in Table 3, on the whole, the identification rates of Rest, AOA1 and AOA2 when using the SVM-based auditory attention classifier are higher than those when using the LDA-based auditory attention classifier. Thus, on the basis of the above results, it is clear that for the identification of auditory object-specific attention the SVM-based auditory attention classifier is more effective than the LDA-based auditory attention classifier.

To investigate which channel of EEG signals was the most sensitive to the identification of Rest, AOA1 and AOA2 of auditory object-specific attention, we exploited the SVM-based auditory attention classifier to identify the auditory object-specific attention using ApEn and CmpMSE in EEG signals per channel as features. Figure 3 shows the identification rates of Rest, AOA1 and AOA2 by the SVM-based auditory attention classifier using ApEn and CmpMSE in EEG signals per channel as features.

The average identification rate is calculated by averaging the identification rates of Rest, AOA1 and AOA2. For ApEn, the P8, P4 and Fz channels are corresponding to the top three of the average identification rates of auditory object-specific attention and the corresponding identification rates are 59.0% ($p = 0.008$), 56.4% ($p = 0.014$) and 56.4% ($p = 0.005$), respectively. For CmpMSE, the T8, P4 and Fz channels are corresponding to the top three of the average identification rates of auditory object-specific attention and the corresponding identification rates are 59.0% ($p = 0.012$), 56.4% ($p < 0.001$), and 48.7% ($p = 0.001$), respectively. It is clearly observed that for ApEn and CmpMSE the performances of the identification of auditory object-specific attention of Rest, AOA1 and AOA2 vary with the different channels, which may because the different informative features were extracted from different channels by the entropy measures.

In order to investigate the influence of the scale factor of entropy measures on the identification rate of auditory object-specific attention, we further studied the identification of auditory object-specific attention of Rest, AOA1 and AOA2 by the SVM-based auditory attention classifier using CmpMSE with different scale factors as features. Figure 4 shows the identification rates of Rest, AOA1 and AOA2 by the adoption of CmpMSE in EEG signals of eight channels with the scale factors $\tau = 1, 5, 10,$

15, 20, 25, 30, 35 and 40, respectively. The average identification rate is calculated by averaging the identification rates of Rest, AOA1 and AOA2. The average identification rates are 56.4% ($p = 0.026$), 56.4% ($p = 0.041$), 69.2% ($p < 0.001$), 46.2% ($p = 0.076$), 56.4% ($p = 0.016$), 56.4% ($p = 0.026$), 53.8% ($p = 0.017$), 56.4% ($p = 0.026$) and 48.7% ($p = 0.146$), respectively. Therefore, for CmpMSE the optimal identification of Rest, AOA1 and AOA2 of auditory object-specific attention is achieved with the scale factor $\tau = 10$, and the corresponding identification rates of Rest, AOA1 and AOA2 are 69.2%, 76.9% and 61.5%, respectively.

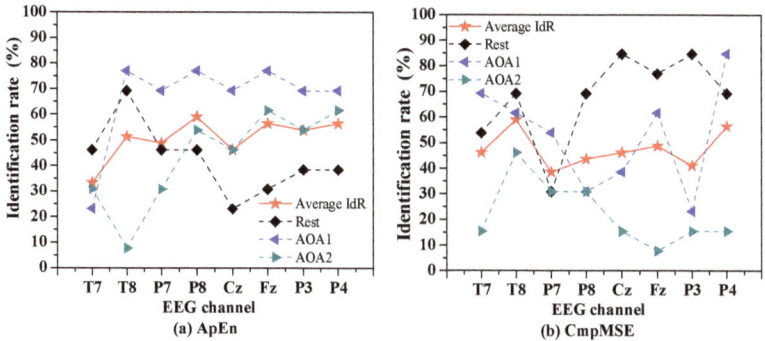

Figure 3. The identification rates of Rest, AOA1 and AOA2 of auditory object-specific attention by the SVM-based auditory attention classifier using (**a**) ApEn and (**b**) CmpMSE in EEG signals per channel as features. The average identification rate (IdR) is calculated by averaging the identification rates of Rest, AOA1 and AOA2. For ApEn the *p* values of chi-squared test are 0.584, 0.001, 0.058, 0.008, 0.048, 0.005, 0.034 and 0.014, which are corresponding to T7, T8, P7, P8, Cz, Fz, P3 and P4, respectively. For CmpMSE the *p* values of chi-squared test are 0.019, 0.012, 0.429, 0.079, 0.003, 0.001, 0.003 and <0.001, which are corresponding to T7, T8, P7, P8, Cz, Fz, P3 and P4, respectively.

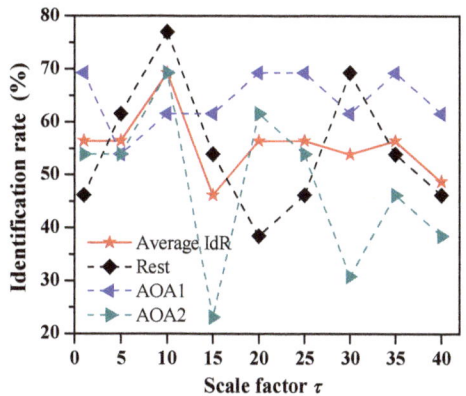

Figure 4. The identification results of auditory object-specific attention by the SVM-based auditory attention classifier using CmpMSE as features. The identification rates of Rest, AOA1 and AOA2 by the adoption of CmpMSE in EEG signals of eight channels with the scale factors $\tau = 1, 5, 10, 15, 20, 25, 30, 35$ and 40, and the *p* values of chi-squared test are 0.026, 0.041, <0.001, 0.076, 0.016, 0.026, 0.017, 0.026 and 0.146, respectively.

In order to investigate the influence of the choice of parameters for the entropy measures on the identification rate of auditory object-specific attention, we carried out a qualitative comparison of the identification results of auditory object-specific attention by the SVM-based auditory attention classifier using CmpMSE with different parameter values of tolerance. For the calculation of CmpMSE the tolerance was selected to r = 0.10, 0.15, 0.20 and 0.25, respectively, and the other parameters were fixed. The qualitative comparison of the identification results of auditory object-specific attention are shown in Table 4. The average identification rates are 59.0% (p = 0.020), 69.2% (p < 0.001), 71.8% (p < 0.001) and 53.8% (p = 0.068), corresponding to the tolerance r = 0.10, 0.15, 0.20 and 0.25, respectively.

Table 4. The qualitative comparison of the identification results of auditory object-specific attention by the SVM-based auditory attention classifier using CmpMSE with different parameter values of tolerance. For the calculation of CmpMSE the tolerance was selected to r = 0.10, 0.15, 0.20 and 0.25, respectively and the other parameters were fixed with the data length N = 30,000, the embedding dimension m = 2 and the scale factor τ = 10.

Tolerance r	Identification Rate (%)			Average IdR (%)
	Rest	**AOA1**	**AOA2**	
0.10	61.5	61.5	53.8	59.0
0.15	69.2	76.9	61.5	69.2
0.20	69.2	76.9	69.2	71.8
0.25	53.8	61.5	46.2	53.8

4. Discussion

In this study, we explored the entropy measures in EEG signals to extract the informative features relating to auditory object-specific attention and then exploited LDA and SVM to construct the auditory attention classifiers. Our proposed method to the identification of auditory object-specific attention is an innovative attempt. Even though the optimal identification rates of Rest, AOA1 and AOA2 of auditory object-specific attention are only 69.2%, 76.9% and 61.5% respectively by the SVM-based auditory attention classifier using CmpMSE in EEG signals of eight channels with the scale factors τ = 10 as features, the identification accuracy is at the same level with the existing studies, for instance, the experimental results reported by Bleichner et al. [38]. We have compared our identification results with the available studies and the comparison results are presented in Table 5.

Table 5. The comparison of our identification results with the existing studies for the identification of auditory attention based on EEG signals.

References	Auditory Task	Number of Channels	Identification Approaches	Identification Results
[38]	The direction of attention (front, left and right)	84	Leave-one-out cross validation template matching approach	70%
[14]	Attending left and attending right	32	ERP templates; Leave-one-out cross-subject validation	71.2%
Our work	Auditory object-specific attention (Rest, AOA1 and AOA2)	8	Composite multiscale entropy; Leave-one-out cross validation	69.2%, 76.9% and 61.5%

EEG signals are a kind of non-stationary, non-linear and often multicomponential dynamic signal and it is challenging to accurately extract the informative features of EEG signals. In this study, based on four well-established entropy measures, i.e., ApEn, SampEn, CmpMSE and FuzzyEn, we demonstrate the use of entropy measures in EEG signals as informative features to reveal the auditory attention states. It is clearly shown that the SVM-based auditory attention classifier using ApEn, SampEn, CmpMSE as features are capable of indicating significant differences in informative features of EEG signals under different auditory object-specific attention. This experiment findings are also in line with existing research findings. Many studies also reported that the physiological and cognitive states of human brain could be determined by the use of entropy measures in EEG signals. For example, discrete wavelet transform and entropy measures were used to identify the focal EEG signals [39]; ApEn, SampEn and multiscale entropy were used to assess the different visual attention levels [40]; ApEn was used to evaluate the wakefulness state [25]. These available studies had suggested that the entropy measures of EEG signals, as a complexity parameters of physiological time-series, could be as an useful indicator to reveal the physiological states of human brain and there was no doubt that the entropy measures of EEG signals had clinical significance. Therefore, the entropy measures in EEG signals could also be as informative features to identify the auditory object-specific attention.

When compared with ApEn, SampEn and FuzzyEn, CmpMSE maybe was regarded as the most informative feature of EEG signals to identify the auditory object-specific attention and the optimal identification was achieved by the SVM-based auditory attention classifier using CmpMSE in EEG signals with the scale factor $\tau = 10$ and the tolerance $r = 0.15$ and 0.20. This might be because the different entropy measures in EEG signals were able to extract the different informative features of EEG signals. As is known to all, ApEn and SampEn can quantify the temporal structure and complexity of time series strictly at a time scale, usually selected to be 1. But CmpMSE can quantify the long-term structures in EEG signals at multiple time scales, which can extract the long-rang informative feature of EEG signals. The main advantages of our proposed method was that the identification of auditory object-specific attention from single-trial EEG signals was achieved without the need to access to the auditory stimulus, when compared with most available studies. For the existing researches, the auditory attention identification from EEG signals usually required the acoustic features of auditory stimuli be known in advance. The main disadvantages of our proposed method was that the entropy measures were usually computationally intensive and it was necessary to perform comprehensive statistical analyses of the optimal parameters of the computation of the entropy measures and the optimization of machine learning algorithms. However, the algorithm complexity and computing time of CmpMSE are lower than those of ApEn, SampEn and FuzzyEn. Therefore, CmpMSE in EEG signals maybe was the most useful indicator to identify the auditory object-specific attention of Rest, AOA1 and AOA2.

There are some limitations in the current study. Firstly, the 60-s duration of EEG signals used to calculate the entropy measures, corresponding to the data length of $N = 30{,}000$, is a slightly long time. It is hard to ensure that the EEG signal is stationary or even weakly stationary, which is especially required for permutation entropy analysis [41–43]. In fact, we had also evaluated the identification results of auditory object-specific attention by the LDA-based and SVM-based auditory attention classifier using permutation entropy in EEG signals of eight channels as features, and the average identification rates are 23.1% ($p = 0.429$) and 30.8% ($p = 0.764$), respectively. if the EEG signals were split into relatively short epochs and the entropy measures in EEG signals which could be deemed stationary were computed for each of the epochs, and then the distributions of values for each EEG signals were obtained [44–47], the experimental data would be better to demonstrate the identification results of auditory object-specific attention via entropy measures and machine learning.

Secondly, only 13 subjects participated in the study, and the number of sample data are not enough, which might lead to the statistical results of the entropy measures not showing significance. In fact, this experiment was not very good to perform. In order to ensure the experimental effects for each subject, EEG signals should be recorded successfully in the first round to avoid the second round of experiments because the subject listening to the same auditory stimulus at the second time might cause adverse effects on the auditory attention. Therefore, in the available studies, the number of subjects who participated in the experiment were usually not many. For example, in the studies of Haghighi [13] and Choi [14], ten subjects' EEG signals were used in the investigations.

Thirdly, for the calculation of entropy measures, it is well-known that some parameters, such as the embedding dimension m and tolerance r, need to be fixed in advance. We did not assess our experimental performances with the different combinations of the parameters. For one thing, if we such did, it would cause the analysis of experiment results to be rather complex. For another, there were no solid methods to obtain the optimal parameters for the entropy measures. Maybe because of these, the identification rates of Rest, AOA1 and AOA2 of auditory object-specific attention were not very high in this study.

Fourthly, in this study the EEG signals were recorded as 8-channel signals and then the 8-dimensional entropy measures (ApEn, SampEn, CmpMSE and FuzzyEn) could be used to identify the auditory object-specific attention. In Table 3 the number of features of ApEn, SampEn, CmpMSE and FuzzyEn were 8. In Figure 4 and Table 4 the number of features of CmpMSE were 8. We believed that the more EEG signal channels used and the more the dimensionality of entropy measures, the more reliable the experimental results may be, because for the entropy measures the more EEG signal channels were used, more informative features of auditory object-specific attention could be extracted. It also should be noted that, however, the identification rate maybe was not always better with the more EEG signal channels. What's more, the current research trend was to identify the auditory attention with less EEG signal channels. For example, O'Sullivan et al. used 128-channel EEG signals [7]; Mirkovic et al. used 25-channel EEG signals [48]; Haghighi et al. used 16-channel EEG signals [13]. This is mainly because using less EEG signal channels to identify auditory attention is more worthy research for practical application.

For future work to improve the study presented here, the identification accuracy of auditory object-specific attention maybe has a great potential for improvement. Firstly, we can adopt other advanced machine learning techniques, such as deep learning [49–51]. In recent years, deep learning has been widely used in physiological signal application analysis (especially EEG signals), such as seizure prediction [51]. Secondly, we can explore the potential of other non-linear feature analysis methods of EEG signals for the identification of auditory object-specific attention, such as higher order spectra [52], phase entropy [53], wavelet transform in conjunction with entropy [39], empirical mode decomposition in conjunction with entropy [54] and so on. In addition, we can adopt several different types of features of EEG signals in conjunction with feature ranking approach [53,55] to further investigate the identification of auditory object-specific attention.

The identification of auditory object-specific attention would undoubtedly have great research value and application potential for the optimization of hearing aids and enhanced listening techniques, which are our main clinical application. For example, the algorithm of identification of auditory object-specific attention would work hand in hand with the algorithms of acoustic scene analysis in hearing aids to form neuro-steered hearing prostheses. With the help of the identification of peoples' attention to a specific auditory object from EEG signals, we can use EEG signals to guide the algorithms of acoustic scene analysis, in effect extending the efferent neural pathways which simulates the top-down cognitive control of auditory attention.

5. Conclusions

In this paper, ApEn, SampEn, CmpMSE and FuzzyEn were used to extract the informative features of EEG signals under three kinds of auditory object-specific attention (Rest, AOA1 and AOA2). The results of statistical analysis of entropy measures indicated that there were significant differences ($p < 0.05$) in the values of ApEn, SampEn, CmpMSE and FuzzyEn in EEG signals under auditory object-specific attention of Rest, AOA1 and AOA2. LDA and SVM were used to construct the auditory attention classifiers respectively and LOOCV was used to evaluate the identification rates of Rest, AOA1 and AOA2 of auditory object-specific attention. Compared with the LDA-based auditory attention classifier, the SVM-based auditory attention classifier was capable of achieving better auditory object-specific attention identification accuracy for Rest, AOA1 and AOA2. According to the identification results, for Rest, AOA1 and AOA2 of auditory object-specific attention, the optimal identification was achieved by the SVM-based auditory attention classifier using CmpMSE with the scale factor $\tau = 10$ and the corresponding identification rates were 69.2%, 76.9% and 61.5%,

Entropy **2018**, *20*, 386

respectively. All results suggest that using the entropy measures in EEG signals as informative features in conjunction with machine learning techniques can provide a novel solution to the identification of auditory object-specific attention from single-trial EEG signals without the need to access to the auditory stimulus.

Author Contributions: Y.L. and M.W. conceived the research theme and designed the experiments; Y.L. and Q.Z. performed the experiments; Y.L. analyzed the experimental data and wrote the manuscript; M.W. contributed materials/analysis tools and assisted to supervise the field activities; Y.H. participated in the revision of the manuscript. All authors read and approved the final manuscript.

Funding: This research was funded in part by the Shenzhen Fundamental Research Project under Grant JCYJ20170412151226061, and Grant JCYJ20170808110410773.

Acknowledgments: The authors would like to thank Wanqing Wu and Rongchao Peng for valuable feedback on this manuscript, they provided much help and suggestions in the revision of the manuscript.

Conflicts of Interest: The authors declare no conflict of interest.

References

1. Griffiths, T.D.; Warren, J.D. What is an auditory object? *Nat. Rev. Neurosci.* **2004**, *5*, 887–892. [CrossRef] [PubMed]
2. Ding, N.; Simon, J.Z. Emergence of neural encoding of auditory objects while listening to competing speakers. *Proc. Natl. Acad. Sci. USA* **2012**, *109*, 11854–11859. [CrossRef] [PubMed]
3. Kaya, E.M.; Elhilali, M. Modelling auditory attention. *Philos. Trans. R. Soc. B Biol. Sci.* **2017**, *372*, 20160101. [CrossRef] [PubMed]
4. Dijkstra, K.; Brunner, P.; Gunduz, A.; Coon, W.; Ritaccio, A.L.; Farquhar, J.; Schalk, G. Identifying the attended speaker using electrocorticographic (ECoG) signals. *Brain Comput. Interfaces* **2015**, *2*, 161–173. [CrossRef] [PubMed]
5. Miran, S.; Akram, S.; Sheikhattar, A.; Simon, J.Z.; Zhang, T.; Babadi, B. Real-time tracking of selective auditory attention from M/EEG: A bayesian filtering approach. *bioRxiv* **2017**. [CrossRef]
6. Akram, S.; Presacco, A.; Simon, J.Z.; Shamma, S.A.; Babadi, B. Robust decoding of selective auditory attention from MEG in a competing-speaker environment via state-space modeling. *Neuroimage* **2016**, *124*, 906–917. [CrossRef] [PubMed]
7. O'Sullivan, J.A.; Power, A.J.; Mesgarani, N.; Rajaram, S.; Foxe, J.J.; Shinn-Cunningham, B.G.; Slaney, M.; Shamma, S.A.; Lalor, E.C. Attentional selection in a cocktail party environment can be decoded from single-trial EEG. *Cereb. Cortex* **2015**, *25*, 1697–1706. [CrossRef] [PubMed]
8. O'Sullivan, J.; Chen, Z.; Herrero, J.; McKhann, G.M.; Sheth, S.A.; Mehta, A.D.; Mesgarani, N. Neural decoding of attentional selection in multi-speaker environments without access to clean sources. *J. Neural. Eng.* **2017**, *14*, 056001. [CrossRef] [PubMed]
9. Gazzaley, A. Influence of early attentional modulation on working memory. *Neuropsychologia* **2011**, *49*, 1410–1424. [CrossRef] [PubMed]
10. Zink, R.; Proesmans, S.; Bertrand, A.; Van Huffel, S.; De Vos, M. Online detection of auditory attention with mobile EEG: Closing the loop with neurofeedback. *bioRxiv* **2017**. [CrossRef]
11. Wu, M.C.K.; David, S.V.; Gallant, J.L. Complete functional characterization of sensory neurons by system identification. *Annu. Rev. Neurosci.* **2006**, *29*, 477–505. [CrossRef] [PubMed]
12. Crosse, M.J.; Di Liberto, G.M.; Bednar, A.; Lalor, E.C. The multivariate temporal response function (mTRF) toolbox: A MATLAB toolbox for relating neural signals to continuous stimuli. *Front. Hum. Neurosci.* **2016**, *10*, 604. [CrossRef] [PubMed]
13. Haghighi, M.; Moghadamfalahi, M.; Akcakaya, M.; Erdogmus, D. EEG-assisted modulation of sound sources in the auditory scene. *Biomed. Signal Process. Control* **2018**, *39*, 263–270. [CrossRef]
14. Choi, I.; Rajaram, S.; Varghese, L.A.; Shinn-Cunningham, B.G. Quantifying attentional modulation of auditory-evoked cortical responses from single-trial electroencephalography. *Front. Hum. Neurosci.* **2013**, *7*, 115. [CrossRef] [PubMed]
15. Horton, C.; Srinivasan, R.; D'Zmura, M. Envelope responses in single-trial EEG indicate attended speaker in a 'cocktail party'. *J. Neural Eng.* **2014**, *11*, 046015. [CrossRef] [PubMed]

16. Chu, W.L.; Huang, M.W.; Jian, B.L.; Cheng, K.S. Analysis of EEG entropy during visual evocation of emotion in schizophrenia. *Ann. Gen. Psychiatry* **2017**, *16*, 34. [CrossRef] [PubMed]

17. Labate, D.; Palamara, I.; Mammone, N.; Morabito, G.; Foresta, F.L.; Morabito, F.C. SVM classification of epileptic EEG recordings through multiscale permutation entropy. In Proceedings of the 2013 International Joint Conference on Neural Networks (IJCNN), Dallas, TX, USA, 4–9 August 2013; pp. 1–5.

18. Huang, J.R.; Fan, S.Z.; Abbod, M.F.; Jen, K.K.; Wu, J.F.; Shieh, J.S. Application of multivariate empirical mode decomposition and sample entropy in EEG signals via artificial neural networks for interpreting depth of anesthesia. *Entropy* **2013**, *15*, 3325–3339. [CrossRef]

19. Aiken, S.J.; Picton, T.W. Human cortical responses to the speech envelope. *Ear Hear.* **2008**, *29*, 139–157. [CrossRef] [PubMed]

20. Boshra, R.; Ruiter, K.; Reilly, J.; Connolly, J. Machine learning based framework for EEG/ERP analysis. *Int. J. Psychophysiol.* **2016**, *108*, 105. [CrossRef]

21. Mu, Z.; Hu, J.; Min, J. Driver fatigue detection system using electroencephalography signals based on combined entropy features. *Appl. Sci.* **2017**, *7*, 150. [CrossRef]

22. Hosseini, S.A.; Naghibi-Sistani, M.B. Emotion recognition method using entropy analysis of EEG signals. *Int. J. Image Graph. Signal Process.* **2011**, *3*, 30–36. [CrossRef]

23. Badie, K. A Comparative investigation of wavelet families for analysis of EEG signals related to artists and nonartists during visual perception, mental imagery, and rest. *J. Neurother.* **2013**, *17*, 248–257.

24. Shourie, N.; Firoozabadi, M.; Badie, K. Analysis of EEG signals related to artists and nonartists during visual perception, mental imagery, and rest using approximate entropy. *Biomed. Res. Int.* **2014**, *2014*, 764382. [CrossRef] [PubMed]

25. Alaraj, M.M.R.; Tadanori, F. Quantification of subject wakefulness state during routine EEG examination. *Int. J. Innov. Comput. Inf. Control* **2013**, *9*, 3211–3223.

26. Pincus, S.M. Approximate entropy as a measure of system complexity. *Proc. Natl. Acad. Sci. USA* **1991**, *88*, 2297–2301. [CrossRef] [PubMed]

27. Richman, J.S.; Moorman, J.R. Physiological time-series analysis using approximate entropy and sample entropy. *Am. J. Physiol. Heart Circ. Physiol.* **2000**, *278*, H2039–H2049. [CrossRef] [PubMed]

28. Wu, S.D.; Wu, C.W.; Lin, S.G.; Wang, C.C.; Lee, K.Y. Time series analysis using composite multiscale entropy. *Entropy* **2013**, *15*, 1069. [CrossRef]

29. Liang, Z.; Duan, X.; Li, X.L. Entropy measures in neural signals. In *Signal Processing in Neuroscience*; Li, X.L., Ed.; Springer: Singapore, 2016; pp. 125–166. ISBN 978-981-10-1822-0.

30. Chen, W.; Wang, Z.; Xie, H.; Yu, W. Characterization of surface EMG signal based on fuzzy entropy. *IEEE Trans. Neural Syst. Rehabil. Eng.* **2007**, *15*, 266–272. [CrossRef] [PubMed]

31. Handojoseno, A.M.A.; Shine, J.M.; Nguyen, T.N.; Tran, Y.; Lewis, S.J.G.; Nguyen, H.T. The detection of freezing of gait in parkinson's disease patients using EEG signals based on wavelet decomposition. In Proceedings of the 2012 Annual International Conference of the IEEE Engineering in Medicine and Biology Society, San Diego, CA, USA, 28 August–1 September 2012; pp. 69–72.

32. Müller, K.R.; Tangermann, M.; Dornhege, G.; Krauledat, M.; Curio, G.; Blankertz, B. Machine learning for real-time single-trial EEG-analysis: From brain–computer interfacing to mental state monitoring. *J. Neurosci. Methods* **2008**, *167*, 82–90. [CrossRef] [PubMed]

33. Park, C.H.; Park, H. A comparison of generalized linear discriminant analysis algorithms. *Pattern Recognit.* **2008**, *41*, 1083–1097. [CrossRef]

34. Alahmadi, N. Classifying children with learning disabilities on the basis of resting state EEG measures using a linear discriminant analysis. *Z. Neuropsychol.* **2015**, *26*, 1–8. [CrossRef]

35. Zhang, R.; Xu, P.; Guo, L.; Zhang, Y.; Li, P.; Yao, D. Z-score linear discriminant analysis for EEG based brain-computer interfaces. *PLoS ONE* **2013**, *8*, e74433. [CrossRef] [PubMed]

36. Multiclass LDA in Matlab. Available online: http://freesourcecode.net/matlabprojects/59485/multiclass-lda-in-matlab (accessed on 10 May 2018).

37. Chih-Chung, C.; Chih-Jen, L. LIBSVM: A library for support vector machines. *ACM Trans. Intell. Syst. Technol.* **2011**, *2*, 1–27. [CrossRef]

38. Bleichner, M.G.; Mirkovic, B.; Debener, S. Identifying auditory attention with ear-EEG: CEEGrid versus high-density cap-EEG comparison. *J. Neural Eng.* **2016**, *13*, 66004. [CrossRef] [PubMed]

39. Acharya, U.R.; Sree, S.V.; Alvin, A.P.; Yanti, R.; Suri, J.S. Application of non-linear and wavelet based features for the automated identification of epileptic EEG signals. *Int. J. Neural Syst.* **2012**, *22*, 1250002. [CrossRef] [PubMed]

40. Li, W.; Ming, D.; Xu, R.; Ding, H.; Qi, H.; Wan, B. Research on visual attention classification based on EEG entropy parameters. In Proceedings of the World Congress on Medical Physics and Biomedical Engineering, Beijing, China, 26–31 May 2012; pp. 1553–1556.

41. Bandt, C.; Pompe, B. Permutation entropy: A natural complexity measure for time series. *Phys. Rev. Lett.* **2002**, *88*, 174102. [CrossRef] [PubMed]

42. Kreuzer, M.; Kochs, E.F.; Schneider, G.; Jordan, D. Non-stationarity of EEG during wakefulness and anaesthesia: Advantages of eeg permutation entropy monitoring. *J. Clin. Monit. Comput.* **2014**, *28*, 573–580. [CrossRef] [PubMed]

43. Keller, K.; Unakafov, A.; Unakafova, V. Ordinal patterns, entropy, and EEG. *Entropy* **2014**, *16*, 6212–6239. [CrossRef]

44. Li, J.; Yan, J.; Liu, X.; Ouyang, G. Using permutation entropy to measure the changes in EEG signals during absence seizures. *Entropy* **2014**, *16*, 3049–3061. [CrossRef]

45. Abásolo, D.; Hornero, R.; Espino, P.; Alvarez, D.; Poza, J. Entropy analysis of the EEG background activity in alzheimer's disease patients. *Physiol. Meas.* **2006**, *27*, 241–253. [CrossRef] [PubMed]

46. Li, X.; Ouyang, G.; Richards, D.A. Predictability analysis of absence seizures with permutation entropy. *Epilepsy Res.* **2007**, *77*, 70–74. [CrossRef] [PubMed]

47. Riedl, M.; Müller, A.; Wessel, N. Practical considerations of permutation entropy. *Eur. Phys. J. Spec. Top.* **2013**, *222*, 249–262. [CrossRef]

48. Mirkovic, B.; Debener, S.; Jaeger, M.; De, V.M. Decoding the attended speech stream with multi-channel EEG: Implications for online, daily-life applications. *J. Neural Eng.* **2015**, *12*, 046007. [CrossRef] [PubMed]

49. Faust, O.; Hagiwara, Y.; Tan, J.H.; Lih, O.S.; Acharya, U.R. Deep learning for healthcare applications based on physiological signals: A review. *Comput. Meth. Programs Biomed.* **2018**, *161*, 1–13. [CrossRef]

50. Schirrmeister, R.T.; Springenberg, J.T.; Ldj, F.; Glasstetter, M.; Eggensperger, K.; Tangermann, M.; Hutter, F.; Burgard, W.; Ball, T. Deep learning with convolutional neural networks for EEG decoding and visualization. *Hum. Brain Mapp.* **2017**, *38*, 5391–5420. [CrossRef] [PubMed]

51. Acharya, U.R.; Oh, S.L.; Hagiwara, Y.; Tan, J.H.; Adeli, H. Deep convolutional neural network for the automated detection and diagnosis of seizure using EEG signals. *Comput. Biol. Med.* **2017**, in press. [CrossRef] [PubMed]

52. Acharya, U.R.; Molinari, F.; Sree, S.V.; Chattopadhyay, S.; Ng, K.H.; Suri, J.S. Automated diagnosis of epileptic EEG using entropies. *Biomed. Signal Process. Control* **2012**, *7*, 401–408. [CrossRef]

53. Sharma, R.; Pachori, R.B.; Acharya, U.R. An integrated index for the identification of focal electroencephalogram signals using discrete wavelet transform and entropy measures. *Entropy* **2015**, *17*, 5218–5240. [CrossRef]

54. Sharma, R.; Pachori, R.B.; Acharya, U.R. Application of entropy measures on intrinsic mode functions for the automated identification of focal electroencephalogram signals. *Entropy* **2015**, *17*, 669–691. [CrossRef]

55. Acharya, U.R.; Ng, E.Y.K.; Eugene, L.W.J.; Noronha, K.P.; Min, L.C.; Nayak, K.P.; Bhandary, S.V. Decision support system for the glaucoma using gabor transformation. *Biomed. Signal Process. Control* **2015**, *15*, 18–26. [CrossRef]

entropy

MDPI

Article

Analog Circuit Fault Diagnosis via Joint Cross-Wavelet Singular Entropy and Parametric t-SNE

Wei He [1], Yigang He [1,2]*, Bing Li [1] and Chaolong Zhang [1,3]

[1] School of Electrical Engineering and Automation, Hefei University of Technology, Hefei 230009, China; wei.he@stud.uni-due.de (W.H.); libinghnu@163.com (B.L.); zhangchaolong@126.com (C.Z.)
[2] School of Electrical Engineering, Wuhan University, Wuhan 430072, China
[3] School of Physics and Electronic Engineering, Anqing Normal University, Anqing 246011, China
* Corresponding: hiway@mail.hfut.edu.cn

Received: 31 May 2018; Accepted: 25 July 2018; Published: 14 August 2018

Abstract: In this paper, a novel method with cross-wavelet singular entropy (XWSE)-based feature extractor and support vector machine (SVM) is proposed for analog circuit fault diagnosis. Primarily, cross-wavelet transform (XWT), which possesses a good capability to restrain the environment noise, is applied to transform the fault signal into time-frequency spectra (TFS). Then, a simple segmentation method is utilized to decompose the TFS into several blocks. We employ the singular value decomposition (SVD) to analysis the blocks, then Tsallis entropy of each block is obtained to construct the original features. Subsequently, the features are imported into parametric t-distributed stochastic neighbor embedding (t-SNE) for dimension reduction to yield the discriminative and concise fault characteristics. Finally, the fault characteristics are entered into SVM classifier to locate circuits' defects that the free parameters of SVM are determined by quantum-behaved particle swarm optimization (QPSO). Simulation results show the proposed approach is with superior diagnostic performance than other existing methods.

Keywords: analog circuit; fault diagnosis; cross wavelet transform; Tsallis entropy; parametric t-distributed stochastic neighbor embedding; support vector machine

---

1. Introduction

With the fast development of electronic science and technology, fault diagnosis and testing as fundamental tasks in preventive maintenance of electronic systems play a vital role in reliability of the product and promoting industrial development [1,2]. It is estimated that testing covers one third of the cost of the product, and majority of the testing is due the testing of the analog parts of the mixed signal circuits [3,4]. Due to continuous parameter and tolerance of analog components, and lack of test nodes, the diagnostics approaches of analog circuits are far less advanced, comparing with well-developed automatic fault diagnosis methodologies for digital circuits. Consequently, there is a pressing need to explore effective fault diagnosis and testing approaches to prevent fault enlargement and guarantee analog electronic system reliable operation.

Faults in analog circuits can be categorized into soft faults and hard faults. Soft faults result in system performance degradation where the parameters of components only deviate from the normal values exceeding the tolerance range. The causes for soft faults mainly include: the aging of an electronic system, fabrication tolerance, electromagnetic interfere and effect of ambient temperature [5]. Conversely, hard faults mainly happen in short- and open- circuit, or they are caused by the larger parameter variation of components [6]. The majority examples of hard faults involve the structural failure in bipolar junction transistor (BJT) and metallic oxide semiconductor field effect transistor (MOSFET) and the parameter deviation of key components in filter circuits.

Currently, there are many diagnosis approaches aiming at the two kinds of analog circuit faults. The vast majority of these methods are only implemented for field failure in factory production processes. However, the implementation of component-level diagnosis is challenging [2,6]. With respect to analog circuits, it is mainly due to the complex and changing operation conditions and external environment, such as strong electromagnetic interference, high-temperature and complicated failure mechanisms. Therefore, it is necessary to investigate an effective diagnosis method for component failure in analog electronic systems.

The rest of this paper is organized as follows. Section 2 contains a survey of the related work. In Section 3, fault feature extraction based on cross-wavelet singular entropy and parametric t-SNE is introduced. In Section 4, the algorithm and implementation procedures of the proposed PSO for parameter selection of SVM are provided. Further, fault diagnosis test in two experimental circuits is performed in Section 5 to verify the effectiveness of the proposed method. In Section 6, a discussion based on Shannon, Rényi and Tsallis entropies is presented. Finally, some conclusions are drawn in Section 7.

2. Related Works

Traditionally, analog circuit fault diagnoses are classified into two broad approaches: Simulation After Test (SAT) and Simulation Before Test (SBT). Compared with SAT approach, the SBT approach is more suitable for diagnostics of analog circuits as it only implements once off-line simulation process, removing on-line computation before testing and running [7]. Among SBT, data-driven diagnostic methods are based on the case that features of the system relatively changed when a fault happens. They extract features from output signals, then apply pattern recognition techniques such as neural networks (NNs) and support vector machines (SVMs) to locate a fault [8]. Meanwhile, the data-driven techniques do not need to construct an explicit model. Hence, the data-driven approaches have been applied to fault diagnosis in many relative works [9,10].

Technically, a data-driven approach can be divided into two phases: feature extraction and classifier application [11–13]. Obviously, feature extraction is the vital steps. To date, increasing numbers of feature extraction tools have been utilized in fault diagnosis, and they can be summarized into three categories: time-domain analysis, frequency-domain analysis, and time-frequency analysis [9,14]. Signals collected from the testing nodes of faulty circuits always carry interference components that probably overwhelm useful information. Thus, it is difficult to effectively recognize the defects of analog electronic systems when only considering the features of time-domain or frequency-domain [15]. As a typical time-frequency domain analysis, wavelet transform (WT) can reveal overlaps in time-frequency domains by decomposing the signal into a set of wavelet coefficients that vary continually over time [10]. Nevertheless, in practice, the measured signals of analog circuits commonly contain random noise, which may lead to misclassification. Therefore, it is necessary to take actions to minimize the impact of random noise. Noise removal can be executed by setting a threshold when computing wavelet coefficients [10]. However, there are some limitations: The threshold needs to be set manually, and the calculation process is time-consuming. Recently, cross-wavelet transforms (XWT) has been employed to handle partial discharge pulses and ECG signal [16,17]. Moreover, XWT has an outstanding ability in extracting time-frequency characteristics of signal and restraining noise. Consequently, XWT is applied to process the fault signals of analog circuits.

However, there are still several open issues that need to be addressed for XWT. In practical application, XWT is limited to being imported into classifiers directly because the transformed result is a high-dimension matrix. Therefore, it is necessary to combine XWT with other feature extraction techniques to reduce information abundance.

As a description of disorder or randomness of matter, entropy is capable of providing rich information about signals, which is fit for feature extraction [18–20]. Many scholars have devoted themselves to the field of feature extraction with use of entropy techniques. Approximative maximum entropy (Apen) has been used to diagnosis faults [9,21]. However, a bad performance could be obtained when processing the short data-set. Moreover, the Apen is sensitive to noise. Because sample entropy (Samp) is insensitive to data length and immune to noise, it can be employed as an input vector of classifiers [20,22]. However, because the Heaviside step function of sample entropy entails discontinuity at the boundary, negative results are possible. In view of this, many scholars adopt Fuzzy entropy (Fen) that vary smoothly and continuously to estimate data complexity [23]. Unfortunately, the membership function in Fen is usually difficult to determine. Some achievements in fault detection have been made using cross entropy and Rényi's entropy [24,25], but the faulty components have not been located. Moreover, none of these techniques are used to extract features with wavelet transform. The utilization of wavelet Shannon entropy (Wse) in feature extraction is proposed, achieving a desirable performance [26]. Nevertheless, the XWT manifests a non-extensive character because of energy leakage and aliasing in the phase of wavelet operation, while Shannon entropy belongs to extensive entropy.

Based on the above, a novel feature extraction technique based on XWT and Tsallis entropy is proposed for fault diagnosis. Owning to its ability of regulating non-extensiveness, Tsallis entropy is employed to construct the feature set, denoting the complexity of fault signals [27,28]. Furthermore, to improve the efficiency of fault pattern recognition, a feasible feature reduction approach needs be implemented. A manifold learning technique is able to unearth intrinsic information embedding in highly dimensional datasets via mapping them into a low-dimensional space and retaining the local neighborhood information. Parametric t-stochastic neighbor embedding (t-SNE) has a good capability in mapping the data with high-dimension into low-dimension representation. It maintains the conditional probability distribution of data associated with the pairwise similarity from the high-dimension space to the feature subspace [29]. Therefore, it is utilized to extract discriminative features between different fault patterns.

To locate the faults, a support vector machine is employed as the classifier. SVM has advantages of high training speed and distinctive generalization ability by finding the optimal hyper-plane [30,31]. However, in practical application, it is difficult to assign the free parameter. To address this issue, various intelligent optimization algorithms, such as genetic algorithm and simulated annealing, have been utilized to determine hyper-parameters of SVM. Owing to high speed of converge and good quality of computation, quantum-behaved particle swarm optimization (QPSO) is adopted to obtain the optimal parameters [32].

3. Feature Extraction

3.1. Cross Wavelet Transform

Given a time domain signal $x(t)$, continuous wavelet transform (CWT) can be defined as:

$$W^x(a, \tau) = a^{-1/2} \int_{-\infty}^{+\infty} \Psi^*(\frac{t-\tau}{a}) dt \tag{1}$$

where Ψ stands for mother wavelet; $*$ denotes complex conjugation; a ($a > 0$) and τ are usual "dilation" and "translation" parameters.

The Morlet wavelet is a commonly used complex valued function, which can reveal the localization property of the signal in the time-frequency domain. The Morlet wavelet function can be described as follows:

$$\Psi(t) = \pi^{-1/4}(e^{-jw_0 t} - e^{-w_0^2/2})e^{-t^2/2} \tag{2}$$

Assuming two time domain signals $x(t)$ and $y(t)$, the cross wavelet transform can be defined as below [33,34]

$$W^{xy}(a, \tau) = W^x(a, \tau) W^{y*}(a, \tau) \tag{3}$$

Accordingly, we can plot the cross-wavelet spectrum by using the magnitude $W^{xy}(a, \tau)$ and phase $\phi = \tan^{-1} \frac{\Im\{W^{xy}(a,\tau)\}}{\Re\{W^{xy}(a,\tau)\}}$.

Via cross wavelet analysis, we can not only estimate the degree of correlation among signals, but also reveal the phase relationship of signals in time-frequency space.

3.2. Singular Value Decomposition (SVD)

On the basis of SVD theory [35], for any $m \times n$ matrix A can be decomposed into a $m \times r$ column-orthogonal matrix U, an $n \times r$ orthogonal matrix V, and a $r \times r$ diagonal matrix Λ, which can be described as below

$$A = U \Lambda V^T \tag{4}$$

where

$$\Lambda = \begin{bmatrix} \lambda_1 & 0 & \cdot & 0 & 0 \\ 0 & \lambda_2 & \cdot & 0 & 0 \\ \cdot & \cdot & \cdot & \cdot & \cdot \\ 0 & 0 & \cdot & \lambda_{r-1} & \cdot \\ 0 & 0 & \cdot & 0 & \lambda_r \end{bmatrix} \tag{5}$$

and its diagonal elements λ_i ($i = 1, 2, \ldots, r$) are called "singular values" of matrix A. The singular values are all nonnegative and arranged in a descending order (i.e., $\lambda_1 \geq \lambda_2 \geq \cdots \geq \lambda_r > 0$).

3.3. Tsallis Entropy

For a uncertain system, the entropy is explored to estimate the uncertainty of the discrete event, which is associated with the probability distribution. Given $p = \{p_i\}$ denotes the probability of the system state i, where $0 \leq p_i \leq 1$ and $\sum_{i=0}^{m} p_i = 1$. Thus, the Shannon entropy can be described as:

$$S = -\sum_{i=1}^{k} p_i \ln(p_i) \tag{6}$$

Besides, Shannon entropy has the extensive property:

$$S(A + B) = S(A) + S(B) \tag{7}$$

Inspired by multi-fractal concepts, Tsallis entropy is investigate to describe non-extensive system [36], which can be expressed as

$$S_q = \frac{1}{q-1}(1 - \sum_{i=1}^{k} (p_i)^q) \tag{8}$$

where q stands for the entropic index, which leads to the non-extensive statistic and k denotes the total number of the system states.

3.4. Definition of XWSE

For a given time domain fault signal $s(t)$ and template signal $e(t)$, the detail about the feature extraction by using XWSE can be described as below:

- First, analyze the $s(t)$ with XWT, where the "morlet" wavelet function is chosen in the process. Then, a XWT spectrum matrix A can be obtained by using Equations (1)~(3).
- Second, the matrix A is divided into eight blocks with the same size as follows:

$$A = \begin{bmatrix} B_1 & B_2 & B_3 & B_4 \\ B_5 & B_6 & B_7 & B_8 \end{bmatrix}$$

- Third, decompose the block $B_n (n = 1, 2, \ldots, 8)$ with SVD, and a singular-value sequence for each block can be obtained as $\{\lambda_1, \lambda_2, \ldots, \lambda_r\}$ where r is the rank of the diagonal matrix Λ.

• Finally, the XWSE of the block B_n is defined by

$$XWSE_n = \frac{1}{q-1}(1 - \sum_{i=1}^{k}(p_i)^q) \ (n = 1, 2, \dots, 8)$$

where the probability p_i associated with λ_i is defined as $p_i = \lambda_i / \sum_{j=1}^{r} \lambda_j$. Thus, the XWSE features of fault signal $s(t)$ can be expressed as $[XWSE_1, XWSE_2, \dots, XWSE_8]$

3.5. Parametric t-Stochastic Neighbor Embedding (Parametric t-SNE)

Given $X = [x_1, x_2, \dots, x_n] \in \Re^{D \times n}$ is the high dimensional data set, where D represents the dimension of x_i $(i = 1, 2, \dots, n)$, and n is the number of samples. Suppose $Y = [y_1, y_2, \dots, y_n] \in \Re^{d \times n}$ $(d < D)$ denotes the low-dimensional map of X. By using t-SNE, the pairwise distance is transformed into the probabilities to measure the similarities between data [37,38]. In the raw space, the pairwise similarities are described as

$$p_{ij} = \frac{\exp(-d_H(x_i, x_j)^2)/2\sigma^2}{\sum_{k \neq l}(-d_H(x_k, x_l)^2/2\sigma^2} \tag{9}$$

where the value of σ is determined by a binary search with a fixed perplexity. Here, the perplexity denotes the effective number of the nearest neighbors of the data x_i, and the pairwise distance $d_H(x_i, x_j)$ represents the Euclidean distance.

In order to solve the "Crowding Problem", the pairwise similarities are employed to described by the long-tailed student t-distribution.

$$q_{ij} = \frac{(1 + d_L(y_i, y_j)^2)^{-1}}{\sum_{k \neq l}(1 + d_L(y_k, y_l)^2)^{-1}} \tag{10}$$

where $d_L(\cdot)$ stands for Euclidean distance.

Via minimizing the Kullback–Leibler divergence between two probability distributions, the cost function $E(Y)$ is obtained to preserve the local structural characteristics of the data.

$$E(Y) = \sum_{i,j} p_{ij} \log(p_{ij}/q_{ij}) \tag{11}$$

However, t-SNE cannot address the out-of-sample extension problem. Accordingly, the parametric t-SNE, an extension of t-SNE technique is proposed [39]. Owing to the excellent capability of the constructed nonlinear projection, Restricted Boltzmann Machines (RBMs) is adopted to construct a pre-trained parametric t-SNE network. The aim is to define a superior initialization for the fine-tuning phase. As the projection is parametric by the deep-forward network f with weight matrix W, q_{ij} can be defined as follows:

$$q_{ij} = \frac{(1 + \| f(x_i|W) - f(x_j|W) \|^2 / \alpha)^{-\frac{\alpha+1}{2}}}{\sum_{k \neq i}(1 + \| f(x_k|W) - f(x_i|W) \|^2 / \alpha)^{-\frac{\alpha+1}{2}}} \tag{12}$$

where α denotes the degrees of freedom of the t-distribution. Then this equation is used as the definition of q_{ij} in Equation (10)

4. SVM and QPSO

4.1. Support Vector Machine (SVM)

Given a training set of N data points $\{(x_i, y_i)\}$, where $x_i \in R^n$ denotes the *i*th data point, and the associated $y_i \in \{+1, -1\}$ represents a class label. Then, the mathematic equation of the classifier by using support vector too can be described as follows:

$$y(x) = \text{sign}(w^T \varphi(x) + b) \tag{13}$$

Here $\varphi(\cdot)$ stands for the kernel function which projects the input samples space into the higher dimensional feature space; *b* denotes the bias parameter, and *w* represents the weight vector of the input features.

The optimal values of *w* and *b* can be obtained by finding the solution of the following optimization problem:

$$\min_{w,b,e} J(w, b, e) = \frac{1}{2} w^T w + \frac{C}{2} \sum_{i=1}^{N} e_i^2 \tag{14}$$
$$\text{s.t. } y_i(w^T \varphi(x_i) + b) = 1 - e_i, i = 1, \dots, N$$

where *C* denotes the regularization parameter which balance the trade-off between complexity and the proportion of non-separable samples; e_i stands for the positive slack term for misclassification.

To address the above problem, Lagrangian function is introduced.

$$L(w, b, e, a) = J(w, b, e) - \sum_{i=1}^{N} a_i\{y_i(w^T \varphi(x_i) + b) - 1 + e_i\} \tag{15}$$

where a_i stands for the Lagrangian multiplier.

Finally, the decision function of the SVM classifier for any test vector $x \in R^N$ can be given as follows:

$$y(x) = \text{sign}(\sum_{i=1}^{N} a_i y_i K(x, x_i) + b) \tag{16}$$

where $K(x, x_i) = \varphi(x_i)^T \varphi(x)$ represents the kernel function. In this work, radial basis function (RBF: $K(x, x_i) = \exp(-\lambda \parallel x_i - x \parallel^2)$) is chosen as the kernel function of the SVM classifier. Here, the term λ plays a important role on the distribution form of the samples in the high dimensional feature space.

After selecting the kernel function, the regularization parameter *C* and the RBF parameter λ should be determined. Thus, QPSO is utilized to find the optimal parameters of *C* and λ in order to improve the classification ability of SVM.

4.2. Quantum-Behaved Particle Swarm Optimization (QPSO)

In 1995, Ederhart and Kennedy came up with the PSO algorithm to search the optimal solutions via imitating the preying behavior of birds [40]. Nevertheless, the algorithm has some drawbacks, such as slow convergence rate and poor search ability. From the view of quantum mechanics, Sun et al. [41] have put forward QPSO. The probability of each particle's next iteration position relies on the potential field of the particle, which is defined as below:

$$X_i(t+1) = P \pm a|nbest - X_i(t)|\ln(1/u) \tag{17}$$

$$nbest = \frac{1}{N} \sum_{i=1}^{N} P_i \tag{18}$$

$$P = sP_i + (1-s)P_g \tag{19}$$

where $i = 1,2,\ldots,N$ and N is swarm size; u and s are uniformly distributed random numbers generated between 0 and 1; Pg is the global optimal position of all particles and Pi is the particle i's optimal position; $X_i(t+1)$ is the position of particle i in iteration $t+1$; *nbest* is the center of all individual optimal positions; a is a contraction expansion coefficient.

4.3. The Procedure of Parameters Optimization

This section introduces the flowchart of the QPSO algorithm-optimized support vector machine for fault diagnosis. The flowchart is shown in Figure 1 and the main steps are described as below:

Step 1: Initialize the QPSO algorithm parameters.

Step 2: For each particle, the fitness is calculated, where the cross-validation testing accuracy is used as the fitness function.

Step 3: Determine each particle optimal position and the global optimal position.

Step 4: Update the velocity and position of each particle in accordance with Equations (17)~(19).

Step 5: Repeat step 2 to step 5 until reaching the stop criterion.

Step 6: Export the optimal 2-dimensional position as the parameters of the SVM.

Step 7: Exit the program.

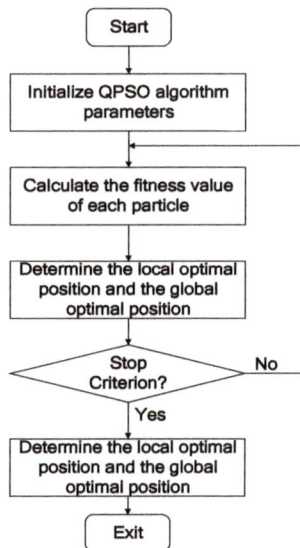

Figure 1. The flow chart of the parameter optimization.

5. Experimental Results and Analysis

The proposed method is investigated on three popular analog circuits in this paper. For the test circuits, each fault class is conducted 60 Monte Carlo analysis. Among these samples, 50% are used for training and the last 50% are used for testing. All testing samples are verified by an SVM classifier, then fault components can be located.

5.1. Example Circuits

(1) CUT 1: The first CUT (circuit under test) shown in Figure 2 is a sallen-key band-pass circuit. In this test, the components R2, R3, C1 and C2 are chosen as fault components. The tolerances of the

resistors and capacitors are all equal to 5%. A total of nine fault classes, including the fault-free (NF) status of circuits, are simulated, and the corresponding fault values and labels are shown in Table 1. In the following Tables 1–3, ↑ and ↓ refer to higher and lower than the nominal value, respectively.

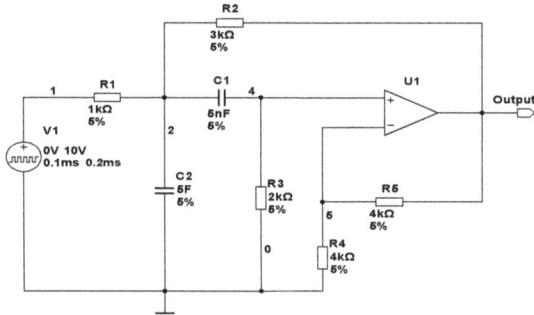

Figure 2. Schematic of a sallen-key band-pass filter.

Table 1. Fault classes for Sallen-key bandpass filter.

Fault Code	Fault Class	Nominal Value	Faulty Value
F0	NF	-	-
F1	R2↓	3 kΩ	2.2 kΩ
F2	R2↑	3 kΩ	3.6 kΩ
F3	R3↓	2 kΩ	1.6 kΩ
F4	R3↑	2 kΩ	2.4 kΩ
F5	C1↓	5 nF	4 nF
F6	C1↑	5 nF	6.5 nF
F7	C2↓	5 nF	4 nF
F8	C2↑	5 nF	6.5 nF

(2) CUT 2: The second CUT, a four-opamps filter circuit, is shown in Figure 3. Thirteen fault classes are all shown in Table 2. The tolerances of the resistors and capacitors are also set to 5%. A pulse signal with 10 V peak, 10 μs duration and 1ms period is considered as the input signal of the circuit.

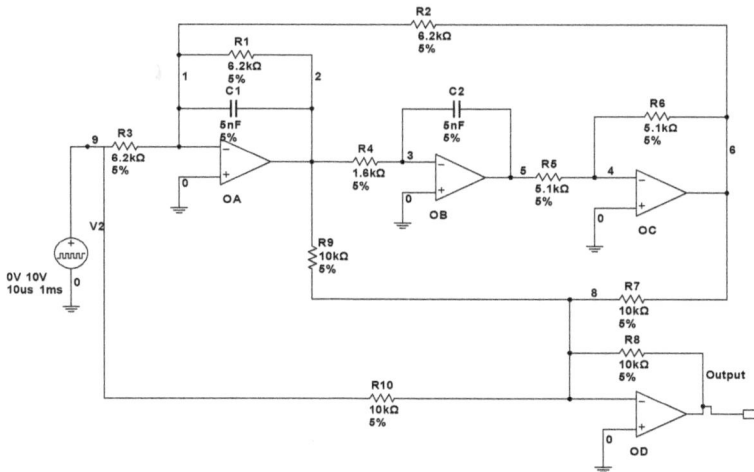

Figure 3. Schematic of a four-opamp filter circuit.

Table 2. Fault classes for four opamp filter circuit.

Fault Code	Fault Class	Nominal Value	Faulty Value
F0	NF	-	-
F1	R1↓	6.2 kΩ	3 kΩ
F2	R1↑	6.2 kΩ	15 kΩ
F3	R2↓	6.2 kΩ	2 kΩ
F4	R2↑	6.2 kΩ	18 kΩ
F5	R3↓	6.2 kΩ	2.7 kΩ
F6	R3↑	6.2 kΩ	12 kΩ
F7	R4↓	6.2 kΩ	0.5 kΩ
F8	R4↑	6.2 kΩ	2.5 kΩ
F9	C1↓	5 nF	2.5 nF
F10	C1↑	5 nF	10 nF
F11	C2↓	5 nF	1.5 nF
F12	C2↑	5 nF	15 nF

(3) CUT 3: To investigate the performance of proposed method in nonlinear circuits, a test of the duffing chaotic circuit shown in Figure 4 is conducted in this section. In this case, an excitation signal with the frequency of 0.155159 Hz and the amplitude of 0.7414148 V is chosen. The normal tolerance of resistor and capacitor is also assumed as 5%. We only collected the signals at the output node, and a 30% deviation of nominal value was considered as a fault condition. The fault modes are listed in Table 3. In this work, the test is denoted as Case 3. After data acquisition, we obtain the original samples set with size of 1080. The size of training samples set and testing samples set are all equal to 540 (30 × 18).

Table 3. Fault classes for duffing chaotic circuit.

Fault Code	Fault Class	Fault Value
F0	-	-
F1	R1↓	7 kΩ
F2	R1↑	13 kΩ
F3	R2↓	7 kΩ
F4	R2↑	13 kΩ
F5	R3↓	7 kΩ
F6	R3↑	13 kΩ
F7	R4↓	14 kΩ
F8	R4↑	26 kΩ
F9	R8↓	7 kΩ
F10	R9↓	0.7 MΩ
F11	R10↑	13 kΩ
F12	C1↓	0.7 μF
F13	C2↑	1.3 μF
F14	R1↑R2↑	(13 kΩ) (13 kΩ)
F15	R1↓R3↓	(7 kΩ) (7 kΩ)
F16	R5↑C1↑	(7 kΩ) (1.3 μF)
F17	R6↓C2↓	(0.7 MΩ) (0.7 μF)

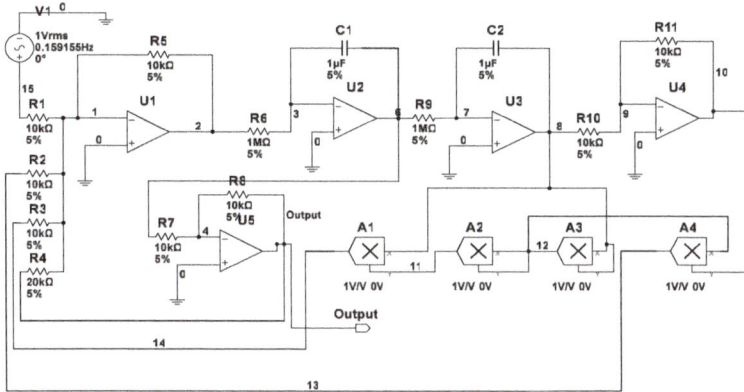

Figure 4. Schematic of a duffing chaotic circuit.

5.2. The Results Analysis of Feature Extraction

First, the sampled signals of CUTs are preprocessed by using XWT to obtain time-frequency spectra (TFS). Owing to the large quantity of fault classes, it is not feasible to list all TFS for all fault classes. Thus, we only present the TFS of F0 and F7 in Figure 5 for CUT1, and the TFS of F0 and F7 in Figure 6 for CUT2. In the figures, the color in the subgraph implies the power in the time-scale plane. And, the black arrow in each sub-image indicates the phase angle. The results from Figures 5 and 6 can be concluded as follows:

(1) As shown in Figure 5, the TFSs between F0 and F7 have tiny differences. It means that the time-frequency distribution only undergos minor changes when faults happen. However, compared with the TFS of F0, the phase distributions in the TFS of F7 has an apparent difference. It indicates that the XWT can fetch phase information effectively.

(2) From Figure 6, compared with the TFS of F0, the phase distribution of F9 in the whole time-frequency plane undergoes dramatic change, and the energy accumulation block in the middle shows a considerable variation.

Consequently, with the application of cross-wavelet transform, the energy and phase characteristics in time-frequency domain can be extracted to analyze the work conditions of analog circuits.

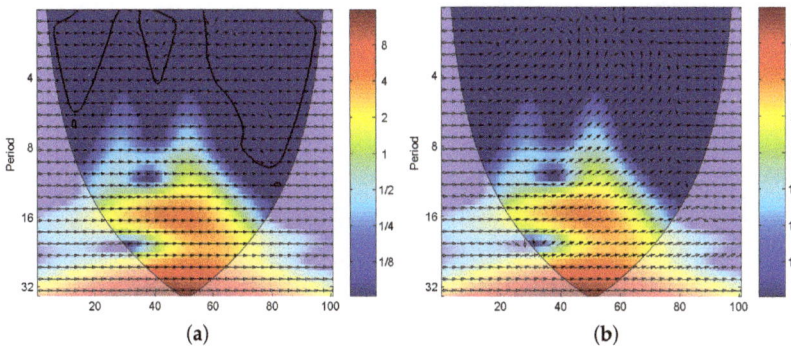

(a)   (b)

Figure 5. The time-frequency spectra obtained by XWT for CUT 1 (a) F0, (b) F7.

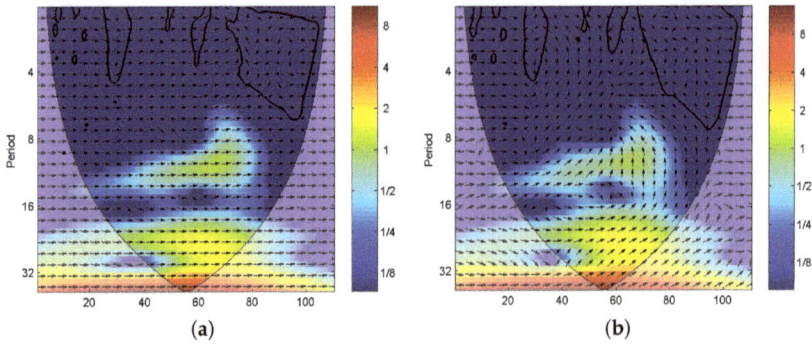

Figure 6. The time-frequency spectra obtained by XWT for CUT 2 (**a**) F0, (**b**) F9.

After calculating singular entropies of blocks in the TFS, Tsallis entropy curves for CUTs are drawn in Figure 7.

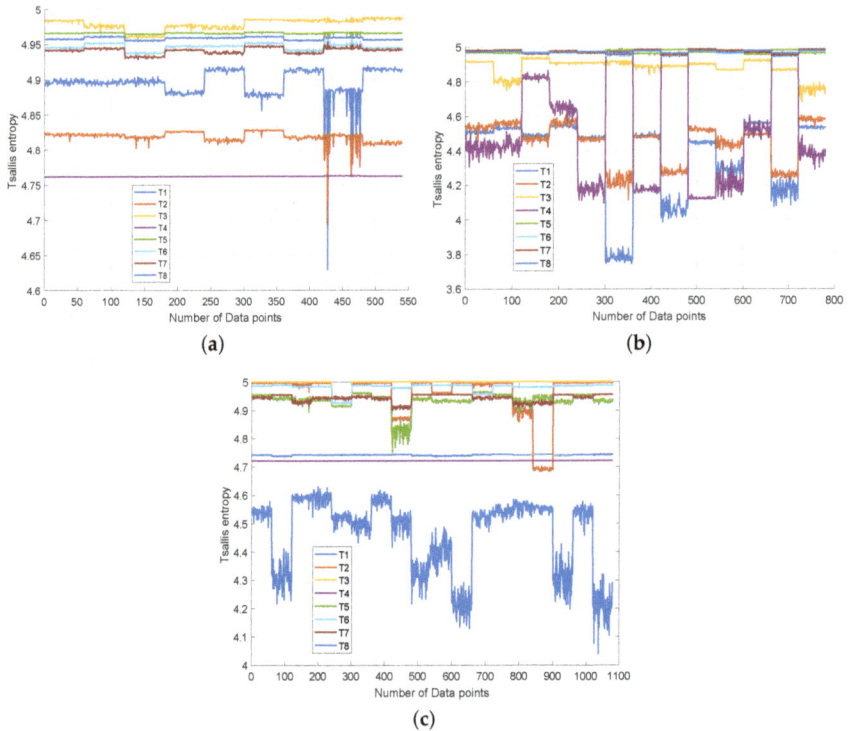

Figure 7. The XWSE features distribution of (**a**) CUT1, (**b**) CUT2, (**c**) CUT3.

As we can see from Figure 7, the eight entropies have apparent difference for all fault modes, although there exist overlapping in some points of different fault classes. It implies that Tsallis entropy can provide some discriminative information for fault recognition.

Here, nine kinds of entropy techniques, including approximate entropy (Apen) [9], sampEn entropy (Samp) [22], fuzzy entropy (Fen) [23], permutation entropy (Per) [42], fuzzy approximate entropy

(Fapen) [43], corrected conditional entropy (Cce) [43], Tsallis entropy [28] and shannon entropy [26], are employed to extract fault features, and these features are directly imported into SVM classifiers. The dimension of features is varied from 1 to 16 and finally, the resultant feature set without feature reduction are employed as the input vectors of SVM classifier. Figure 8 shows the classification rates for CUT1 and CUT 2 varying from the first features to all features. It can be observed that the recognition rate of Tsallis entropy increases steadily and achieves the highest accuracy in whole scale. Hence, it can be concluded from Figure 8 that Tsallis entropy is superior to the other entropy techniques on feature extraction.

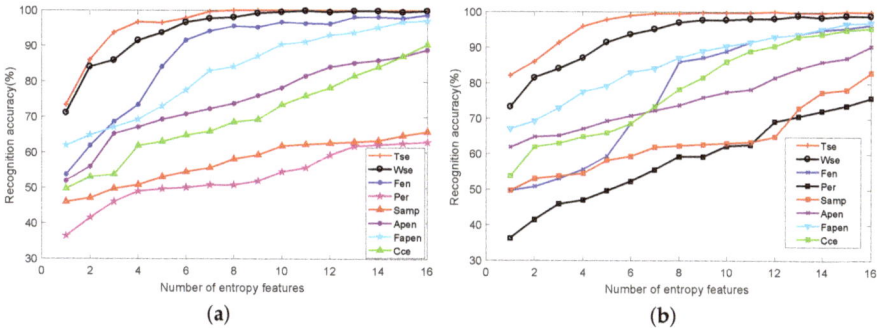

Figure 8. The plots of classification accuracy versus the number of features for various entropy techniques (**a**) CUT1, (**b**) CUT2.

Finally, we apply the parametric t-SNE to obtain the optimal low-dimensional representation. It not only requires less training and processing time, but also leads to a smaller structure and better generalization performance for the adopted SVM. The 220-600-600-2500-2 parametric t-SNE network structure is utilized on the fault data.

The 2-D scatter plots for the whole fault classes in CUT1 and CUT2 are shown in Figure 9. Meanwhile, the visualization of the fault data using locality preserving projection (LPP) [44] and linear local tangent space alignment (LLTSA) [45] are reported in Figures 10 and 11. From Figure 9, it can be concluded that the proposed algorithm can substantially improves the separability degree of different fault classes. On the contrary, there are strong overlapping between different fault classes in Figures 10 and 11. Therefore, it can be concluded that the optimal low-dimensional features can be obtained by using the Parametric t-SNE.

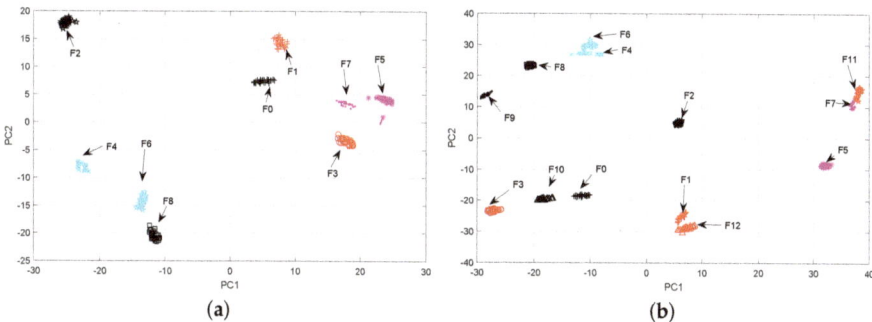

Figure 9. *Cont.*

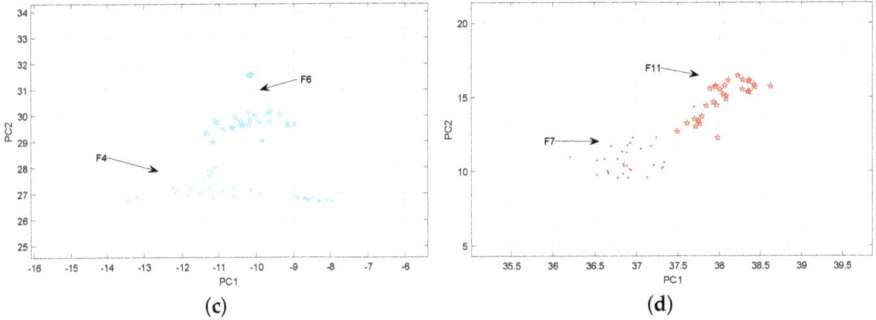

Figure 9. The scatter plots of two-dimensional features obtained by parametric t-SNE (**a**) CUT1, (**b**) CUT2, (**c**) F4 and F6 of CUT 2, (**d**) F7 and F11 of CUT 2.

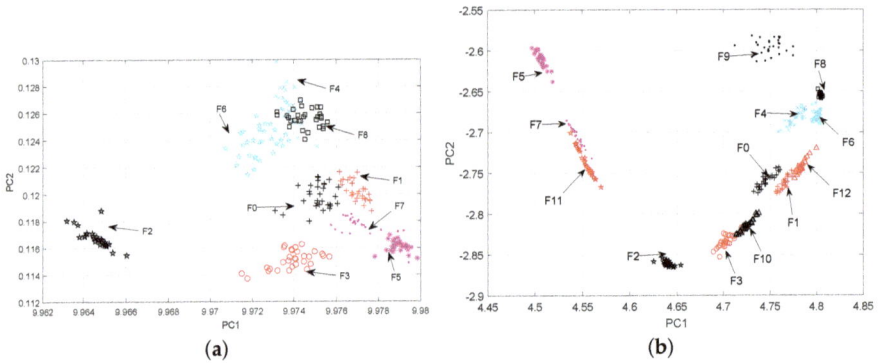

Figure 10. The scatter plots of two-dimensional features obtained by LPP (**a**) CUT1, (**b**) CUT2.

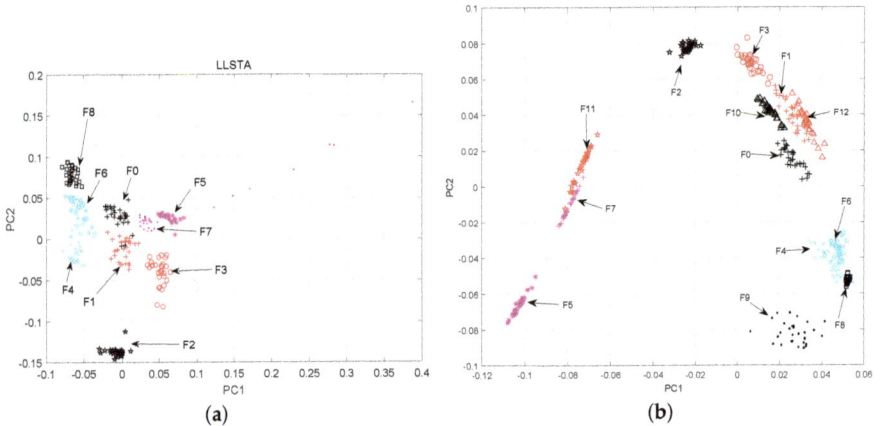

Figure 11. The scatter plots of two-dimensional features obtained by LLSTA(**a**) CUT1, (**b**) CUT2.

5.3. Classification Result by Using QPSO-SVM Model

In this study, the QPSO-based SVM is used as a classifier. After z-score normalization, the optimal features obtained by using parametric t-SNE are imported into the classifier to locate the faults.

Because 60 Monte-Carlo runs are implemented for each fault class, there are 540 samples for CUT 1780 samples for CUT 2, and 1080 samples for CUT3. Each samples set is divided into two subsets with the same size. The two subsets are used as training and testing data sets, respectively. Figure 12 shows the parameter optimization procedures of these three cases. As illustrated in the figures, the presented optimization algorithm achieves desirable performances during the training stage with consuming much few time. Thus, it can be concluded that the characters in different fault classes of the circuits tend to separate obviously, and the proposed QPSO-SVM. have excellent classification ability. The optimal solutions $[C, \lambda]$ for the three CUTs are [0.01, 4.076], [0.01, 86.15] and [6.5152, 0.1595] respectively.

Subsequently, the test samples are used as the input vectors of the SVM model to recognize the states. The classification accuracy comparisons with other current works for CUT1 and CUT2 are given in Table 4. Additionally, the diagnosis result of the proposed method for CUT3 is shown in Figure 13.

Table 4. Recongnition performance comparision of the proposed method with other existing methods.

| Works | Approach | Accuracy (%) | |
		CUT 1	CUT 2
Aminian et al. [1]	WT + PCA + NN	97	95
Xiao et al. [4]	FrWT + KPCA + Ridgelet−NN	100	98.52
Yuan et al. [9]	Entropy + Kurtosis + NN	100	99
Vasan et al. [10]	WT + entropy, Kurtosis + SVM	99.70	95.69
Song et al. [46]	FrFT statistical feature + SVM	98.41	95.12
Chen et al. [47]	WPT + DCQGA−SVM	97.41	98.72
Proposed	XWSE + Pt−SNE + QPSO-SVM	99.26	99.74

As shown in Table 4, it can be observed that our proposed method achieves a better result than that of other listed works, with other exceptions [4,9,10]. However, the fault components in our work have smaller parametric deviation. When fault components occur with smaller parametric deviation, the features of different fault classes tend to overlap, which results in a lower diagnosis accuracy. For the second CUT, the proposed method achieves the highest diagnostic accuracy. Therefore, with the diagnosis performance of CUT1 and CUT2, it can be summed up that the proposed scheme can effectively and accurately diagnose the soft faults in analog circuits.

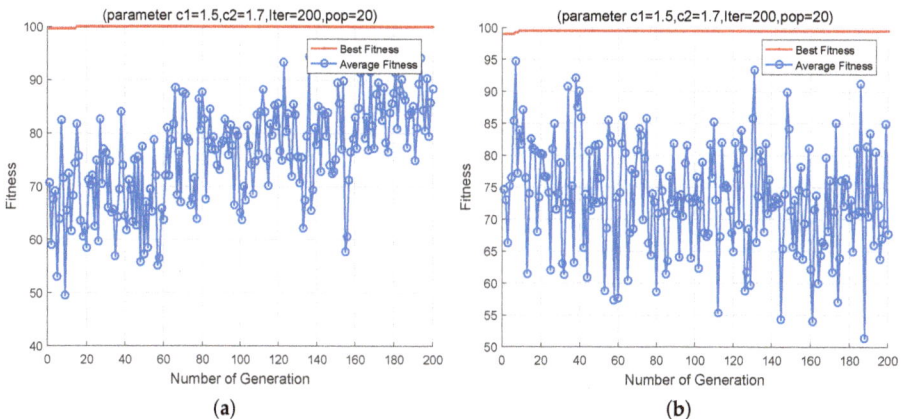

Figure 12. *Cont.*

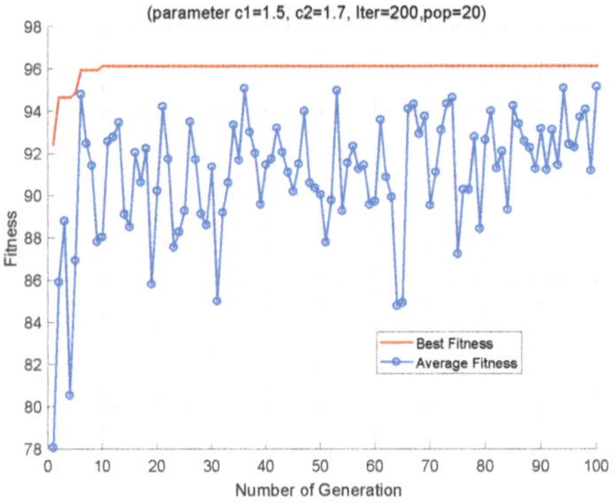

(c)

Figure 12. Best and the average fitness values versurs number of iterations for (**a**) CUT1, (**b**) CUT2, (**c**) CUT3.

As shown in Figure 13, it can be observed that some diagonal elements in the confusion matrix are close to 1. It means that the proposed algorithm has a good ability in classifying fault patterns into its actual class. However, the proposed algorithm gets unsatisfied results when dealing with some fault samples in F2 and F11. It implies that the proposed approach still needs to be improved further to fulfil the task of fault diagnosis in complex nonlinear circuits.

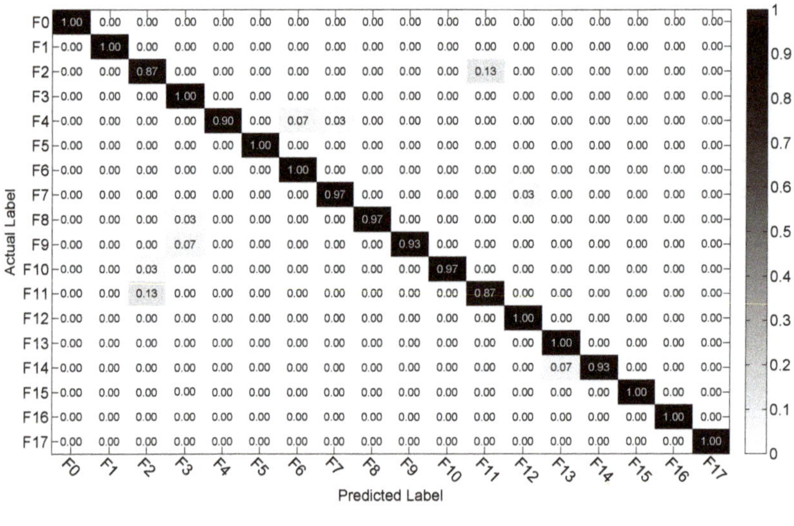

Figure 13. The diagnosis results of the proposed method for CUT 3.

6. Discussion

Compared with Tsallis entropy, other entropy techniques, such as Rényi and Shannon entropies have already been applied to many diverse practical problems [48,49]. Therefore, a discussion based on Shannon, Rényi and Tsallis entropy is described in this section.

For given two probabilities p_1 and $p_2(p_2 = 1 - p_1)$, the plots of Shannon, Rényi and Tsallis entropies are shown in Figures 14–16. Here, the Rényi entropy is defined as $I_q = \frac{1}{1-q} \log(\sum_{i=1}^{n} p_i^q)$.

As shown in the figures, with the increase of q, the statistical range of Rényi entropy and Tsallis entropy will change, and the entropy values of the probability events will decrease correspondingly. However, with Shannon entropy, the statistical value of the probability events remains unchanged due to the equal weights in the entropy computation. For Tsallis entropy, the events with high probability contribute more than lower probabilities. The Rényi entropy with higher q parameter is determined by events with higher probabilities and the lower values of q coefficients weigh the events more equally.

For a signal containing noise components, the low energy components which can be used to characterize may be corrupted by the background noise that is relative to the events with small probability. In this context, Rényi and Tsallis entropies can achieve better results in extracting features by selecting appropriate q parameter to minimize noise as compared to Shannon entropy. Furthermore, Tsallis entropy is a much more sensitive function than Rényi entropy with respect to changes in q value, which is conducive to determine the proper q parameter. Besides, Tsallis entropy has been found to possess non-extensive property, which is helpful to deal with non-extensive character of XWT transform [50]. Based on the above advantages of the Tsallis entropy, it is applied to the fault feature extraction of analog circuits in this work.

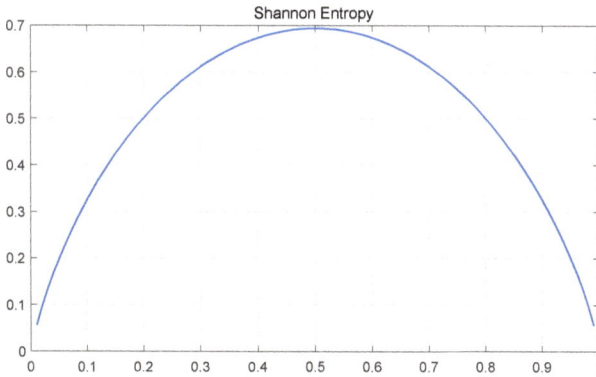

Figure 14. Plot of the Shannon entropy.

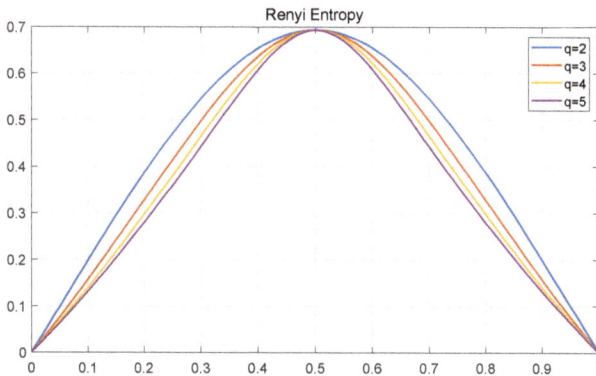

Figure 15. Plots of the Rényi entropy for several values of *p*.

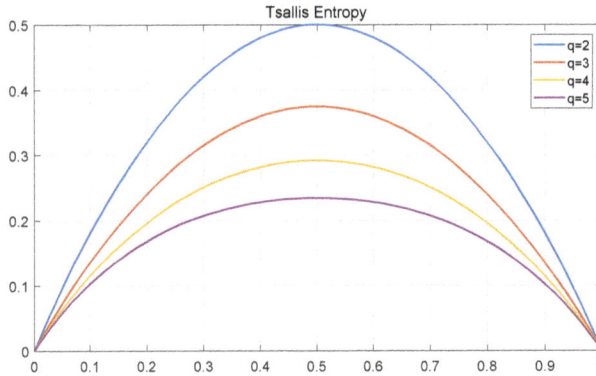

Figure 16. Plots of the Tsallis entropy for several values of p.

7. Conclusions

In this work, a new feature extraction technique based on XWSE and parametric t-SNE is put forward, and a PSO-SVM classifier is presented to locate faults as well. The conclusions validated by the simulation experiments are drawn as blew.

- Via making full use of the time-frequency distribution characteristics and entropy description, the XWSE method has a better ability to effectively extract essential features of the analyzed fault signals, and the experimental results lead us to believe that the proposed algorithm offers great potential in revealing the difference between different fault classes.
- For the sake of eliminating useless information, the parametric t-SNE is implemented to provide a nonlinear projection from the input space to the reduced space for enhancing the feature separation degree of the fault classes. The comparisons with other dimensionality reduction methods have demonstrated its feasibility and effectiveness.
- Moreover, this work also proposes a promising means for the optimization of SVM classifier by using QPSO, which is an bionic heuristic algorithm that shows faster and better convergence rate than other methods. Simulation tests have been conducted to validate that the presented QPSO-SVM model can achieve a desirable classification performance in linear circuits as well as nonlinear circuits.

In addition to all the above achievements, several issues also need to be investigated in subsequent studies. For instance, the method of extracting features effectively under incipient and multiple faults conditions should be explored, the problem of integrating the advantages of other semi-supervised dimensionality reduction methods and parametric t-SNE needs to be studied and the performance of the proposed scheme for actual circuits fault diagnosis should be further analyzed.

Author Contributions: W.H. wrote the draft and designed the research method. Y.H. provided important guidance. B.L. and C.Z. gave a detailed revision. All authors have read and approved the final manuscript.

Funding: This work was supported by the National Natural Science Foundation of China under Grant No. 51577046 and 51777050, the State Key Program of National Natural Science Foundation of China under Grant No. 51637004, the national key research and development plan "important scientific instruments and equipment development" Grant No. 2016YFF0102200, Equipment research project in advance Grant No. 41402040301, Natural Science Foundation of Hunan Province under Grant No. 2017JJ2080, Basic Research Service Fee Project of Central University under Grant No. JDK16TD01.

Conflicts of Interest: The authors declare no conflict of interest.

References

1. Aminian, F.; Member, S.; Aminian, M.; Collins, H.W. Analog Fault Diagnosis of Actual Circuits Using Neural Networks. *IEEE Trans. Instrum. Meas.* **2002**, *51*, 544–550. [CrossRef]
2. Liu, Z.; Liu, T.; Han, J.; Bu, S.; Tang, X.; Pecht, M. Signal Model-based Fault Coding for Diagnostics and Prognostics of Analog Electronic Circuits. *IEEE Trans. Ind. Electron.* **2016**, *46*, 605–614. [CrossRef]
3. Kumar, A.; Singh, A.P. Fuzzy classifier for fault diagnosis in analog electronic circuits. *ISA Trans.* **2013**, *52*, 816–824. [CrossRef] [PubMed]
4. Xiao, Y.; Feng, L. A novel linear ridgelet network approach for analog fault diagnosis using wavelet-based fractal analysis and kernel PCA as preprocessors. *Measurement* **2012**, *45*, 297–310. [CrossRef]
5. Liu, Z.; Jia, Z.; Vong, C.M.; Member, S.; Bu, S. Capturing High-Discriminative Fault Features for Electronics-Rich Analog System via Deep Learning. *IEEE Trans. Ind. Inf.* **2017**, *13*, 1213–1226. [CrossRef]
6. Binu, D.; Kariyappa, B. A survey on fault diagnosis of analog circuits: Taxonomy and state of the art. *AEU Int. J. Electron. Commun.* **2017**, *73*, 68–83. [CrossRef]
7. Catelani, M.; Fort, A. Soft fault detection and isolation in analog circuits: Some results and a comparison between a fuzzy approach and radial basis function networks. *IEEE Trans. Instrum. Meas.* **2002**, *51*, 196–202. [CrossRef]
8. Tian, S.; Yang, C.; Chen, F.; Liu, Z. Circle Equation-Based Fault Modeling Method for Linear Analog Circuits. *IEEE Trans. Instrum. Meas.* **2014**, *63*, 2145–2159. [CrossRef]
9. Yuan, L.; He, Y.; Huang, J.; Sun, Y. A new neural-network-based fault diagnosis approach for analog circuits by using kurtosis and entropy as a preprocessor. *IEEE Trans. Instrum. Meas.* **2010**, *59*, 586–595. [CrossRef]
10. Vasan, A.S.S.; Long, B.; Pecht, M. Diagnostics and Prognostics Method for Analog Electronic Circuits. *IEEE Trans. Ind. Electron.* **2013**, *60*, 5277–5291. [CrossRef]
11. Wang, Z.; Wang, J.; Zhao, Z.; Wang, R. A Novel Method for Multi-Fault Feature Extraction of a Gearbox under Strong Background Noise. *Entropy* **2018**, *20*, 10. [CrossRef]
12. Cui, Y.; Shi, J.; Wang, Z. Analog circuits fault diagnosis using multi-valued Fisher's fuzzy decision tree (MFFDT). *Int. J. Circuit Theory Appl.* **2016**, *44*, 240–260. [CrossRef]
13. Kumar, S.; Chow, T.W.S.; Pecht, M. Approach to Fault Identification for Electronic Products Using Mahalanobis Distance. *IEEE Trans. Instrum. Meas.* **2010**, *59*, 2055–2064. [CrossRef]
14. Kang, H.J.; Van, M. Bearing-fault diagnosis using non-local means algorithm and empirical mode decomposition-based feature extraction and two-stage feature selection. *IET Sci. Meas. Technol.* **2015**, *9*, 671–680.
15. Roh, J.; Abraham, J.A. Subband filtering for time and frequency analysis of mixed-signal circuit testing. *IEEE Trans. Instrum. Meas.* **2004**, *53*, 602–611. [CrossRef]
16. Dey, D.; Chatterjee, B.; Chakravorti, S.; Munshi, S. Cross-wavelet transform as a new paradigm for feature extraction from noisy partial discharge pulses. *IEEE Trans. Dielectr. Electr. Insul.* **2010**, *17*, 157–166. [CrossRef]
17. Banerjee, S.; Member, S.; Mitra, M. Application of Cross Wavelet Transform for ECG Pattern Analysis and Classification. *IEEE Trans. Instrum. Meas.* **2014**, *63*, 326–333. [CrossRef]
18. Gao, Y.; Villecco, F.; Li, M.; Song, W. Multi-Scale Permutation Entropy Based on Improved LMD and HMM for Rolling Bearing Diagnosis. *Entropy* **2017**, *19*, 176. [CrossRef]
19. Yang, Q.; Wang, J. Multi-Level Wavelet Shannon Entropy-Based Method for Single-Sensor Fault Location. *Entropy* **2015**, *17*, 7101–7117. [CrossRef]
20. Widodo, A.; Shim, M.C.; Caesarendra, W.; Yang, B.S. Intelligent prognostics for battery health monitoring based on sample entropy. *Expert Syst. Appl.* **2011**, *38*, 11763–11769. [CrossRef]
21. Zhang, Z.; Duan, Z.; Long, Y.; Yuan, L. A new swarm-SVM-based fault diagnosis approach for switched current circuit by using kurtosis and entropy as a preprocessor. *Analog Integr. Circuits Signal Process.* **2014**, *81*, 289–297. [CrossRef]
22. Han, M.; Pan, J. A fault diagnosis method combined with LMD, sample entropy and energy ratio for roller bearings. *Measurement* **2015**, *76*, 7–19. [CrossRef]
23. Li, Y.; Xu, M.; Zhao, H.; Huang, W. Hierarchical fuzzy entropy and improved support vector machine based binary tree approach for rolling bearing fault diagnosis. *Mech. Mach. Theory* **2016**, *98*, 114–132. [CrossRef]
24. Li, X.; Xie, Y. Analog circuits fault detection using cross-entropy approach. *J. Electron. Test. Theory Appl.* **2013**, *29*, 115–120. [CrossRef]

25. Xie, X.; Li, X.; Bi, D.; Zhou, Q.; Xie, S.; Xie, Y. Analog Circuits Soft Fault Diagnosis Using Rényi's Entropy. *J. Electron. Test.* **2015**, *31*, 217–224. [CrossRef]

26. Kankar, P.K.; Sharma, S.C.; Harsha, S.P. Fault diagnosis of ball bearings using continuous wavelet transform. *Appl. Soft Comput. J.* **2011**, *11*, 2300–2312. [CrossRef]

27. Zhang, Y.; Wu, L. Optimal Multi-Level Thresholding Based on Maximum Tsallis Entropy via an Artificial Bee Colony Approach. *Entropy* **2011**, *13*, 841–859. [CrossRef]

28. Dong, S.; Tang, B.; Chen, R. Bearing running state recognition based on non-extensive wavelet feature scale entropy and support vector machine. *Measurement* **2013**, *46*, 4189–4199. [CrossRef]

29. Maaten, L.V.D. Learning a Parametric Embedding by Preserving Local Structure. *J. Mach. Learn. Res.* **2009**, *5*, 384–391.

30. Boser, E.; Vapnik, N.; Guyon, I.M.; Laboratories, T.B. A Training Algorithm Margin for Optimal Classifiers. In Proceedings of the Fifth Annual Workshop on Computational Learning Theory, Pittsburgh, PA, USA, 27–29 July 1992; pp. 144–152.

31. Saitta, L. Support-Vector Networks. *Mach. Learn.* **1995**, *297*, 273–297.

32. Soliman, M.M.; Hassanien, A.E.; Onsi, H.M. An adaptive watermarking approach based on weighted quantum particle swarm optimization. *Neural Comput. Appl.* **2016**, *27*, 469–481. [CrossRef]

33. Grinsted, A.; Moore, J.C.; Jevrejeva, S.; Grinsted, A.; Moore, J.C.; Application, S.J. Application of the cross wavelet transform and wavelet coherence to geophysical time series. *Nonlinear Process. Geophys.* **2004**, *11*, 561–566. [CrossRef]

34. Ruessink, B.G.; Coco, G.; Ranasinghe, R.; Turner, I.L. A cross-wavelet study of alongshore nonuniform nearshore sandbar behavior. In Proceedings of the 2006 IEEE International Joint Conference on Neural Network Proceedings, Vancouver, BC, Canada, 16–21 July 2006; pp. 4310–4317.

35. Series, H.; Algebra, L. Singular Value Decomposition and Least Squares Solutions. *Numer. Math.* **1970**, *14*, 403–420.

36. Physics, S.; November, R. Possible Generalization of Boltzmann-Gibbs Statistics. *J. Stat. Phys.* **1988**, *52*, 479–487.

37. Maaten, L.V.D.; Hinton, G. Visualizing Data using t-SNE. *J. Mach. Learn. Res.* **2008**, *9*, 2579–2605.

38. Pan, M.; Jiang, J.; Kong, Q.; Shi, J.; Sheng, Q.; Zhou, T. Radar HRRP Target Recognition Based on t-SNE Segmentation and Discriminant Deep Belief Network. *IEEE Geosci. Remote Sens. Lett.* **2017**, *14*, 1609–1613. [CrossRef]

39. Li, M.; Luo, X.; Yang, J. Extracting the nonlinear features of motor imagery EEG using parametric t-SNE. *Neurocomputing* **2016**, *218*, 371–381. [CrossRef]

40. Eberhart, R.; Kennedy, J. A New Optimizer Using Particle Swarm Theory. In Proceedings of the Sixth International Symposium on International Symposium on Micro Machine and Human Science, Nagoya, Japan, 4–6 October 1995; pp. 39–43.

41. Sun, J. Quantum-Behaved Particle Swarm Optimization: Analysis of Individual Particle Behavior and Parameter Selection. *Evol. Comput.* **2012**, *20*, 349–393. [CrossRef] [PubMed]

42. Zhao, L.Y.; Wang, L.; Yan, R.Q. Rolling Bearing Fault Diagnosis Based on Wavelet Packet Decomposition and Multi-Scale Permutation Entropy. *Entropy* **2015**, *17*, 6447–6461. [CrossRef]

43. Pang, Q.; Liu, X.; Sun, B.; Ling, Q. Approximate Entropy Based Fault Localization and Fault Type Recognition for Non-solidly Earthed Network. *Meas. Sci. Rev.* **2012**, *12*, 309–313. [CrossRef]

44. Su, Z.; Tang, B.; Liu, Z.; Qin, Y. Multi-fault diagnosis for rotating machinery based on orthogonal supervised linear local tangent space alignment and least square support vector machine. *Neurocomputing* **2015**, *157*, 208–222. [CrossRef]

45. Tang, G.; Wang, X.; He, Y. A Novel Method of Fault Diagnosis for Rolling Bearing Based on Dual Tree Complex Wavelet Packet Transform and Improved Multiscale Permutation Entropy. *Math. Probl. Eng.* **2016**, *2016*, 5432648. [CrossRef]

46. Song, P.; He, Y.; Cui, W. Statistical property feature extraction based on FRFT for fault diagnosis of analog circuits. *Analog Integr. Circuits Signal Process.* **2016**, *87*, 427–436. [CrossRef]

47. Chen, P.; Yuan, L.; He, Y.; Luo, S. An improved SVM classifier based on double chains quantum genetic algorithm and its application in analogue circuit diagnosis. *Neurocomputing* **2016**, *211*, 202–211. [CrossRef]

48. Gajowniczek, K.; Zabkowski, T.; Orlowski, A. Comparison of decision trees with Rényi and Tsallis entropy applied for imbalanced churn dataset. In Proceedings of the 2015 Federated Conference on Computer Science and Information Systems, Lodz, Poland, 13–16 September 2015; pp. 39–44.

49. Chen, J.; Dou, Y.; Wang, Z.; Li, G. A novel method for PD feature extraction of power cable with Rényi entropy. *Entropy* **2015**, *17*, 7698–7712. [CrossRef]

50. Johal, R.S.; Tirnakli, U. Tsallis versus Rényi entropic form for systems with q-exponential behaviour: The case of dissipative map. *Phys. A Stat. Mech. Appl.* **2004**, *331*, 487–496. [CrossRef]

![entropy logo] *entropy*

MDPI

Concept Paper

Learning Entropy as a Learning-Based Information Concept

Ivo Bukovsky [1,*], Witold Kinsner [2] and Noriyasu Homma [3]

[1] Department of Mechanics, Biomechanics, and Mechatronics, Research Centre for Low-Carbon Energy Technologies, Faculty of Mechanical Engineering, Czech Technical University in Prague, Technicka 4, 166 07 Prague 6, Czech Republic

[2] Department of Electrical and Computer Engineering, University of Manitoba, Winnipeg, MB R3T 5V6, Canada; witold.kinsner@umanitoba.ca

[3] Department of Radiological Imaging and Informatics, Tohoku University Graduate School of Medicine, Intelligent Biomedical System Engineering Laboratory, Graduate School of Biomedical Engineering, Tohoku University, Sendai 980-8575, Japan; homma@ieee.org

* Correspondence: ivo.bukovsky@fs.cvut.cz; Tel.: +420-2-2435-7300

Received: 30 December 2018; Accepted: 5 February 2019; Published: 11 February 2019

Abstract: Recently, a novel concept of a non-probabilistic novelty detection measure, based on a multi-scale quantification of unusually large learning efforts of machine learning systems, was introduced as learning entropy (LE). The key finding with LE is that the learning effort of learning systems is quantifiable as a novelty measure for each individually observed data point of otherwise complex dynamic systems, while the model accuracy is not a necessary requirement for novelty detection. This brief paper extends the explanation of LE from the point of an informatics approach towards a cognitive (learning-based) information measure emphasizing the distinction from Shannon's concept of probabilistic information. Fundamental derivations of learning entropy and of its practical estimations are recalled and further extended. The potentials, limitations, and, thus, the current challenges of LE are discussed.

Keywords: learning; information; novelty detection; non-probabilistic entropy; learning systems

1. Introduction

Complexity measures and novelty detection measures, which are based on Shannon's entropy [1], are probabilistic measures that do not consider the governing laws of systems explicitly. On the contrary, computational learning systems can approximate at least the contemporary governing laws of dynamical behavior. Novelty detection in dynamical systems is approached either by probabilistic approaches (e.g., [2]) or by utilization of learning systems, e.g., [3].

As the representative examples of probability-based novelty detection approaches, i.e., the statistical novelty measures and probabilistic entropy measures, we should mention sample entropy (SampEn) and approximate entropy (ApEn) [4,5]; SampEn and ApEn relate to fractal measures and thus to multi-scale evaluation [6–9] that is based on the concept of power-law [10]. The benefits of these multi-scale techniques were also shown via works on coarse-graining extensions to SampEn in [11,12] and recently also in [13]. Further, compensated transfer entropy [14] is another probabilistic technique for entropy evaluation via the conditional mutual information between present and past states. The probabilistic entropy approach for fault detection was published in [15] and probabilistic technique for sensor data concept drift (also concept shift) appeared in [16].

Among the probabilistic novelty approaches, we shall also mention the currently popular concepts of generalized entropies, especially, the extensively studied Tsallis and Rényi entropies and their potentials, e.g., [17] and references therein. The example of their application, e.g., to probabilistic anomaly detection in cybersecurity, can be found in [18]. In learning systems such as neural networks, the generalized entropies are also naturally studied to improve the learning process, e.g., [19]. As regards the proposed learning entropy (LE), it could also be used in the difficult problem of anomaly detection during a cyberattack, e.g., a system has been developed to detect such anomalies using convolutional neural networks and multi-scale and poly-scale measures [20]; adding LE to such a system could enhance the real-time detection of DDOS and other attacks. However, LE is a non-probabilistic measure that evaluates unusually large learning efforts of a learning system, so it is also different from the studied applications of generalized entropies.

The second direction of novelty detection in dynamical systems, i.e., the direction of non-probabilistic novelty measures, is based on learning systems that employ machine learning, and we apply this direction in our research too. As some more recent survey works on non-probabilistic detection methods with learning systems, we may refer to [21–24] and to [25] as to an example that involves incremental learning. The usage of residuals of learning system output for fault detection with nonlinear estimators was studied in [22–24]. For recent works on adaptive concept drift detection, we may refer to works [25–28] and to newer works in area of neural networks [29–31] and [32] as a work on a cognitive system for sensor fault diagnosis. A certain similarity with our proposed concept can also be found in the adaptive resonance theory [33].

While the probabilistic approaches do not explicitly reflect the governing law of data, the learning-system-based methods rely on the evaluation of model residuals and thus on certain accuracy of models. However, there has been a missing concept for novelty detection that would utilize the learning process without probability computations and that would not rely on the accuracy of learning systems, and learning entropy is such a novel concept in this sense.

The motivation of this brief paper is to discuss and extend the recently introduced concept of LE [34,35] in the sense of a (machine) learning-based information measure as a founding concept of the cognitive novelty detection based on quantification of unusual learning efforts of learning systems. Novelty detection via LE is based on real-time learning of systems after they had been pretrained on an initial pretraining data set. Aside from the founding work [34], other examples of works that indicate the usefulness of LE in biomedical or technical data and for novelty detection in data with concept drift can be found in [36–39].

Section 2 first discusses a loose parallel between entropy in the sense of thermodynamics and importantly, the distinction from the concept of Shannon's information theory, with the proposal of the concept of (machine) learning-based information measure. Secondly, the original multiscale LE algorithm [34,35] based on unusually large learning efforts is reviewed and followed by its approximate version. Thirdly, it provides an alternative (more direct) formula for practical computations.

In the following text, terms such as learning system, neural network, model, observer, and predictor are used interchangeably, unless it is stated otherwise. To simplify mathematical notations, the discrete time index k is dropped from notations unless it is necessary for clarity.

2. Concept of Learning Information Measure

A loose parallel between entropy concepts of informatics and thermodynamics can be drawn regarding learning systems and training data. More novel data carries more information from the point of view of a learning system. When novel training data are presented to the learning system (after its pretraining), then the learning algorithm responds via its learning activity with its adaptive parameters. Hence, the information (novelty) that training data means to the learning system changes the activity of learning system, similar to how heat changes the energy of thermodynamical systems. For incrementally learning systems with a vector of all adaptive parameters $\mathbf{w}$, the novelty in data can change the actual learning effort, so the actual weight updates $\Delta \mathbf{w}$ or at least some of them, indicate the novelty that the data provides to the contemporary trained learning systems. The weight updates represent additional information for a better description of real systems by neural networks (or learning systems in general). These loose analogies between the novelty of training data, learning effort, and weight updates are the necessary information elements to improve system description, as well as to draw connotations from the concept and meaning of entropy in a general sense, including those of thermodynamics and information theory (a review on meanings of entropy can be found in [40]).

According to Shannon's probabilistic approach, the amount of self-information *I* that the value $y(k)$ can provide to an observer, depends on its inverse probability as follows:

$$I(y(k)) = -log(p(y(k))). \tag{1}$$

where $p(y(k))$ is the probability of value $y(k)$ that is in fact independent of the discrete time index k, and the less frequent value of $y(k)$ the larger information it provides to an observer. However, if the observer is a learning system that learns the governing law of data, then the statistically new data do not necessarily provide new information (i.e., as with the non-repeating, yet deterministic chaos). The statistically new data can still comply with the temporarily learned governing law, so the learning system is not "surprised" by its appearance. This points us to the essence of calculating the novelty (information) that data provide to an observer in a different way than that established via the Shannon probabilistic sense as in (1). While the probabilistic information measure is based on clustering that utilizes a distance between vectors of data, we may quantify the familiarity of a learning-system with data because the learning system considers data to be novel if the data do not comply with the contemporary learned governing law via the following:

- Supervised learning (as for given input–output patterns with supervised learning), or via
- unsupervised learning (such as learned by clustering methods, SOMs, or auto-encoders).

The most straightforward way to quantify novelty with supervised learning is to use a model (e.g., prediction) error that indicates the expectancy of the actual data from the governing-law viewpoint. However, this assumes to have a correctly designed learning model that is not trivial to obtain for the real-world data. In fact, the (prediction) error is not the most straightforward quantity that either tells us how much information the learning system needs, or how much learning effort it is going to spend to become more familiar with the new data. The higher error does not necessarily mean that the actually presented data are novel because the model can be limited in the quality of approximation, and its generalization capability is unknown for data that never occurred before. Also, the model error is only one component of the learning algorithm and each model parameter can be updated with different significance and magnitude, depending on other factors including inputs. During sample-by-sample or sliding-window batch pretraining on the initial training dataset, the weights become updated with smaller and smaller updates in each consecutive training epoch, so the parameters of a learning system converge up to a certain pattern of behavior. Thus, for the pretraining dataset, the average update magnitudes of individual weights finally become constant, i.e., $\overline{|\Delta \mathbf{w}|}$ = const.. If retraining continued for further data that comply with the pretrained governing law, then in principle, further weight updates of a pretrained learning system would not be larger than those during pretraining (even if the model could not learn the governing law properly). However, if the retraining data involve data samples that do not comply with the temporarily learned governing law, the weight update behavior changes as the learning system tends to get adapted to novel data and weight updates can be larger (see middle axes in Figure 1 for $k \geq 400$).

Thus, the learning updates $\Delta \mathbf{w}$ represent learning effort and they are suitable for evaluation of how much information the new data convey to a learning system in terms of the contemporary learned governing law. In particular, if all weights are updated within the usually large magnitudes, then the retraining data do not bring any new information to the learning system. However, if more weights are updated with unusual updates, the data appear to be more unexpected, thus leading to a more unusual learning effort. This also means that data convey more information to the learning system. Thus, the detection of unusual weight updates can be used to detect novel data, and naturally the higher count of unusual updates the more information the retraining data convey to the already pretrained model. Then, a (machine) learning-based information measure can be generally proposed via a suitable aggregation of unusual learning increments as follows:

$$L(k) = A(f(\Delta \mathbf{w}(k))) \tag{2}$$

where $A(.)$ represents a general aggregation function and $f(.)$ denotes a function that quantifies the unusuality of the learning effort via learning increments (assuming the learning system has been pretrained on the training data). So far in our research of LE [34,36,38,41–43], a summation has been applied as the aggregation function $A(.)$ as follows:

$$L(k) = \sum_{\forall \Delta w \in \Delta w} f(\Delta w(k)) \; [/] \tag{3}$$

and $f(.)$ for detection of unusually large learning effort has been defined via unusually large weight increments as follows

$$f(\Delta w) = \begin{cases} 1 & \text{if } |\Delta w| \text{ is unusually large} \\ 0 & \text{if } |\Delta w| \text{ else} \end{cases} \tag{4}$$

In reality, it is practically impossible to choose the best bias that determines the unusually large weight update magnitudes for proper evaluation of (4), so the detection sensitivity for unusually large weight updates was resolved via a power-law based multi-scale approach as in [34,43] and that is reviewed and modified in later sections.

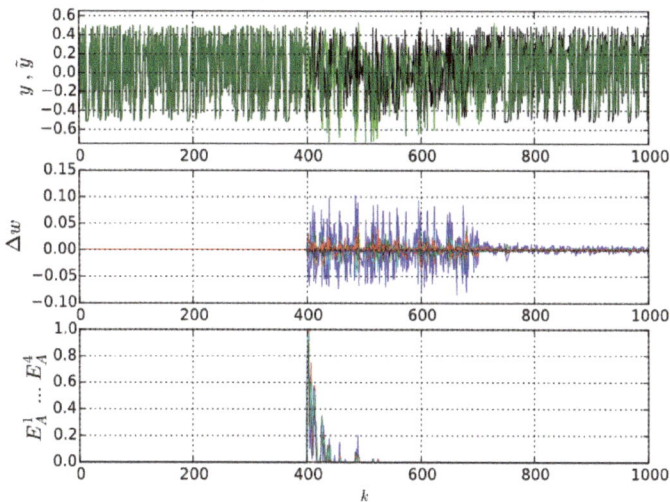

Figure 1. (**Top**) Chaotic (deterministic) time series with a sudden occurrence of white noise (k > 400) superimposed on the output of its real-time sample-by-sample learning predictor. (**Middle**) The weight updates cannot converge to noise. (**Bottom**) Approximate Learning Entropies (of various orders) via (19) detect the noise as the novelty immediately at its occurrence at k > 400 and then LE decreases as the large variance of learning increments becomes a new usual learning pattern (details on LE and its orders can be found in Sections 4.1 and 4.2).

3. Shannon Entropy versus Learning Entropy

Until now, we have discussed the Shannon entropy, i.e., the probabilistic, information measure I (1) vs. the learning-system-based concept of information measure L (3) and (4). Both $I(y(k))$ and

$L(y(k))$ represents the quantity of how unusual data sample $y(k)$ is. However, we cannot think about L in the sense of histogram-bin clustered data, because while for the Shannon concept I it holds that

$$y(k_1) = y(k_2) \Rightarrow I(y(k_1)) = I(y(k_2)) = I_i \tag{5}$$

where i denotes the bin index, the learning measure L is likely to be different for two identical values of data at different times because of the learning process; i.e.,

$$y(k_1) = y(k_2) \nRightarrow (y(k_1)) = L(y(k_2)) \tag{6}$$

Thus, it is apparent from (5) and (6) that the Shannon entropy definition, i.e., the probability-weighted average of the information measure

$$H = \sum_i p_i \cdot I_i \tag{7}$$

where i denotes the normalized histogram bin index, and cannot be used in the same way for the learning-based measure L. In light of the learning-based information measure L and its distinction from the Shannon measure I, a multiscale extension of L via (3) and (4) was introduced as the approximate individual sample learning entropy (AISLE) in [34] (for more details, see Section 4 below). AISLE reflects the amount of the unusually large learning effort that learning system spends on updating to novel data, and thus it reflects the amount of new information that data means to a learning system (or loosely such as the energy with which novel data boosts the learning engine).

The most straightforward measure based on AISLE is the learning entropy profile (LEP) that was defined in [34] as the cumulative sum of LE in time over the whole interval of data as follows

$$LEP = \sum_{k=1}^{N} LE(k) \approx \sum_{k=1}^{N} L(k) \tag{8}$$

Thus, the LEP is a function that quantifies the novelty that a pretrained learning system is able to find in a new dataset in terms of its unusual learning effort. The last point of LEP is called the learning entropy of a model (LEM)

$$LEM = LEP(k = N) \tag{9}$$

In other words, LE characterizes how pretrained neural network is unfamiliar with each new data point (in time), while the LEP quantifies the total amount of novelty that the interval of data has conveyed to the pretrained learning system, and it also gives a notion about the novelty (learning information) in data from the point of learning effort for the used mathematical structure and its particularly used learning algorithm. Based on incremental learning (11), we can see from (3), (4), (8) and (9) that the learning entropy of a model is always increasing.

4. Algorithms for Learning Entropy Estimation

The previous sections recalled the concept of LE and discussed it with connotation to a (machine) learning-based information concept. Further, the theoretical multiscale algorithm for the estimation of LE is reviewed in Section 4.1, followed with practical formula in Section 4.2 with new direct formula in Section 4.3.

4.1. The Multiscale Approach

A general form of a learning system (LS) is as follows:

$$\tilde{y} = F(\mathbf{w}, \mathbf{u}) \tag{10}$$

where $\tilde{y}$ is the vector of actual outputs, $\mathbf{u}$ is the vector of inputs (including feedbacks in case of a recurrent learning system), $F(.)$ is the general mapping function of LS, and $\mathbf{w}$ represents the vector of all adaptable parameters (weights). Further derivations apply when the learning entropy considers all neural weights in **w**; however, customization of the algorithm for individual weights may be an interesting research challenge, particularly for deep neural networks. Further for simplicity, let us assume that all neural weights are updated at the same time according to the incremental scheme as follows

$$\mathbf{w}(k+1) = \mathbf{w}(k) + \Delta\mathbf{w}(k) \tag{11}$$

where $\Delta\mathbf{w}(k)$ is the vector of actual weight updates that depend on a particularly chosen learning algorithm and its potential modification. The concept of learning entropy is based on the evaluation of unusual weight updates as the unusual learning pattern can indicate novelty in training data; i.e., the new information that new samples of data carry in respect to what the NN contemporary has learned already [34]. This methodology to evaluate the learning entropy through the unusually large weight updates was recently introduced [34] and then reviewed with some simplifications recently in [35,36,38]. The first important parameters here are as follows:

- α is the relative detection sensitivity parameter that defines the crisp border between usual weight updates and unusually large ones (since the optimal α is never known in advance, the multi-scale evaluation has to be adopted).
- M is the length of the floating window over which the average magnitudes of the recent weight updates are calculated (for periodical data, there is also the lag m between the actual time and the end of the window, see p. 4179 in [34]),

Then the unusual learning effort of LS can be evaluated at each learning update (through (11)) as the count of unusually large weight increments for all weights of the LS as follows:

$$L(\alpha) = \sum_{\forall \Delta w \in \Delta w} f(\Delta w(k), \alpha) \tag{12}$$

where $f(.)$ is the detection function defined for every individual weight increment as follows:

$$f(\Delta w, \alpha) = \begin{cases} 1 \text{ if } \left(\left| |\Delta w| - \overline{|\Delta\mathbf{w}|} \right| \right) > \alpha \cdot \sigma_{\Delta\mathbf{w}} \\ 0 \text{ else} \end{cases} \tag{13}$$

where the detection sensitivity α is defined above, $\sigma_{\Delta\mathbf{w}}$ is the standard deviation of recently usual weight update magnitude, and the average weight-update magnitude can be calculated as follows:

$$\overline{|\Delta\mathbf{w}|} = \overline{|\Delta\mathbf{w}^M|} = \frac{1}{M} \sum_{j=k-M-m}^{k-1-m} |\Delta w(j)| \tag{14}$$

where M is the length of the floating window and m is the optional lag for data with features of periodicity (as indicated in Equation (27) in [34]. Notice, we should calculate $\overline{|\Delta\mathbf{w}|}$ when a learning system had been already pretrained in such a way so learning does not display any more convergence (LE is attractive also for that it is principally independent of any model accuracy [43], while the pretraining and further learning are the key principles of LE).

Since the count of all unusual weight updates $L(k, \alpha)$ depends on detection sensitivity α, and since we do not know the optimal sensitivity for the particular learning system (i.e., for the particular LS, or the learning algorithm used, or for the data) we shall overcome this single scale issue by using a multi-scale approach that evaluates the unusual learning effort over the whole interval of detection sensitivities $\alpha \in \boldsymbol{\alpha}$. Considering that the real-word quantities non-linearly depend on parameters and being inspired by the use of the power-law from fractal analysis, we can assume that the dependence of the count of unusual weight updates on the detection sensitivity can be characterized via exponent H in the power-law relationship as follows:

$$L(\alpha) \cong (\alpha)^{-H} \Rightarrow log(L(\alpha)) \cong -H \cdot log(\alpha) \tag{15}$$

and the characterizing exponent H then can be estimated as the slope of the log-log plot as

$$H = \lim_{\alpha \to \alpha^-_{max}} \left(-\frac{log(L(\alpha))}{log(\alpha)} \right) \tag{16}$$

where α_{max} was defined in [34] as the value where first unusual weight updates can be detected within all data. Alternatively, α_{max} is defined as follows:

$$\alpha_{max}: \begin{cases} \text{if } \alpha > \alpha_{max} \Rightarrow \sum_{\forall k} L(\alpha, k) = 0 \\ \text{else} \qquad \sum_{\forall k} L(\alpha, k) \geq 1 \end{cases} \tag{17}$$

Finally, we arrive at the definition of the learning entropy E as the normalized measure of unusually large learning effort at every weight update as follows

$$E(k) = \frac{2}{\pi} \cdot \arctan(H(k)) \Rightarrow E(k) \in [0,1) \tag{18}$$

where $E = 0$ means that no learning updates of all parameters are unusually large $\forall \alpha \in \alpha$ and $E \to 1$ as all learning updates of all parameters are unusually large $\forall \alpha \in \alpha$. In fact, the learning entropy E in (18) is considered to be the first-order learning entropy because the detection function (13) is calculated with the first difference ($\approx$ first-order derivative) of weights (as it results from (11)). It appeared useful to practically enhance LE computation with higher-order differences of weight updates that contribute to more reliable novelty detection as the higher order weight difference terms indicates useful noise filtering [34–36,38]. To compute the LE of various orders, the corresponding weight differences can be used in formulas (12)–(14) as in Table 1.

Table 1. Order of learning entropy (OLE) is determined by the difference in the order of weight increments in (12)–(14).

E^r r^{th}OLE	Detection Function Modifications for Varying Orders of LE
E^0	$L(\alpha) = \sum_{\forall \omega} f\left(\left|\|\Delta\omega\| - \overline{\|\Delta w\|}\right| > \alpha\sigma_{\Delta\omega}\right)$
E^1	$L(\alpha) = \sum_{\forall \omega} f\left(\left|\|\Delta\omega\| - \overline{\|\Delta w\|}\right| > \alpha\sigma_{\Delta\omega}\right)$ $\Delta w(k) = w(k) - w(k-1)$
E^2	$L(\alpha) = \sum_{\forall \omega} f\left(\left|\|\Delta^2\omega\| - \overline{\|\Delta^2 w\|}\right| > \alpha\sigma_{\Delta\omega}\right)$ $\Delta^2 w(k) = \Delta w(k) - \Delta w(k-1)$
E^r	$L(\alpha) = \sum_{\forall \omega} f\left(\left|\|\Delta^r\omega\| - \overline{\|\Delta^r w\|}\right| > \alpha\sigma_{\Delta\omega}\right)$ $\Delta^r w(k) = \Delta^{r-1} w(k) - \Delta^{r-1} w(k-1)$

It should be emphasized, that the first important factor that affects the quality of the use of LE for novelty detection (i.e., for detecting data samples or intervals that carry new information that the neural network is not yet familiar with), is the proper pretraining of the neural network (e.g., an initial data set for further use online). In this case, the proper pretraining can be defined as such a long or repeated training as long as a learning performance index decreases, i.e., until the learning system tends to learn from data. In general, of course, the quality of adaptive novelty detection using the above derived LE further depends on the particularly chosen type of learning system, on the selected learning rule, on other setups that can be optimized the better we understand LS, the process, and its data.

This section recalled the theoretical derivation of learning entropy based on the fractal characterization of the power-law relationship of increased learning effort with a multiscale setup of detection parameter sensitivity α. The next section recalls a practical algorithm for the estimation of LE via cumulative sums and then a new direct algorithm based on the z-scoring of the temporal matrix of learning increments is introduced.

4.2. Practical Algorithm for Learning Entropy

The theoretical derivation of learning entropy (18) in Section 4.1 is based on estimating the characterizing exponent H as the slope of the log-log plot. In works [34,43], the calculation of characterizing exponent H of a log–log plot was replaced by the sum of quantities $L(\alpha)$ calculated for multiple values of detection sensitivities α and for all neural weight, so the learning entropy can be approximated as follows

$$E \approx E_A = \frac{1}{n_\alpha \cdot n_w} \sum \{L(\alpha); \ \alpha \in \boldsymbol{\alpha}\}$$
$$\alpha \in \boldsymbol{\alpha} = [\alpha_1 < \alpha_2 < \ldots < \alpha_{n_\alpha}], E_A \in \langle 0, 1 \rangle \tag{19}$$

where $E = 0$ means that no learning updates of all parameters are unusually large for any sensitivity α, and $E = 1$ means that all learning updates of all parameters are unusually large for all sensitivities α, and where the sum is normalized for the length of vector $\boldsymbol{\alpha}$ and for the total number of neural weights n_w, and thus (19) represents an approximation of LE. Particularly in [34], it is shown that the sum of $L(\alpha)$ along given by formula (12) in principle correlates to the log–log plot slope H calculated by formula (16). In particular, the steeper slope H is in a log–log plot, the more the $L(\alpha)$ counts increase along sensitivities $\alpha \in \boldsymbol{\alpha}$, and that naturally results in the largest sum for most novel samples in data because $L(\alpha)$ starts increasing as soon as the neural network is learning more novel data. It is not necessary to find the exact value of α_{max} (see (16) and (17)), because α can in principle contain even larger values of α when calculated by (19). Thus, E_A in (19) was introduced as approximate individual sample learning entropy (AISLE) when the sample-by-sample adaptation learning rule is used; e.g., the gradient descent learning in [34] and it was used also in works [35,36,38]. An example of AISLE of various orders is shown in Figure 1.

4.3. A Direct Algorithm

With mathematical symbols for the mean such as $\bar{x}$, for standard deviation as $\sigma(x)$, and considering (14) introduces a special Z-scoring as follows:

$$z(|\Delta w_i(k)|) = \frac{|\Delta w_i(k)| - \overline{|\Delta \mathbf{w}_i^M(k-1)|}}{\sigma(|\Delta \mathbf{w}_i^M(k-1)|)} \tag{20}$$

then a new formula for the estimation of LE can be introduced as an alternative to AISLE from (19) as follows:

$$E(k) = \sum_{i=1}^{n_w} z(|\Delta w_i(k)|) ; \ E \in \mathbb{R} \tag{21}$$

In contrast to previously proposed formulas for LE (18) and (19) that involved only the occurrences of unusually large learning efforts, the new direct formula (21) has the potential to quantify both unusually large learning efforts as well as unusually small ones; i.e., when the novelty in data makes weights become rapidly converging so their updates yield is unusually small in time and thus (21) results in unusually small values (see Figure 3 and the discussion there). Nevertheless, the novelty in data may be potentially detected even when only very few weight updates (or even a single one) unusually largely increase, and this makes LE be a very sensitive method. However, in principle, this is not well detectable by the LE formula (21) because the other weight updates may result in a negative contribution to E as $z(|\Delta w_i(k)|) < 0$ for some i. Since (21) can also result in negative values of E (when the majority of weights are usually updated, or with even smaller updates), it would not provide a sharp border between usual and unusual learning effort. Thus, we can enhance (21) as follows

$$E(k) = \sum_{\forall w} \max\{0, z(|\Delta w(k)|) - \beta\};$$
$$E(k) \in \langle 0, +\infty)$$
(22)

that both

- detects only unusually large weight-update increments, larger than their recent mean plus $\beta\times$ standard deviation, and
- also directly computes their absolute significance (due to Z-scoring) for each weight while it was calculated in the previous concept of LE (18) and (19) via the multiscale evaluation over sensitivity setups (as recalled in Section 4.1 and (19)).

In order to achieve a normalized value of E in (22) as well as to cope with the single-scale issue of selection β, we propose to estimate the r-th order LE with this direct approach using a threshold function $f(.)$ as well as multiple setups of sensitivity β as follows

$$E_A(k) = \frac{1}{n_\beta \cdot n_w} \sum_{\forall w} \sum_{\forall \beta} f(\max\{0, z(|\Delta w(k)|) - \beta\});$$
$$\beta \in \beta = \left[\beta_1 \beta_2 \cdots \beta_{n_\beta}\right], f(.) = \begin{cases} 1 \text{ if } (.) = True \\ 0 \text{ if } (.) = False \end{cases}$$
$$E(k) \in \langle 0, 1 \rangle$$
(23)

where again $E_A = 0$ means that no learning updates of all parameters are unusually large $\forall \beta \in \beta$, and $E_A = 1$ means that all learning updates of all parameters are unusually large $\forall \beta \in \beta$. Furthermore, β represents a parameter of detection sensitivity that is related to the standard deviation of recent weight-update magnitudes and it causes formulas (22) and (23) work as follows:

- if $\beta = 0$, the weight-update magnitudes larger than their recent mean are summed in (22) or counted in (23), i.e., the detection of unusual learning effort is the most sensitive one,
- if $\beta = 1$, only the weight-update magnitudes larger than their recent mean plus one standard deviation are summed in (22) or counted in (23), i.e., the detection of unusual learning effort is less sensitive,
- if $\beta = 2$, only the weight-update magnitudes larger than their recent mean plus two standard deviations are summed in (22) or counted in (23), i.e., the detection of unusual learning effort is even less sensitive, and, similarly, the detection of unusually large learning effort is less sensitive with the increasing parameter β while the vector of detection sensitivities β must not necessarily be a vector of integers.

The performance of the direct algorithm (23) is demonstrated in Figures 2 and 3, and we found the performance fairly comparable to the previously introduced estimation of LE (19) as we compared it with the same learning system and the same (gradient descent) learning algorithm and similar data as in Figure 1.

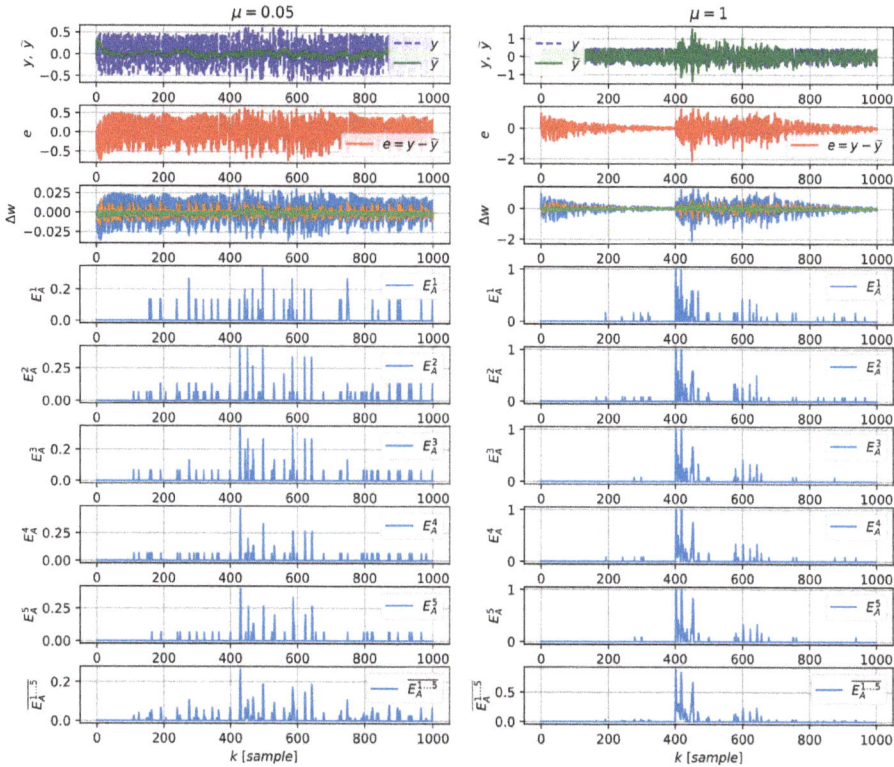

Figure 2. The performance of the direct algorithm for estimation of learning entropy of various orders (23) for not pretrained adaptive predictor with a too low learning rate $\mu = 0.05$ (left graphs) and for reasonable learning rate $\mu = 1$ (right graphs); normally distributed noise is within $k \in< 400, 750 >$ (same as in Figures 1 and 3).

5. Limitations and Further Challenges

At first, the main considerations for both the power and weakness of learning entropy are the choice of a proper learning system, the learning algorithm, and its setups. Thus, background knowledge and relevant skills with machine learning can crucially affect the performance of LE.

Second, the fundamental assumption for learning entropy is that a learning system that adapts its weights via $\Delta\mathbf{w}$ is already pretrained.

This is demonstrated both in Figure 1 and also in Figure 3 for $k > 700$, where the adaptive predictor was pretrained on initial deterministic time series that suddenly changed to white noise for $k \in \langle 400, 700 \rangle$. Then, the noisy data samples at $k > 400$ results in the immediate increase of learning effort, so the LE increases immediately after $k = 400$. However; the complexity of pure white noise for $k \in \langle 400, 700 \rangle$ disables the learning system for retraining, so the weight increments do not converge at all yet the LE decreases because the learning pattern is usually found within $k \in \langle 400, 700 \rangle$. Thus, when the data changes back to deterministic ones for $k > 700$ (Figure 3), the LE fails to detect this novelty (bottom axis Figure 3) because the adaptive predictor was not retrained due to extreme complexity of the preceding signal (noise) and thus the new data for $k > 700$ do not induce the increased learning effort.

The previously demonstrated limitation of LE (18,19,23) is based on a theoretical example, and so far, we have not encountered this issue in our research with deterministic systems or with real-world data. However, this theoretical case certainly demonstrates the challenge for further enhancement of algorithms for estimation of LE.

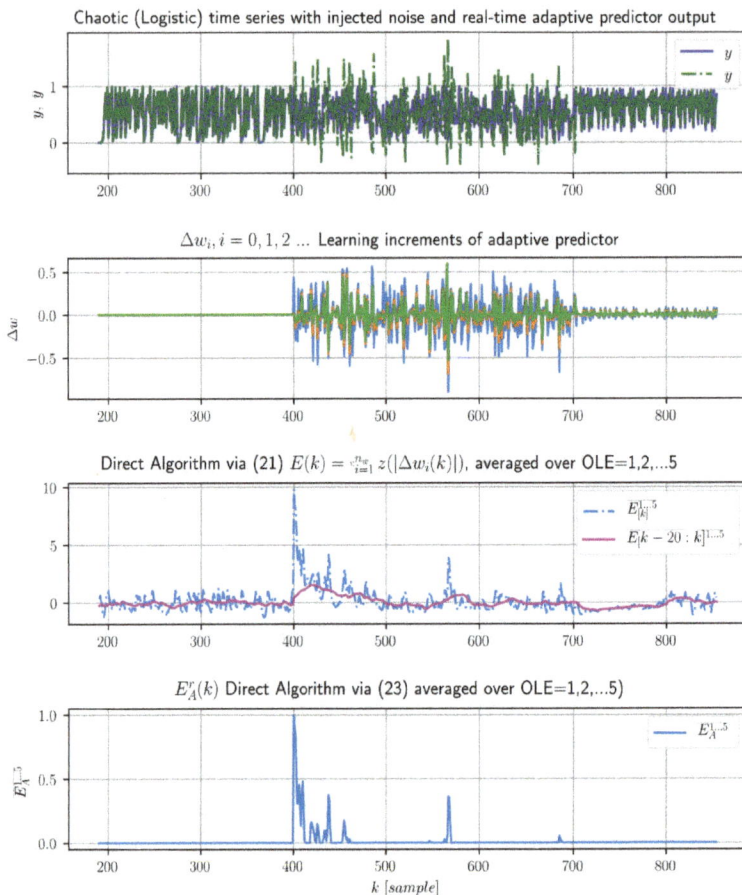

Figure 3. The limitation and challenges: The alternative LE estimation (21) displays capability to capture both unusually large learning effort as well as unusually small one, while the currently proposed algorithms of LE (18,19,23) are based on capturing unusually large learning effort and the novelty detection when the noise ($400 < k \leq 700$) changes back to deterministic signal for $k > 700$ is still a challenge. So far, we found the Direct Algorithm (23) (bottom axes) be practically comparable to the original LE estimation (19) (see Figure 1 with a similar type of data).

In future research, the intermediate alternative of the direct algorithm (21) shall be investigated as it can capture both suddenly increased learning effort and, with some latency due to floating averaging, also the immediate decrease of learning effort (Figure 3, middle axes, $k > 700$). Also, it is important to study LE with learning systems with more powerful learning criteria such as the ones employing generalized entropies (square-error based learning criteria have been investigated so far).

6. Conclusions

The main finding is that the learning effort of a pretrained learning model is quantifiable as a new (machine) learning-oriented information measure for each individually observed sample of data of otherwise complex dynamic systems while the model accuracy is not a necessary requirement for novelty detection. The method and the obtained results present LE as a cognitive concept of real-time novelty detection, where new information in data can be quantified via unusual learning effort, while, in principle, the error of the learning systems itself is not substantial. Being relieved from the assumption that model errors and novelty in data must be correlated, LE has the potential to detect novelty in complex behavior even with the use of imprecise learning systems. Thus, LE establishes a novel concept for research of new cognitive information measures with prospects to adaptive signal processing and cognitive computational intelligence methods with the very essence of learning systems.

Author Contributions: Conceptualization, I.B., W.K., and N.H.; methodology, I.B., W.K., and N.H.; software, I.B.; investigation, I.B., W.K. and N.H.; writing—original draft preparation, I.B.; writing—review and editing, I.B.; visualization, I.B.; supervision, W.K. and N.H.

Funding: Ivo Bukovsky was supported by the Ministry of Education, Youth and Sports of the Czech Republic under OP RDE grant number CZ.02.1.01/0.0/0.0/16_019/0000753 "Research Centre for Low-Carbon Energy Technologies. Witold Kinsner was supported by the Natural Sciences and Engineering Research Council of Canada (NSERC) and by the Mathematics of Information Technology and Complex Systems of Canada (MITACS). Noriyasu Homma was supported and by the Japanese project Smart aging research center grant and JSPS Kakenhi #17H04117 and #18K19892.

Acknowledgments: Ivo would like to acknowledge and thank Jan Vrba (currently with University of Chemistry and Technology, Prague) for mutually enriching discussions that we had on learning entropy during Jan's stay at CTU in Prague.

Conflicts of Interest: The authors declare no conflict of interest.

References

1. Shannon, C.E. A mathematical theory of communication. *Bell Syst. Tech. J.* **1948**, *27*, 379–423. [CrossRef]
2. Markou, M.; Singh, S. Novelty detection: A review—Part 1: Statistical approaches. *Signal Process.* **2003**, *83*, 2481–2497. [CrossRef]
3. Markou, M.; Singh, S. Novelty detection: A review—Part 2: Neural network based approaches. *Signal Process.* **2003**, *83*, 2499–2521. [CrossRef]
4. Pincus, S.M. Approximate entropy as a measure of system complexity. *Proc. Natl. Acad. Sci. USA* **1991**, *88*, 2297–2301. [CrossRef] [PubMed]
5. Richman, J.S.; Moorman, J.R. Physiological time-series analysis using approximate entropy and sample entropy. *Am. J. Physiol. Heart Circ. Physiol.* **2000**, *278*, H2039–H2049. [CrossRef] [PubMed]
6. Kinsner, W. Towards cognitive machines: Multiscale measures and analysis. *Int. J. Cogn. Inf. Nat. Intel. (IJCINI)* **2007**, *1*, 28–38. [CrossRef]
7. Kinsner, W. A Unified Approach To Fractal Dimensions. *Int. J. Cogn. Inf. Nat. Intel. (IJCINI)* **2007**, *1*, 26–46. [CrossRef]
8. Kinsner, W. Is Entropy Suitable to Characterize Data and Signals for Cognitive Informatics? *Int. J. Cognit. Inform. Nat. Int. (IJCINI)* **2007**, *1*, 34–57. [CrossRef]
9. Zurek, S.; Guzik, P.; Pawlak, S.; Kosmider, M.; Piskorski, J. On the relation between correlation dimension, approximate entropy and sample entropy parameters, and a fast algorithm for their calculation. *Phys. A Stat. Mech. Appl.* **2012**, *391*, 6601–6610. [CrossRef]
10. Schroeder, M.R. *Fractals, Chaos, Power Laws: Minutes from an Infinite Paradise*; W. H. Freeman: New York, NY, USA, 1991; ISBN 0-7167-2136-8.
11. Costa, M.; Goldberger, A.L.; Peng, C.-K. Multiscale Entropy Analysis of Complex Physiologic Time Series. *Phys. Rev. Lett.* **2002**, *89*, 068102. [CrossRef]
12. Costa, M.; Goldberger, A.L.; Peng, C.-K. Multiscale entropy analysis of biological signals. *Phys. Rev. E* **2005**, *71*, 021906. [CrossRef] [PubMed]
13. Wu, S.-D.; Wu, C.-W.; Lin, S.-G.; Wang, C.-C.; Lee, K.-Y. Time series analysis using composite multiscale entropy. *Entropy* **2013**, *15*, 1069–1084. [CrossRef]
14. Faes, L.; Nollo, G.; Porta, A. Compensated transfer entropy as a tool for reliably estimating information transfer in physiological time series. *Entropy* **2013**, *15*, 198–219. [CrossRef]

15. Yin, L.; Zhou, L. Function based fault detection for uncertain multivariate nonlinear non-gaussian stochastic systems using entropy optimization principle. *Entropy* **2013**, *15*, 32–52. [CrossRef]

16. Vorburger, P.; Bernstein, A. Entropy-based Concept Shift Detection. In Proceedings of the Sixth International Conference on Data Mining (ICDM'06), Hong Kong, China, 18–22 December 2006; pp. 1113–1118.

17. Amigó, J.; Balogh, S.; Hernández, S. A Brief Review of Generalized Entropies. *Entropy* **2018**, *20*, 813. [CrossRef]

18. Bereziński, P.; Jasiul, B.; Szpyrka, M. An Entropy-Based Network Anomaly Detection Method. *Entropy* **2015**, *17*, 2367–2408. [CrossRef]

19. Gajowniczek, K.; Orłowski, A.; Ząbkowski, T. Simulation Study on the Application of the Generalized Entropy Concept in Artificial Neural Networks. *Entropy* **2018**, *20*, 249. [CrossRef]

20. Ghanbari, M.; Kinsner, W. Extracting Features from Both the Input and the Output of a Convolutional Neural Network to Detect Distributed Denial of Service Attacks. In Proceedings of the 2018 IEEE 17th International Conference on Cognitive Informatics & Cognitive Computing (ICCI*CC), Berkeley, CA, USA, 16–18 July 2018; pp. 138–144.

21. Willsky, A.S. A survey of design methods for failure detection in dynamic systems. *Automatica* **1976**, *12*, 601–611. [CrossRef]

22. Gertler, J.J. Survey of model-based failure detection and isolation in complex plants. *IEEE Control Syst. Mag.* **1988**, *8*, 3–11. [CrossRef]

23. Isermann, R. Process fault detection based on modeling and estimation methods—A survey. *Automatica* **1984**, *20*, 387–404. [CrossRef]

24. Frank, P.M. Fault diagnosis in dynamic systems using analytical and knowledge-based redundancy: A survey and some new results. *Automatica* **1990**, *26*, 459–474. [CrossRef]

25. Widmer, G.; Kubat, M. Learning in the presence of concept drift and hidden contexts. *Mach. Learn.* **1996**, *23*, 69–101. [CrossRef]

26. Polycarpou, M.M.; Helmicki, A.J. Automated fault detection and accommodation: A learning systems approach. *IEEE Trans. Syst. Man Cybern.* **1995**, *25*, 1447–1458. [CrossRef]

27. Demetriou, M.A.; Polycarpou, M.M. Incipient fault diagnosis of dynamical systems using online approximators. *IEEE Trans. Autom. Control* **1998**, *43*, 1612–1617. [CrossRef]

28. Trunov, A.B.; Polycarpou, M.M. Automated fault diagnosis in nonlinear multivariable systems using a learning methodology. *IEEE Trans. Neural Netw.* **2000**, *11*, 91–101. [CrossRef] [PubMed]

29. Alippi, C.; Roveri, M. Just-in-Time Adaptive Classifiers—Part I: Detecting Nonstationary Changes. *IEEE Trans. Neural Netw.* **2008**, *19*, 1145–1153.

30. Alippi, C.; Roveri, M. Just-in-Time Adaptive Classifiers—Part II: Designing the Classifier. *IEEE Trans. Neural Netw.* **2008**, *19*, 2053–2064.

31. Alippi, C.; Boracchi, G.; Roveri, M. Just-In-Time Classifiers for Recurrent Concepts. *IEEE Trans. Neural Netw. Learn. Syst.* **2013**, *24*, 620–634. [CrossRef]

32. Alippi, C.; Ntalampiras, S.; Roveri, M. A Cognitive Fault Diagnosis System for Distributed Sensor Networks. *IEEE Trans. Neural Netw. Learn. Syst.* **2013**, *24*, 1213–1226. [CrossRef]

33. Grossberg, S. Adaptive Resonance Theory: How a Brain Learns to Consciously Attend, Learn, and Recognize a Changing World. *Neural Netw.* **2013**, *37*, 1–47. [CrossRef]

34. Bukovsky, I. Learning Entropy: Multiscale Measure for Incremental Learning. *Entropy* **2013**, *15*, 4159–4187. [CrossRef]

35. Bukovsky, I.; Oswald, C.; Cejnek, M.; Benes, P.M. Learning entropy for novelty detection a cognitive approach for adaptive filters. In Proceedings of the Sensor Signal Processing for Defence (SSPD), Edinburgh, UK, 8–9 September 2014; pp. 1–5.

36. Bukovsky, I.; Homma, N.; Cejnek, M.; Ichiji, K. Study of Learning Entropy for Novelty Detection in lung tumor motion prediction for target tracking radiation therapy. In Proceedings of the 2014 International Joint Conference on Neural Networks (IJCNN), Beijing, China, 6–11 July 2014; pp. 3124–3129.

37. Bukovsky, I.; Cejnek, M.; Vrba, J.; Homma, N. Study of Learning Entropy for Onset Detection of Epileptic Seizures in EEG Time Series. In Proceedings of the 2016 International Joint Conference on Neural Networks (IJCNN), Vancouver, BC, Canada, 24–29 July 2016.

38. Bukovsky, I.; Oswald, C. Case Study of Learning Entropy for Adaptive Novelty Detection in Solid-fuel Combustion Control. In *Intelligent Systems in Cybernetics and Automation Theory (CSOC 2015)*; Advances in Intelligent Systems and Computing; Silhavy, R., Senkerik, R., Oplatkova, Z., Prokopova, Z., Silhavy, P., Eds.; Springer: Cham, Switzerland, 2015.

39. Cejnek, M.; Bukovsky, I. Concept drift robust adaptive novelty detection for data streams. *Neurocomputing* **2018**, *309*, 46–53. [CrossRef]

40. Brissaud, J.-B. The meanings of entropy. *Entropy* **2005**, *7*, 68–96. [CrossRef]

41. Bukovsky, I. Modeling of Complex Dynamic Systems by Nonconventional Artificial Neural Architectures and Adaptive Approach to Evaluation of Chaotic Time Series. PhD thesis (in English), CTU in Prague, Prague, Czech Republic. Available online: https://aleph.cvut.cz/F?func=direct&doc_number=000674522&local_base=DUPL&format=999 (accessed on 11 February 2019).

42. Bukovsky, I.; Bila, J. Adaptive Evaluation of Complex Dynamical Systems Using Low-Dimensional Neural Architectures. In *Advances in Cognitive Informatics and Cognitive Computing. Studies in Computational Intelligence*; Wang, Y., Zhang, D., Kinsner, W., Eds.; Springer: Berlin/Heidelberg, Germany, 2010; Volume 323, pp. 33–57. ISBN 978-3-642-16082-0.

43. Bukovsky, I.; Kinsner, W.; Bila, J. Multiscale analysis approach for novelty detection in adaptation plot. In Proceedings of the Sensor Signal Processing for Defence (SSPD 2012), London, UK, 25–27 September 2012; pp. 1–6.

MDPI

Article

Assessing Information Transmission in Data Transformations with the Channel Multivariate Entropy Triangle

Francisco J. Valverde-Albacete [†] **and Carmen Peláez-Moreno** [*,†]

Department of Signal Theory and Communications, Universidad Carlos III de Madrid, Leganés 28911, Spain; fva@tsc.uc3m.es
* Correspondence: carmen@tsc.uc3m.es; Tel.: +34-91-624-8771
† These authors contributed equally to this work.

Received: 3 May 2018; Accepted: 20 June 2018; Published: 27 June 2018

Abstract: Data transformation, e.g., feature transformation and selection, is an integral part of any machine learning procedure. In this paper, we introduce an information-theoretic model and tools to assess the quality of data transformations in machine learning tasks. In an unsupervised fashion, we analyze the transformation of a discrete, multivariate source of information $\overline{X}$ into a discrete, multivariate sink of information $\overline{Y}$ related by a distribution $P_{\overline{XY}}$. The first contribution is a decomposition of the maximal potential entropy of $(\overline{X}, \overline{Y})$, which we call a balance equation, into its (a) non-transferable, (b) transferable, but not transferred, and (c) transferred parts. Such balance equations can be represented in (de Finetti) entropy diagrams, our second set of contributions. The most important of these, the aggregate channel multivariate entropy triangle, is a visual exploratory tool to assess the effectiveness of multivariate data transformations in transferring information from input to output variables. We also show how these decomposition and balance equations also apply to the entropies of $\overline{X}$ and $\overline{Y}$, respectively, and generate entropy triangles for them. As an example, we present the application of these tools to the assessment of information transfer efficiency for Principal Component Analysis and Independent Component Analysis as unsupervised feature transformation and selection procedures in supervised classification tasks.

Keywords: entropy, entropy visualization; entropy balance equation; Shannon-type relations; multivariate analysis; machine learning evaluation; data transformation

---

1. Introduction

Information-related considerations are often cursorily invoked in many machine learning applications, sometimes to suggest why a system or procedure is seemingly better than another at a particular task. In this paper, we set out to ground our work on measurable evidence phrases such as "this transformation retains more information from the data" or "this learning method uses the information from the data better than this other".

This has become particularly relevant with the increase of complexity of machine learning methods, such as deep neuronal architectures [1], which prevents straightforward interpretations. Nowadays, these learning schemes almost always become black-boxes, where the researchers try to optimize a prescribed performance metric without looking inside. However, there is a need to assess what the deep layers are actually accomplishing. Although some answers have started to appear [2,3], the issue is by no means settled.

In this paper, we put forward that framing the previous problem in a generic information-theoretical model can shed light on it by exploiting the versatility of information theory. For instance, a classical end-to-end example of an information-based model evaluation can be observed in Figure 1a. In this supervised scheme introduced in [4], the evaluation of the performance of the classifier involves only the comparison of the true labels K vs. the predicted labels $\hat{K}$. This means that all the complexity enclosed in the classifier box cannot be accessed, measured or interpreted.

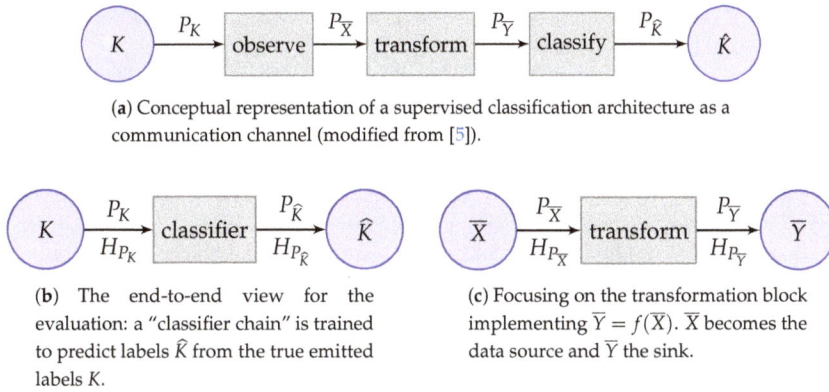

(a) Conceptual representation of a supervised classification architecture as a communication channel (modified from [5]).

(b) The end-to-end view for the evaluation: a "classifier chain" is trained to predict labels $\hat{K}$ from the true emitted labels K.

(c) Focusing on the transformation block implementing $\overline{Y} = f(\overline{X})$. $\overline{X}$ becomes the data source and $\overline{Y}$ the sink.

Figure 1. Different views of a supervised classification task as an information channel: (**a**) as individualized blocks; (**b**) for end-to-end evaluation; and (**c**) focused on the transformation.

In this paper, we want to expand the previous model into the scheme of Figure 1a, which provides a more detailed picture of the contents of the black-box where:

- A random source of classification labels K is subjected to a measurement process that returns random observations $\overline{X}$. The n instances of pairs $(k_i, \overline{x}_i), 1 \leq i \leq n$ is often called the (task) dataset.
- Then, a generic data transformation block may transform the available data, e.g., the observations in the dataset $\overline{X}$, into other data with "better" characteristics, the transformed feature vectors $\overline{Y}$. These characteristics may be representational power, independence among individual dimensions, reduction of complexity offered to a classifier, etc. The process is normally called feature transformation and selection.
- Finally, the $\overline{Y}$ are the inputs to an actual classifier of choice that obtains the predicted labels $\hat{K}$.

This would allow us to better understand the flow of information in the classification process with a view toward assessing and improving it.

Note the similarity between the classical setting of Figure 1b and the transformation block of Figure 1a reproduced in Figure 1c for convenience. Despite this, the former represents a Single-Input Single-Output (SISO) block with $(K, \hat{K}) \sim P_{K\hat{K}}$, whereas the latter represents a multivariate Multiple-Input Multiple-Output (MIMO) block described by the joint distribution of random vectors $(\overline{X}, \overline{Y}) \sim P_{\overline{X}\overline{Y}}$.

This MIMO kind of block may represent an unsupervised transformation method—for instance, a Principal Component Analysis (PCA) or Independent Component Analysis (ICA)—in which case, the "effectiveness" of the transformation is supplied by a heuristic principle, e.g., least reconstruction error on some test data, maximum mutual information, etc. However, it may also represent a supervised transformation method—for instance, $\overline{X}$ are the feature instances, and $\overline{Y}$ are the (multi-)labels or classes in a classification task, or $\overline{Y}$ may be the activation signals of a convolutional neural network trained using an implicit target signal— in which case, the "effectiveness" should measure the conformance to the supervisory signal.

In [4], we argued for carrying out the evaluation of classification tasks that can be modeled by Figure 1b with the new framework of entropy balance equations and their related entropy triangles [4–6]. This has provided a means of quantifying and visualizing the end-to-end information transfer for SISO architectures. The gist of this framework is explained in Section 2.1: if a classifier working on a certain dataset obtained a confusion matrix $P_{K\hat{K}}$, then we can information-theoretically assess the classifier by analyzing the entropies and information in the related distribution $P_{K\hat{K}}$ with the help of a balance equation [6]. However, looking inside the black-box poses a challenge since $\overline{X}$ and $\overline{Y}$ are random vectors and most information-theoretic quantities are not readily available in their multivariate version.

If we want to extend the same framework of evaluation to random vectors in general, we need the multivariate generalizations of the information-theoretic measures involved in the balance equations, an issue that is not free of contention. With this purpose in mind, we review the best-known multivariate generalizations of mutual information in Section 2.2.

We present our contributions finally in Section 3. As a first result, we develop a balance equation for the joint distribution $P_{\overline{XY}}$ and related representation in Sections 3.1 and 3.2, respectively. However we are also able to obtain split equations for the input and output multivariate sources only tied by one multivariate extension of mutual information, much as in the SISO case. As an instance of use, in Section 3.3, we analyze the transfer of information in PCA and ICA transformations applied to some well-known UCI datasets. We conclude with a discussion of the tools in light of this application in Section 3.4.

2. Methods

In Section 3, we will build a solution to our problem by finding the minimum common multiple, so to speak, of our previous solutions to the SISO block we describe in Section 2.1 and the multivariate source cases, to be described in Section 2.2.

2.1. The Channel Bivariate Entropy Balance Equation and Triangle

A solution to conceptualizing and visualizing the transmission of information through a channel where input and output are reduced to a single variable, that is with $|\overline{X}| = 1$ and $|\overline{Y}| = 1$, was presented in [6] and later extended in [4]. For this case, we use simply X and Y to describe the random variables. Notice that in the Introduction, and later in the example application, these are called K and $\hat{K}$, but here, we want to present this case as a simpler version of the one we set out to solve in this paper. Figure 2a, then, depicts a classical information-diagram (i-diagram) [7,8] of an entropy decomposition around P_{XY} in which we have included the exterior boundaries arising from the entropy balance equation, as we will show later. Three crucial regions can be observed:

- The (normalized) redundancy ([9], Section 2.4), or divergence with respect to uniformity (yellow area), $\Delta H_{P_X \cdot P_Y}$, between the joint distribution where P_X and P_Y are independent and the uniform distributions with the same cardinality of events as P_X and P_Y,

$$\Delta H_{P_X \cdot P_Y} = H_{U_X \cdot U_Y} - H_{P_X \cdot P_Y}. \tag{1}$$

- The mutual information, $MI_{P_{XY}}$ [10] (each of the green areas), quantifies the force of the stochastic binding between P_X and P_Y, "towards the outside" in Figure 2a,

$$MI_{P_{XY}} = H_{P_X \cdot P_Y} - H_{P_{XY}} \tag{2}$$

but also "towards the inside",

$$MI_{P_{XY}} = H_{P_X} - H_{P_{X|Y}} = H_{P_Y} - H_{P_{Y|X}}. \tag{3}$$

- The variation of information (the sum of the red areas), $VI_{P_{XY}}$ [11], embodies the residual entropy, not used in binding the variables,

$$VI_{P_{XY}} = H_{P_{X|Y}} + H_{P_{Y|X}}. \tag{4}$$

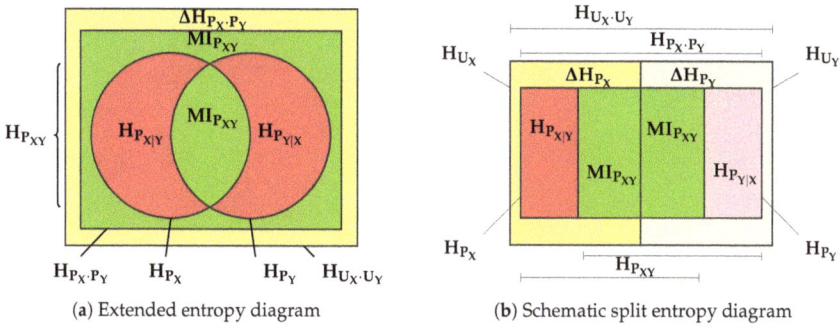

(a) Extended entropy diagram   (b) Schematic split entropy diagram

Figure 2. Extended entropy diagram related to a bivariate distribution, from [4].

Then, we may write the following entropy balance equation between the entropies of X and Y:

$$H_{U_X \cdot U_Y} = \Delta H_{P_X \cdot P_Y} + 2 \cdot MI_{P_{XY}} + VI_{P_{XY}} \tag{5}$$
$$0 \leq \Delta H_{P_X \cdot P_Y}, MI_{P_{XY}}, VI_{P_{XY}} \leq H_{U_X \cdot U_Y}$$

where the bounds are easily obtained from distributional considerations [6]. If we normalize (5) by the overall entropy $H_{U_X \cdot U_Y}$, we obtain:

$$1 = \Delta' H_{P_X \cdot P_Y} + 2 \cdot MI'_{P_{XY}} + VI'_{P_{XY}} \qquad\qquad 0 \leq \Delta' H_{P_X \cdot P_Y}, MI'_{P_{XY}}, VI'_{P_{XY}} \leq 1 \tag{6}$$

Equation (6) is the 2-simplex in normalized $\Delta H'_{P_X \cdot P_Y} \times 2MI'_{P_{XY}} \times VI'_{P_{XY}}$ space. Each joint distribution P_{XY} can be characterized by its joint entropy fractions, $F(P_{XY}) = [\Delta H'_{P_{XY}}, 2 \cdot MI'_{P_{XY}}, VI'_{P_{XY}}]$, whose projection onto the plane with director vector $(1, 1, 1)$ is its de Finetti or compositional diagram [12]. This diagram of the 2-simplex is an equilateral triangle, the coordinates of which are $F(P_{XY})$, so every bivariate distribution is shown as a point in the triangle, and each zone in the triangle is indicative of the characteristics of distributions, the coordinates of which fall in it. This is what we call the Channel Bivariate Entropy Triangle (CBET) whose schematic is shown in Figure 3.

We can actually decompose (5) and the quantities in it into two split balance equations,

$$H_{U_X} = \Delta H_{P_X} + MI_{P_{XY}} + H_{P_{X|Y}} \qquad\qquad H_{U_Y} = \Delta H_{P_Y} + MI_{P_{XY}} + H_{P_{Y|X}}. \tag{7}$$

with the obvious limits. These can be each normalized by H_{U_X}, respectively H_{U_Y}, leading to the 2-simplex equations:

$$1 = \Delta' H_{P_X} + MI'_{P_{XY}} + H'_{P_{X|Y}} \qquad\qquad 1 = \Delta' H_{P_Y} + MI'_{P_{XY}} + H'_{P_{Y|X}}. \tag{8}$$

Since these are also equations on a 2-simplex, we can actually represent the coordinates $F_X(P_{XY}) = [\Delta H'_{P_X}, MI'_{P_{XY}}, H'_{P_{X|Y}}]$ and $F_Y(P_{XY}) = [\Delta H'_{P_Y}, MI'_{P_{XY}}, H'_{P_{Y|X}}]$ in the same triangle side by side the original $F(P_{XY})$, whereby the representation seems to split in two.

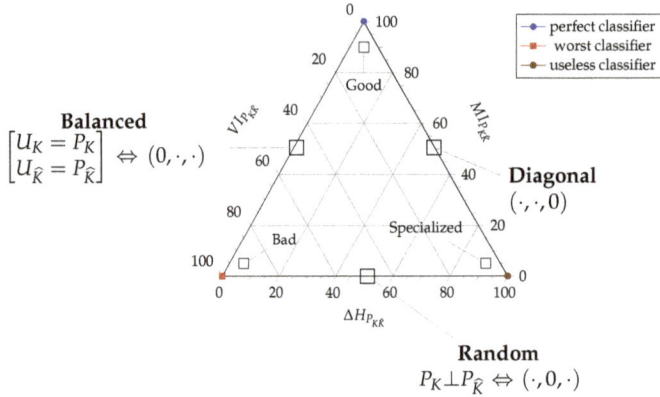

Figure 3. Schematic CBET as applied to supervised classifier assessment. An actual triangle shows dots for each classifier (or its split coordinates see Figure 6 for example) and none of the callouts for specific types of classifiers (from [4]). The callouts situated in the center of the sides of the triangle apply to the whole side.

2.1.1. Application: The Evaluation of Multiclass Classification

The CBET can be used to visualize the performance of supervised classifiers in a straightforward manner as announced in the Introduction: Consider the confusion matrix $N_{K\widehat{K}}$ of a classifier chain on a supervised classification task given the random variable of true class labels $K \sim P_K$ and that of predicted labels $\widehat{K} \sim P_{\widehat{K}}$ as depicted in Figure 1a, which now play the role of P_X and P_Y. From this confusion matrix, we can estimate the joint distribution $P_{K\widehat{K}}$ between the random variables, so that the entropy triangle for $P_{K\widehat{K}}$ produces valuable information about the actual classifier used to solve the task [6,13] and even the theoretical limits of the task; for instance, whether it can be solved in a trustworthy manner by classification technology and with what effectiveness.

The CBET acts, in this case, as an exploratory data analysis tool for visual assessment, as shown in Figure 3.

The success of this approach in the bivariate, supervised classification case is a strong hint that the multivariate extension will likewise be useful for other machine learning tasks. See [4] for a thorough explanation of this procedure.

2.2. Quantities around the Multivariate Mutual Information

The main hurdle for a multivariate extension of the balance Equation (5) and the CBET is the multivariate generalization of binary mutual information, since it quantifies the information transport from input to output in the bivariate case and is also crucial for the decoupling of (5) into the split balance Equation (7). For this reason, we next review the different "flavors" of information measures describing sets of more than two variables looking for these two properties. We start from very basic definitions both in the interest of self-containment and to provide a script of the process of developing future analogues for other information measures.

To fix notation, let $\overline{X} = \{X_i \mid 1 \leq i \leq m\}$ be a set of discrete random variables with joint multivariate distribution $P_{\overline{X}} = P_{X_1...X_m}$ and the corresponding marginals $P_{X_i}(x_i) = \sum_{j \neq i} P_{\overline{X}}(\overline{x})$ where $\overline{x} = x_1 \ldots x_m$ is a tuple of m elements; likewise for $\overline{Y} = \{Y_j \mid 1 \leq j \leq l\}$, with $P_{\overline{Y}} = P_{Y_1...Y_l}$ and the marginals P_{Y_j}. Furthermore, let $P_{\overline{XY}}$ be the joint distribution of the $(m+l)$-length tuples $\overline{XY}$. Note that two different situations can be clearly distinguished:

Situation 1: All the random variables form part of the same set $\overline{X}$, and we are looking at information transfer within this set, or

Situation 2: They are partitioned into two different sets $\overline{X}$ and $\overline{Y}$, and we are looking at information transfer between these sets.

An up-to-date review of multivariate information measures in both situations is [14], which follows the interesting methodological point from [15] of calling information those measures that involve amounts of entropy shared by multiple variables and entropies those that do not—although, this poses a conundrum for the entropy written as the self-information $H_{P_X} = MI_{P_{XX}}$.

Since i-diagrams are a powerful tool to visualize the interaction of distributions in the bivariate case, we will also try to use them for sets of random variables. For multivariate generalizations of mutual information as seen in the i-diagrams, the following caveats apply:

- Their multivariate generalization is only warranted when signed measures of probability are considered, since it is well known that some of these "areas" can be negative, contrary to the geometric intuitions in this respect.
- We should retain the bounding rectangles that appear when considering the most entropic distributions with similar support to the ones being graphed [6]. This is the sense of the bounding rectangles in Figure 4a,b.

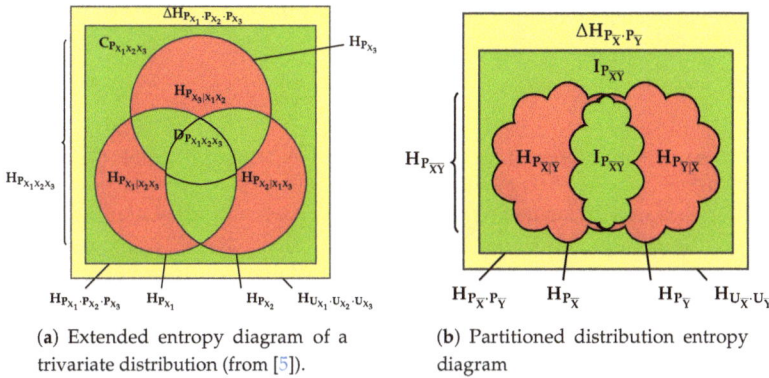

(a) Extended entropy diagram of a trivariate distribution (from [5]).

(b) Partitioned distribution entropy diagram

Figure 4. (Color online) Extended entropy diagram of multivariate distributions for (**a**) a trivariate distribution (from [5]) as an instance of Situation 1; and (**b**) a joint distribution where a partitioning of the variables is made evident (Situation 2). The color scheme follows that of Figure 2, to be explained in the text.

With great insight, the authors of [15] point out that some of the multivariate information measures stem from focusing on a particular property of the bivariate mutual information and generalizing it to the multivariate setting. The properties in question—including already stated (2) and (3)—are: The properties in question are:

$$MI_{P_{XY}} = H_{P_X} + H_{P_Y} - H_{P_{XY}}$$
$$MI_{P_{XY}} = H_{P_X} - H_{P_{X|Y}} = H_{P_Y} - H_{P_{Y|X}}$$
$$MI_{P_{XY}} = \sum_{x,y} P_{XY}(x,y) \log \frac{P_{XY}(x,y)}{P_X(x)P_Y(y)} \tag{9}$$

Regarding the first situation of a vector of random variables $\overline{X} \sim P_{\overline{X}}$, let $\Pi_{\overline{X}} = \prod_{i=1}^n P_{X_i}$ be the (jointly) independent distribution with similar marginals to $P_{\overline{X}}$. To picture this (virtual) distribution consider Figure 4a depicting an i-diagram for $\overline{X} = [X_1, X_2, X_3]$. Then, $\Pi_{\overline{X}} = P_{X_1} \cdot P_{X_2} \cdot P_{X_3}$ is the inner rectangle containing both green areas. The different extensions of mutual information that concentrate on different properties are:

- the total correlation [16], integration [17] or multi-information [18], which is a generalization of (2), represented by the green area outside $H_{P_{\overline{X}}}$.

$$C_{P_{\overline{X}}} = H_{\Pi_{\overline{X}}} - H_{P_{\overline{X}}} \tag{10}$$

- the dual total correlation [19,20] or interaction complexity [21] is a generalization of (3), represented by the green area inside $H_{P_{\overline{X}}}$:

$$D_{P_{\overline{X}}} = H_{P_{\overline{X}}} - VI_{P_{\overline{X}}} \tag{11}$$

- the interaction information [22], multivariate mutual information [23] or co-information [24] is the generalization of (9), the total amount of information to which all variables contribute.

$$MI_{P_{\overline{X}}} = \sum P_{\overline{X}}(\overline{x}) \log \frac{P_{\overline{X}}(\overline{x})}{\Pi_{\overline{X}}(\overline{x})} \tag{12}$$

It is represented by the inner convex green area (within the dual total correlation), but note that it may in fact be negative for $n > 2$ [25].
- the local exogenous information [15] or the bound information [26] is the addition of the total correlation and the dual total correlation:

$$M_{P_{\overline{X}}} = C_{P_{\overline{X}}} + D_{P_{\overline{X}}}. \tag{13}$$

Some of these generalizations of the multivariate case were used in [5,26] to develop a similar technique as the CBET, but applied to analyzing the information content of data sources. For this purpose, it was necessary to define for every random variable a residual entropy $H_{P_{X_i | X_i^c}}$, where $X_i^c = \overline{X} \setminus \{X_i\}$, which is not explained by the information provided by the other variables. We call residual information [15] or (multivariate) variation of information [11,26] the generalization of the same quantity in the bivariate case, i.e., the sum of these quantities across the set of random variables:

$$VI_{P_{\overline{X}}} = \sum_{i=1}^{n} H_{P_{X_i | X_i^c}}. \tag{14}$$

Then, the variation of information can easily be seen to consist of the sum of the red areas in Figure 4a and amounts to information particular to each variable.

The main question regarding this issue is which, if any, of these generalizations of bivariate mutual information are adequate for an analogue of the entropy balance equations and triangles. Note that all of these generalizations consider $\overline{X}$ as a homogeneous set of variables, that is Situation 1 described at the beginning of this section, and none consider the partitioning of the variables in $\overline{X}$ into two subsets (Situation 2), for instance to distinguish between input and output ones, so the answer cannot be straightforward. This issue is clarified in Section 3.1.

3. Results

Our goal is now to find a decomposition of the entropies around characterizing a joint distribution $P_{\overline{XY}}$ between random vectors $\overline{X}$ and $\overline{Y}$ in ways analogous to those of (5) but considering multivariate input and output.

Note that it provides no advantage trying to do this on continuous distributions, as the entropic measures used are basic. Rather, what we actually capitalize on is in the outstanding existence of a balance equation between these apparently simple entropic concepts, and what their intuitive meanings afford to the problem of measuring the transfer of information in data processing tasks. As we set out to demonstrate in this section, our main results are in complete analogy to those of the binary case, but with the flavour of the multivariate case.

3.1. The Aggregate and Split Channel Multivariate Balance Equation

Consider the modified information diagram of Figure 4b highlighting entropies for some distributions around $P_{\overline{XY}}$. When we distinguish two random vectors in the set of variables $\overline{X}$ and $\overline{Y}$, a proper multivariate generalization of the variation of information in (4) is

$$VI_{P_{\overline{XY}}} = H_{P_{\overline{X}|\overline{Y}}} + H_{P_{\overline{Y}|\overline{X}}}. \tag{15}$$

and we will also call it the *variation of information*. It represents the addition of the information in $\overline{X}$ not shared with $\overline{Y}$ and vice-versa, as captured by the red area in Figure 4b. Note that this is a non-negative quantity, since its is the addition of two entropies.

Next, consider

- $U_{\overline{XY}}$, the uniform distribution over the supports of $\overline{X}$ and $\overline{Y}$, and
- $P_{\overline{X}} \times P_{\overline{Y}}$, the distribution created with the marginals of $P_{\overline{XY}}$ considered independent.

Then, we may define a *multivariate divergence with respect to uniformity*—in analogy to (1)—as

$$\Delta H_{P_{\overline{X}} \times P_{\overline{Y}}} = H_{U_{\overline{XY}}} - H_{P_{\overline{X}} \times P_{\overline{Y}}}. \tag{16}$$

This is the yellow area in Figure 4b representing the divergence of the virtual distribution $P_{\overline{X}} \times P_{\overline{Y}}$ with respect to uniformity. The virtuality comes from the fact that this distribution does not properly exist in the context being studied. Rather, it only appears in the extreme situation that the marginals of $P_{\overline{XY}}$ are independent.

Furthermore, recall that both the total entropy of the uniform distribution and the divergence from uniformity factor into individual equalities $H_{U_{\overline{X}} U_{\overline{Y}}} = H_{U_{\overline{X}}} + H_{U_{\overline{Y}}}$—since uniform joint distributions always have independent marginals—and $H_{P_{\overline{X}} \times P_{\overline{Y}}} = H_{P_{\overline{X}}} + H_{P_{\overline{Y}}}$. Therefore (16) admits splitting as $\Delta H_{P_{\overline{X}} \times P_{\overline{Y}}} = \Delta H_{P_{\overline{X}}} + \Delta H_{P_{\overline{Y}}}$ where

$$\Delta H_{P_{\overline{X}}} = H_{U_{\overline{X}}} - H_{P_{\overline{X}}} \qquad\qquad \Delta H_{P_{\overline{Y}}} = H_{U_{\overline{Y}}} - H_{P_{\overline{Y}}}. \tag{17}$$

Now, both $U_{\overline{X}}$ and $U_{\overline{Y}}$ are the most entropic distributions definable in the support of $\overline{X}$ and $\overline{Y}$ whence both $\Delta H_{P_{\overline{X}}}$ and $\Delta H_{P_{\overline{Y}}}$ are non-negative, as is their addition. These generalizations are straightforward and intuitively mean that *we expect them to agree with the intuitions developed in the CBET*, which is an important usability concern.

The problem is finding a quantity that fulfills the same role as the (bivariate) mutual information. The first property that we would like to have is for this quantity to be a "transmitted information" after conditioning away any of the entropy of either partition, so we propose the following as a definition:

$$I_{P_{\overline{XY}}} = H_{P_{\overline{XY}}} - VI_{P_{\overline{XY}}} \tag{18}$$

represented by the inner green area in the i-diagram of Figure 4b. This can easily be "refocused" on each of the subsets of the partition:

Lemma 1. *Let $P_{\overline{XY}}$ be a discrete joint distribution. Then*

$$H_{P_{\overline{X}}} - H_{P_{\overline{X}|\overline{Y}}} = H_{P_{\overline{Y}}} - H_{P_{\overline{Y}|\overline{X}}} = I_{P_{\overline{XY}}} \tag{19}$$

Proof. Recalling that the conditional entropies are easily related to the joint entropy by the chain rule $H_{P_{\overline{XY}}} = H_{P_{\overline{X}}} + H_{P_{\overline{Y}|\overline{X}}} = H_{P_{\overline{Y}}} + H_{P_{\overline{X}|\overline{Y}}}$, simply subtract $VI_{P_{\overline{XY}}}$. □

This property introduces the notion that this information is *within each of $\overline{X}$ and $\overline{Y}$ independently but mutually induced*. It is easy to see that this quantity appears once again in the i-diagram:

Lemma 2. *Let $P_{\overline{XY}}$ be a discrete joint distribution. Then*

$$I_{P_{\overline{XY}}} = H_{P_{\overline{X}} \times P_{\overline{Y}}} - H_{P_{\overline{XY}}} . \tag{20}$$

Proof. Considering the entropy decomposition of $P_{\overline{X}} \times P_{\overline{Y}}$:

$$H_{P_{\overline{X}} \times P_{\overline{Y}}} - H_{P_{\overline{XY}}} = H_{P_{\overline{X}}} + H_{P_{\overline{Y}}} - \left(H_{P_{\overline{Y}}} + H_{P_{\overline{X}|\overline{Y}}} \right) = H_{P_{\overline{X}}} - H_{P_{\overline{X}|\overline{Y}}} = I_{P_{\overline{XY}}}$$

□

In other words, this is the quantity of information required to bind $P_{\overline{X}}$ and $P_{\overline{Y}}$; equivalently, it is the amount of information *lost* from $P_{\overline{X}} \times P_{\overline{Y}}$ to achieve the binding in $P_{\overline{XY}}$. Pictorially, this is the outermost green area in Figure 4b, and *it must be non-negative*, since $P_{\overline{X}} \times P_{\overline{Y}}$ is more entropic than $P_{\overline{XY}}$. Notice that (18) and (19) are the analogues of (10) and (11), respectively, but with the flavor of (2) and (3). Therefore, this quantity must be the multivariate mutual information of $P_{\overline{XY}}$ as per the Kullback-Leibler divergence definition:

Lemma 3. *Let $P_{\overline{XY}}$ be a discrete joint distribution. Then*

$$I_{P_{\overline{XY}}} = \sum_{i,j} P_{\overline{XY}}(x_i, y_j) \log \frac{P_{\overline{XY}}(x_i, y_j)}{P_{\overline{X}}(x_i) P_{\overline{Y}}(y_j)} \tag{21}$$

Proof. This is an easy manipulation.

$$\sum_{i,j} P_{\overline{XY}}(x_i, y_j) \log \frac{P_{\overline{XY}}(x_i, y_j)}{P_{\overline{X}}(x_i) P_{\overline{Y}}(y_j)} = \sum_{i,j} P_{\overline{XY}}(x_i, y_j) \log \frac{P_{\overline{X}|\overline{Y}=y_j}(x_i|y_j)}{P_{\overline{X}}(x_i)} = \sum_{i} P_{\overline{X}}(x_i) \log \frac{1}{P_{\overline{X}}(x_i)} -$$

$$- \sum_{j} P_{\overline{Y}}(y_j) \sum_{i} P_{\overline{X}|\overline{Y}=y_j}(x_i|y_j) \log \frac{1}{P_{\overline{X}|\overline{Y}=y_j}(x_i|y_j)} =$$

$$= H_{P_{\overline{X}}} - H_{P_{\overline{X}|\overline{Y}}} = I_{P_{\overline{XY}}},$$

after a step of marginalization and considering (3). □

With these relations we can state our first theorem:

Theorem 1. *Let $P_{\overline{XY}}$ be a discrete joint distribution. Then the following decomposition holds:*

$$H_{U_{\overline{X}} \times U_{\overline{Y}}} = \Delta H_{P_{\overline{X}} \times P_{\overline{Y}}} + 2 \cdot I_{P_{\overline{XY}}} + VI_{P_{\overline{XY}}} \tag{22}$$

$$0 \le \Delta H_{P_{\overline{X}} \times P_{\overline{Y}}}, I_{P_{\overline{XY}}}, VI_{P_{\overline{XY}}} \le H_{U_{\overline{X}} \times U_{\overline{Y}}}$$

Proof. From (16) we have $H_{U_{\overline{X}} \times U_{\overline{Y}}} = \Delta H_{P_{\overline{X}} \times P_{\overline{Y}}} + H_{P_{\overline{X}} \times P_{\overline{Y}}}$ whence by introducing (18) and (20) we obtain:

$$H_{U_{\overline{X}} \times U_{\overline{Y}}} = \Delta H_{P_{\overline{X}} \times P_{\overline{Y}}} + I_{P_{\overline{XY}}} + H_{P_{\overline{XY}}} = \Delta H_{P_{\overline{X}} \times P_{\overline{Y}}} + I_{P_{\overline{XY}}} + I_{P_{\overline{XY}}} + VI_{P_{\overline{XY}}}. \tag{23}$$

Recall that each quantity is non-negative by (15), (16) and (21), so the only things left to be proven are the limits for each quantity in the decomposition. For that purpose, consider the following clarifying *conditions*,

1. **$\overline{X}$ marginal uniformity** when $H_{P_{\overline{X}}} = H_{U_{\overline{X}}}$, **$\overline{Y}$ marginal uniformity** when $H_{P_{\overline{Y}}} = H_{U_{\overline{Y}}}$ and **marginal uniformity** when both conditions coocur.
2. **Marginal independence**, when $P_{\overline{XY}} = P_{\overline{X}} \times P_{\overline{Y}}$.

3. $\overline{Y}$ **determines** $\overline{X}$ when $H_{P_{\overline{X}|\overline{Y}}} = 0$, $\overline{X}$ **determines** $\overline{Y}$ when $H_{P_{\overline{Y}|\overline{X}}} = 0$ and **mutual determination**, when both conditions hold.

Notice that these conditions are *independent of each other* and that *each fixes the value of one of the quantities in the balance*:

- For instance, in case $H_{P_{\overline{X}}} = H_{U_{\overline{X}}}$ then $\Delta H_{P_{\overline{X}}} = 0$ after (17). Similarly, if $H_{P_{\overline{Y}}} = H_{U_{\overline{Y}}}$ then $\Delta H_{P_{\overline{Y}}} = 0$. Hence when marginal uniformity holds, we have $\Delta H_{P_{\overline{XY}}} = 0$.
- Similarly, when marginal independence holds, we see that $I_{P_{\overline{XY}}} = 0$ from (20). Otherwise stated, $H_{P_{\overline{X}|\overline{Y}}} = H_{P_{\overline{X}}}$ and $H_{P_{\overline{Y}|\overline{X}}} = H_{P_{\overline{Y}}}$.
- Finally, if mutual determination holds—that is to say the variables in either set are deterministic functions of those of the other set—by the definition of the multivariate variation of information, we have $VI_{P_{\overline{X}|\overline{Y}}} = 0$.

Therefore, these three conditions fix the lower bounds for their respectively related quantities. Likewise, the upper bounds hold when *two* of the conditions hold at the same time. This is easily seen invoking the previously found balance equation (23):

- For instance, if marginal uniformity holds, then $\Delta H_{P_{\overline{XY}}} = 0$. But if marginal independence also holds, then $I_{P_{\overline{X}|\overline{Y}}} = 0$ whence by (23) $VI_{P_{\overline{XY}}} = H_{U_{\overline{X}} \times U_{\overline{Y}}}$.
- But if both marginal uniformity and mutual determination hold, then we have $\Delta H_{P_{\overline{XY}}} = 0$ and $VI_{P_{\overline{XY}}} = 0$ so that $I_{P_{\overline{XY}}} = H_{U_{\overline{X}} \times U_{\overline{Y}}}$.
- Finally, if both mutual determination and marginal indepence holds, then a fortiori $\Delta H_{P_{\overline{XY}}} = H_{U_{\overline{X}} \times U_{\overline{Y}}}$.

This concludes the proof. □

Notice how the bounds also allow an interpretation similar to that of (5). In particular, the interpretation of the conditions for actual joint distributions will be taken again in Section 3.2.

The next question is whether the balance equation also admits splitting.

Theorem 2. *Let $P_{\overline{XY}}$ be a discrete joint distribution. Then the Channel Multivariate Entropy Balance equation can be split as:*

$$H_{U_{\overline{X}}} = \Delta H_{P_{\overline{X}}} + I_{P_{\overline{XY}}} + H_{P_{\overline{X}|\overline{Y}}} \qquad 0 \le \Delta H_{P_{\overline{X}}}, I_{P_{\overline{XY}}}, H_{P_{\overline{X}|\overline{Y}}} \le H_{U_{\overline{X}}} \qquad (24)$$

$$H_{U_{\overline{Y}}} = \Delta H_{P_{\overline{Y}}} + I_{P_{\overline{XY}}} + H_{P_{\overline{Y}|\overline{X}}} \qquad 0 \le \Delta H_{P_{\overline{Y}}}, I_{P_{\overline{XY}}}, H_{P_{\overline{Y}|\overline{X}}} \le H_{U_{\overline{Y}}} \qquad (25)$$

Proof. We prove (24): the proof of (25) is similar *mutatis mutandis*.

In a similar way as for (22), we have that $H_{U_{\overline{X}}} = \Delta H_{P_{\overline{X}}} + H_{P_{\overline{X}}}$. By introducing the value of $H_{P_{\overline{X}}}$ from (19) we obtain the decomposition of $H_{U_{\overline{X}}}$ of (24).

These quantities are non-negative, as mentioned. Next consider the $\overline{X}$ marginal uniformity condition applied to the input vector introduced in the proof of Theorem 1. Clearly, $\Delta H_{\overline{X}} = 0$. Marginal independence, again, is the condition so that $I_{\overline{XY}} = 0$. Finally, if $\overline{Y}$ determines $\overline{X}$ then $H_{P_{\overline{X}|\overline{Y}}} = 0$. These conditions individually provide the lower bounds on each quantity.

On the other hand, when we put together any two of these conditions, we obtain the upper bound for the unspecified variable: so, if $\Delta H_{P_{\overline{X}}} = 0$ and $I_{P_{\overline{XY}}} = 0$ then $H_{P_{\overline{X}|\overline{Y}}} = H_{P_{\overline{X}}} = H_{U_{\overline{X}}}$. Also, if $I_{P_{\overline{XY}}} = 0$ and $H_{P_{\overline{X}|\overline{Y}}} = 0$, then $H_{P_{\overline{X}}} = H_{P_{\overline{X}|\overline{Y}}} = 0$ and $\Delta H_{P_{\overline{X}}} = H_{U_{\overline{X}}} - 0$. Finally, if $H_{P_{\overline{X}|\overline{Y}}} = 0$ and $\Delta H_{P_{\overline{X}}} = 0$, then $I_{P_{\overline{XY}}} = H_{P_{\overline{X}}} - H_{P_{\overline{X}|\overline{Y}}} = H_{U_{\overline{X}}} - 0$. □

3.2. Visualizations: From i-Diagrams to Entropy Triangles

3.2.1. The Channel Multivariate Entropy Triangle

Our next goal is to develop an exploratory analysis tool similar to the CBET introduced in Section 2.1. As in that case, we need the equation of a simplex to represent the information balance of a multivariate transformation. For that purpose, as in (6) we may normalize by the overall entropy $H_{U_{\overline{X}} \times U_{\overline{Y}}}$ to obtain the equation of the 2-simplex in multivariate entropic space,

$$1 = \Delta' H_{P_{\overline{X}} \times P_{\overline{Y}}} + 2 \cdot I'_{P_{\overline{X}\overline{Y}}} + VI'_{P_{\overline{X}\overline{Y}}} \tag{26}$$

$$0 \leq \Delta' H_{P_{\overline{X}} \times P_{\overline{Y}}}, I'_{P_{\overline{X}\overline{Y}}}, VI'_{P_{\overline{X}\overline{Y}}} \leq 1.$$

The de Finetti diagram of this equation then provides the aggregated *Channel Multivariate Entropy Triangle, CMET*.

A *formal* graphical assessment of multivariate joint distribution with the CMET is fairly simple using the schematic in Figure 5a and the conditions of Theorem 1:

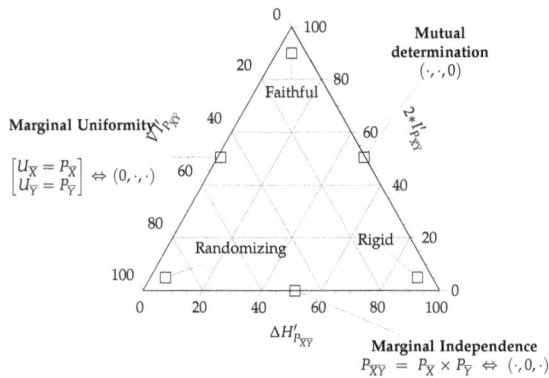

(a) Schematic CMET with a formal interpretation.

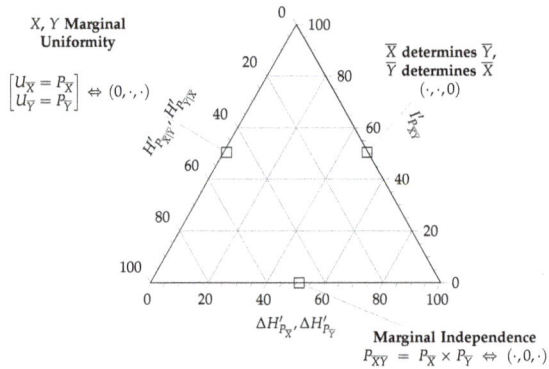

(b) Schematic *split CMETs* with formal interpretations. Note that there are **two types** of overimposed entropy triangles in this figure.

Figure 5. Schematic Channel Multivariate Entropy Triangles (CMET) showing interpretable zones and extreme cases using formal conditions. The annotations on the center of each side are meant to hold for that whole side, those for the vertices are meant to hold in their immediate neighborhood too.

- The lower side of the triangle with $I'_{P_{\overline{X}\overline{Y}}} = 0$, affected of *marginal independence* $P_{\overline{X}\overline{Y}} = P_{\overline{X}} \times P_{\overline{Y}}$, is the locus of partitioned joint distributions who do not share information between the two blocks $\overline{X}$ and $\overline{Y}$.

- The right side of the triangle with $VI'_{P_{XY}} = 0$, described with *mutual determination* $H'_{P_{X|Y}} = 0 = H'_{P_{Y|X}}$, is the locus of partitioned joint distributions whose groups do not carry supplementary information to that provided by the other group.
- The left side with $\Delta H'_{P_{XY}} = 0$, describing distributions with *uniform marginals* $P_{\overline{X}} = U_{\overline{X}}$ and $P_{\overline{Y}} = U_{\overline{Y}}$, is the locus of partitioned joint distributions that offer as much potential information for transformations as possible.

Based on these characterizations we can attach interpretations to other regions of the CMET:

- If we want a transformation from $\overline{X}$ to $\overline{Y}$ to be *faithful*, then we want to maximize the information used for mutual determination $I'_{P_{XY}} \to 1$, equivalently, minimize at the same time the divergence from uniformity $\Delta H'_{P_{XY}} \to 0$ and the information that only pertains to each of the blocks in the partition $VI'_{P_{XY}} \to 0$. So the coordinates of a faithful partitioned joint distribution will lay close to the apex of the triangle.
- However, if the coordinates of a distribution lay close to the left vertex $VI'_{P_{XY}} \to 1$, then it shows marginal uniformity $\Delta H'_{P_{XY}} \to 0$ but shares little or no information between the blocks $I'_{P_{XY}} \to 0$, hence it must be a *randomizing* transformation.
- Distributions whose coordinates lay close to the right vertex $\Delta H'_{P_{XY}} \to 1$ are essentially deterministic and in that sense carry no information $I'_{P_{XY}} \to 0, VI'_{P_{XY}} \to 0$. Indeed in this instance there does not seem to exist a transformation, whence we call them *rigid*.

These qualities are annotated on the vertices of the schematic CMET of Figure 5a. Note that different applications may call for partitioned distributions with different qualities and the one used above is pertinent when the partitioned joint distributions models a transformation of $\overline{X}$ into $\overline{Y}$ or vice-versa.

3.2.2. Normalized Split Channel Multivariate Balance Equations

With a normalization similar to that from (7) to (8), (24) and (25) naturally lead to 2-simplex equations normalizing by $H_{U_{\overline{X}}}$ and $H_{U_{\overline{Y}}}$, respectively

$$1 = \Delta' H_{P_{\overline{X}}} + I'_{P_{XY}} + H'_{P_{X|Y}} \qquad\qquad 0 \le \Delta' H_{P_{\overline{X}}}, I'_{P_{XY}}, H'_{P_{X|Y}} \le 1 \qquad (27)$$

$$1 = \Delta' H_{P_{\overline{Y}}} + I'_{P_{XY}} + H'_{P_{Y|X}} \qquad\qquad 0 \le \Delta' H_{P_{\overline{Y}}}, I'_{P_{XY}}, H'_{P_{Y|X}} \le 1 \qquad (28)$$

Note that the quantities $\Delta H'_{P_{\overline{X}}}$ and $\Delta H'_{P_{\overline{Y}}}$ have been independently motivated and named *redundancies* ([9], Section 2.4).

These are actually two different representations for each of the two blocks in the partitioned joint distribution. Using the fact that they share one coordinate—$I'_{P_{XY}}$—and the rest are analogues—$\Delta' H_{P_{\overline{X}}}$ and $\Delta' H_{P_{\overline{Y}}}$ on one side, and $H'_{P_{X|Y}}$ and $H'_{P_{Y|X}}$ on the other—we can represent both equations *at the same time* in a single de Finetti diagram. We call this representation the *split Channel Multivariate Entropy Triangle*, a schema of which can be seen in Figure 5b. The qualifying "split" then refers to the fact that each partitioned joint distribution appears as *two points* in the diagram. Note the double annotation in the left and bottom coordinates implying that there are *two* different diagrams overlapping.

Conventionally, the point referring to the $\overline{X}$ block described by (27) is represented with a cross, while the point referring to the $\overline{Y}$ block described by (28) is represented with a circle as will be noted in Figure 6.

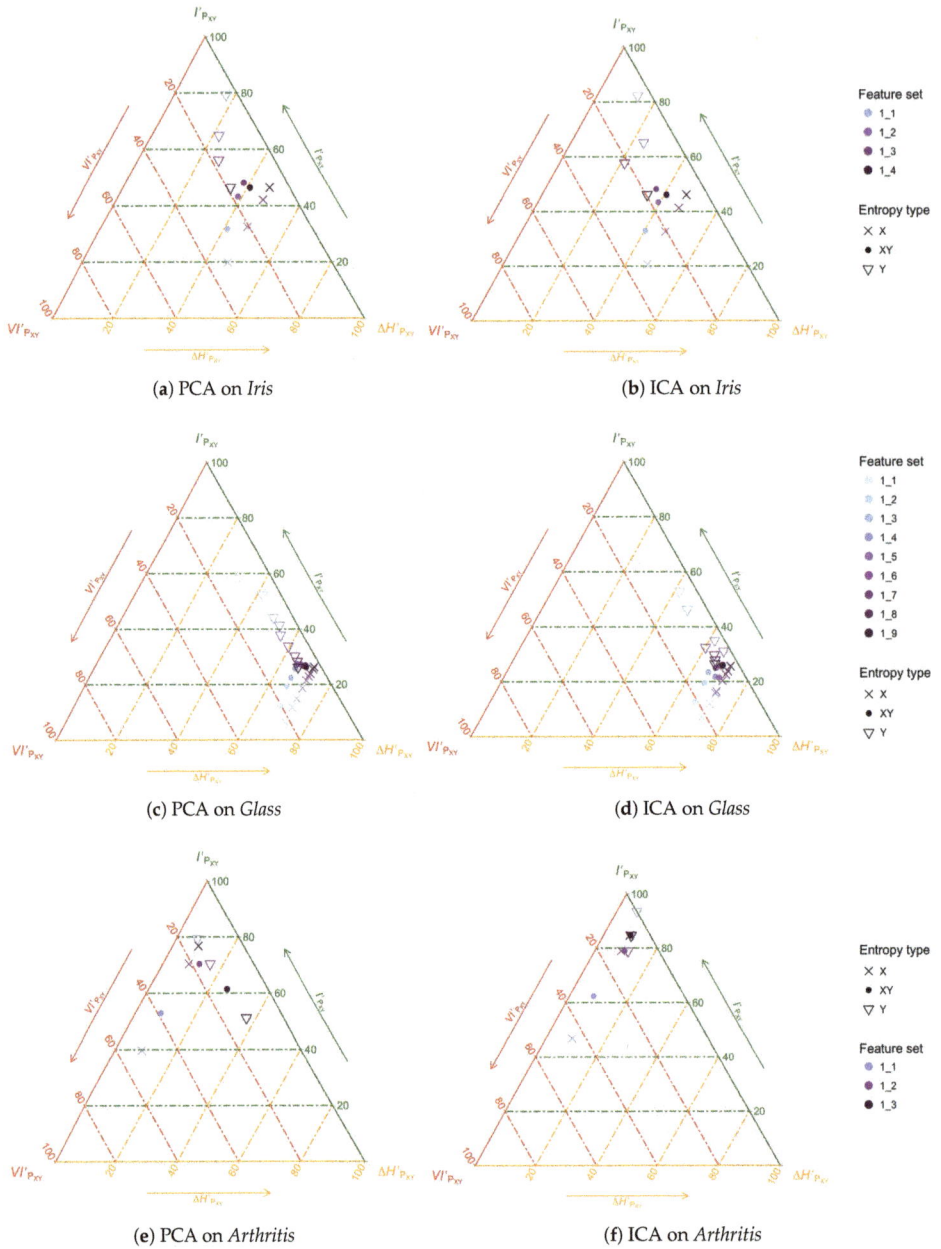

Figure 6. (Color online) Split CMET exploration of feature transformation and selection with PCA (left) and ICA (right) on *Iris*, *Glass* and *Arthritis* when selecting the first *n* ranked features as obtained for each method. The colors of the axes have been selected to match those of Figure 4.

The formal interpretation of this split diagram with the conditions of Theorem 1 follows that of the aggregated CMET but considering only one block at a time, for instance, for $\overline{X}$:

- The lower side of the triangle is interpreted as before.
- The right side of the triangle is the locus of the partitioned joint distribution whose $\overline{X}$ block is completely determined by the $\overline{Y}$ block, that is, $H'_{P_{\overline{X}|\overline{Y}}} = 0$.
- The left side of the triangle $\Delta H'_{P_{\overline{X}}} = 0$ is the locus of those partitioned joint distributions whose $\overline{X}$ marginal is uniform $P_{\overline{X}} = U_{\overline{X}}$.

The interpretation is analogue for $\overline{Y}$ *mutatis mutandis*.

The purpose of this representation is to investigate the formal conditions separately on each block. However, for this split representation we have to take into consideration that the normalizations may not be the same, that is $H_{P_{\overline{X}}}$ and $H_{P_{\overline{Y}}}$ are, in general, different.

A full example of the interpretation of both types of diagrams, the CMET and the split CMET is provided in the next Section in the context of feature transformation and selection.

3.3. Example Application: The Analysis of Feature Transformation and Selection with Entropy Triangles

In this Section we present an application of the results obtained above to a machine learning subtask: the transformation and selection of features for supervised classification.

The task. An extended practice in supervised classification is to explore different transformations of the observations and then evaluate such different approaches on different classifiers for a particular task [27]. Instead of this "in the loop" evaluation—that conflates the evaluation of the transformation and the classification—we will use the CMET to evaluate *only* the transformation block using the information transferred from the original to the transformed features as heuristic. As specific instances of transformations, we will evaluate the use of Principal Component Analysis (PCA) [28] and Independent Component Analysis (ICA) [29] which are often employed for dimensionality reduction.

Note that we may evaluate feature transformation and dimensionality reduction at the same time with the techniques developed above: the transformation procedure in the case of PCA and ICA may provide the $\overline{Y}$ as a ranking of features, so that we may carry out *feature selection* afterwards by selecting subsets $\overline{Y}_j$ spanning from the first-ranked to the j-th feature.

The tools. PCA is a staple technique in statistical data analysis and machine learning based in the Singular Value Decomposition of the data matrix to obtain projections along the singular vectors that account for its variance in decreasing amount, so PCA ranks the transformed features by this order. The implementation used in our examples are those of the publicly available R packages `stats` (v. 3.3.3) (https://stat.ethz.ch/R-manual/R-devel/library/stats/html/00Index.html, accessed on 11 June 2018).

While PCA aims at the orthogonalization of the projections, ICA finds the projections, also known as *factors*, by maximimizing their statistical independence, in our example by minimizing a cost term related to their mutual information [30]. However, this does not result in a ranking of the transformed features, hence we have created a pseudo-ranking by carrying an ICA transformation obtaining j transformed features for all sensible values of $1 \leq j \leq l$ using independent runs of the ICA algorithm. The implementation used in our examples is that of fastICA [30] as implemented in the R package `fastICA` (v. 1.2-1) (https://cran.r-project.org/package=fastICA, accessed on 11 June 2018, with standard parameter values (`alg.typ`="parallel", `fun`="logcosh", `alpha`=1, `method`="C", `row.norm`= FALSE, `maxit`=200, `tol`=0.0001).

The entropy diagrams and calculations were carried out with the open-source `entropies` experimental R package that provides an implementation of the present framework (available at https://github.com/FJValverde/entropies.git, accessed on 11 June 2018). The analysis carried out in this section is part of an illustrative vignette for the package and will remain so in future releases.

Analysis of results. We analyzed in this way some UCI classification datasets [31], whose number of classes k, features m, and observations n are listed in Table 1.

Table 1. Datasets analyzed.

	Name	k	m	n
1	Ionosphere	2	34	351
2	Iris	3	4	150
3	Glass	7	9	214
4	Arthritis	3	3	84
5	BreastCancer	2	9	699
6	Sonar	2	60	208
7	Wine	3	13	178

For simplicity issues, we decided to illustrate our new techniques on three datasets: *Iris, Glass* and *Arthritis. Ionosphere, BreastCancer, Sonar* and *Wine* have a similar pattern to *Glass*, but less interesting, as commented below. Besides, both *Ionosphere* and *Wine* have too many features for the kind of neat visualization we are trying to use in this paper. We have also used a slightly modified entropy triangles in which the colors of the axes are related to those of the information diagrams of Figure 4b.

For instance, Figure 6a presents the results of the PCA transformation on the logarithm of the features of Anderson's Iris. Crosses represent the information decomposition of the input features $\overline{X}$ using (27) while circles represent the information decomposition of transformed features $\overline{Y}_j$ using (28) and filled circles the aggregate decomposition of (26). We represent several possible features sets $\overline{Y}_j$ as output where each is obtained selecting the first *j* features in the ranking provided by PCA. For example, since Iris has four features we can make four different feature sets of 1 to *j* features, named in the Figure as "1_*j*", that is, "1_1" to "1_4". The figure then explores how the information in the whole database $\overline{X}$ is transported to different, nested candidate feature sets $\overline{Y}_j$ as per the PCA recipe: choose as many ranked features as required to increase the transmitted information.

We first notice that all the points for $\overline{X}$ lie on a line parallel to the left side of the triangle and their average transmitted information is increasing, parallel to a decrease in remanent information. Indeed, the redundancy $\Delta H'_{\overline{X}} = \frac{\Delta H_{\overline{X}}}{H U_{\overline{X}}}$ is the same regardless of the choice of $\overline{Y}_j$. The monotonic increase with the number of features selected *j* in *average transmitted information* $I'_{P_{\overline{XY}_j}} = \frac{I_{P_{\overline{XY}_j}}}{H U_{\overline{X}}}$ in (27) corresponds to the monotonic increase in absolute transmitted information $I_{P_{\overline{XY}_j}}$: for a given input set of features $\overline{X}$, the more output features are selected, the higher the mutual information between input and output. This is the basis of the effectiveness of the feature-selection procedure.

Regarding the points for $\overline{Y}_j$, note that the *absolute* transmitted information also appears in the *average* transmitted information (with respect to $\overline{Y}_j$) as $I'_{P_{\overline{XY}_j}} = \frac{I_{P_{\overline{XY}_j}}}{H U_{\overline{Y}_j}}$ in (28). While $I_{P_{\overline{XY}_j}}$ increases with *j*, as mentioned, we actually see a monotonic *decrease* in $I'_{P_{\overline{XY}_j}}$. The reason for this is the rapidly increasing value of the denominator $H U_{\overline{Y}_j}$ as we select more and more features.

Finally, notice how these two tendencies are conflated in the aggregate plot for the $\overline{XY}_j$ in Figure 6a that shows a lopsided, inverted U pattern, peaking before *j* reaches its maximum. This suggests that if we balance aggregated transmitted information against number of features selected—the complexity of the representation—in the search for a *faithful* representation, the average transmitted information is the quantity to optimize, that is, the *mutual determination* between the two feature sets.

Figure 6b presents similar results on the ICA transformation on the logarithm of the features of Anderson's Iris with the same glyph convention as before, but with a ranking resulting from carrying the ICA method *in full* for each value of *j*. That is, we first work out $\overline{Y}_1$ which is a single component, then we calculate $\overline{Y}_2$ which the two best ICA components, and so on. The reason for this is that ICA does not rank the features it produces, so we have to create this ranking by carrying the ICA algorithm for all values of *j* to obtain each $\overline{Y}_j$. Note that the transformed features produce by PCA and ICA are, in principle, very different, but the phenomena described for PCA are also apparent here: an increase in *aggregate* transmitted information, checked by the increase of the denominator represented by $H_{U_{\overline{Y}_j}}$ which implies a decreasing transmitted information *per feature* for $\overline{Y}_j$.

With the present framework the question of which transformation is "better" for this dataset can be given content and rephrased as *which transformation transmits more information on average on this dataset*, and also, importantly, *whether the aggregate information available in the dataset is being transmitted* by either of these methods. This is explored in Figure 7 for *Iris*, *Glass* and *Arthritis*, where, for reference, we have included a point for the (deterministic) transformation of the logarithm, the cross, giving an idea of what a lossless information transformation can achieve.

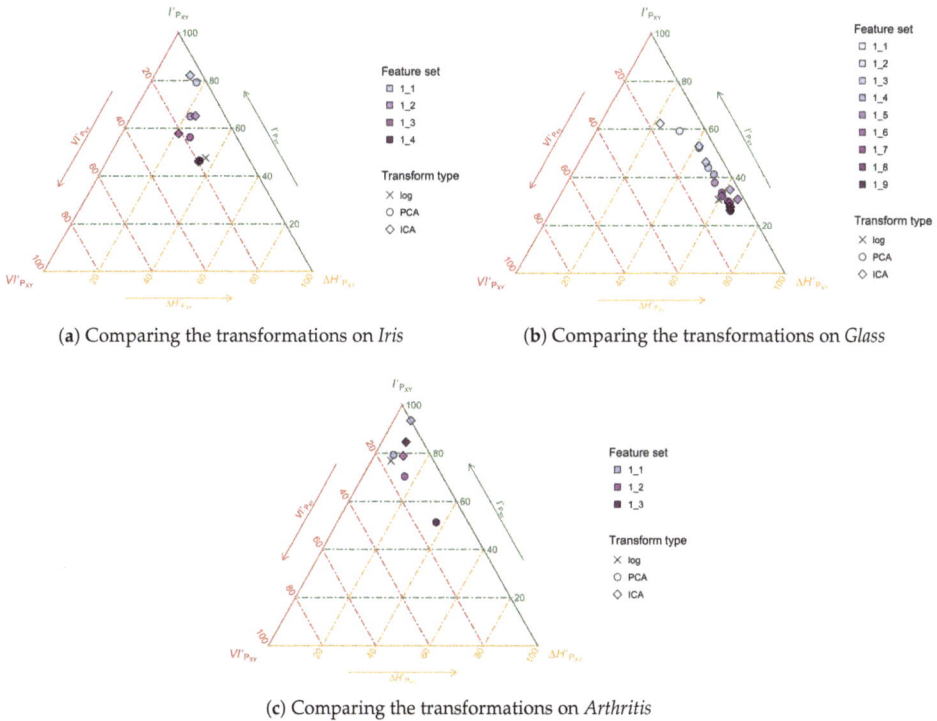

(a) Comparing the transformations on *Iris*

(b) Comparing the transformations on *Glass*

(c) Comparing the transformations on *Arthritis*

Figure 7. (Color online) Comparison of PCA and ICA as data transformations using the CMET on *Iris*, *Glass* and *Arthritis*. Note that these are the same positions represented as inverted triangles in Figure 6a,b.

Consider Figure 7a for *Iris*. The first interesting observation is that neither technique is transmitting all of the information in the database, which can be gleaned from the fact that both feature sets "1_4"—when all the features available have been selected—are below the cross. This clearly follows the data processing inequality, but is still surprising since transformations like ICA and PCA are extensively used and considered to work well in practice. In this instance it can only be explained by the advantages of the achieved dimensionality reduction. Actually, the observation in the CMET suggests that we can *improve on the average transmitted information per feature* by retaining the three first features for each PCA and ICA.

The analysis of *Iris* turns out to be an intermediate case between that of *Arthritis* and *Glass*, the latter being the most typical in our analysis. This is the case with a lot of original features $\overline{X}$ which transmit very little private, distinctive information per feature. The typical behavior, both for PCA and

ICA is to select at first, features that carry very little average information $\overline{Y}_1$. As we select more and more transformed features, information accumulates but at a very slow pace as shown in Figure 6c,d. Typically, the transformed features chosen last are very redundant. In the case of *Glass*, specifically, there is no point in retaining features beyond the sixth (out of 9) for either PCA or ICA as shown in Figure 7b. As to comparing the techniques, in some similarly-behaving datasets PCA is better, while in others ICA is. In the case of *Glass*, it is better to use ICA when retaining up to two transformed features, but it is better to use PCA when retaining between 2 and 6.

The case of *Arthritis* is quite different, perhaps due to the small number of original features $n = 3$. Our analyses show that just choosing the first ICA component $\overline{Y}_1$—perhaps the first two—provides an excellent characterization of the dataset, being extremely efficient in what regards information transmission. This phenomenon is also seen in the first PCA component, but is lost as we aggregate more PCA components. Crucially, taking the 3 ICA components amounts to taking all of the original information in the dataset, while taking the 3 components in the case of PCA is rather inefficient, as confirmed by Figure 7c.

All in all, our analyses show that the unsupervised transformation and selection of features in datasets can be assessed using an information-theoretical heuristic: maximize the average mutual information accumulated by the transformed features. And we have also shown how to carry out this assessment with entropic balance equations and entropy triangles.

3.4. Discussion

The development of the multivariate case is quite parallel to the bivariate case. An important point to realize is that the multivariate transmitted information between two different random vectors $I_{P_{\overline{X}\overline{Y}}}$ is the proper generalization for the usual mutual information $MI_{P_{XY}}$ in the bivariate case, rather than the more complex alternatives used in multivariate sources (see Section 2.2 and [5,14]). Indeed properties (18) and (20) are crucial in transporting the structure and intuitions built from the bivariate channel entropy triangle to the multivariate one, of which the former is a proper instance. This was not the case with balance equations and entropy triangles for stochastic sources of information [5].

The crucial quantities in the balance equation and the triangle have been independently motivated in other works. First, multivariate mutual information is fundamental in Information Theory, and we have already mentioned the redundancy ΔH_{P_X} [9]. We also mentioned the input-entropy normalized $I'_{P_{\overline{X}\overline{Y}}}$ used as a standalone assessment measure in intrusion detection [32]. Perhaps the least known quantity in the paper was the variation of information. Despite being inspired by the concept proposed by Meila [11], to the best of our knowledge it is completely new in the multivariate setting. However, the underlying concepts of conditional or remanent entropies have proven their usefulness time and again. All of the above is indirect proof that the quantities studied in this paper are significant, and the existence of a balance equation binding them together important.

The paragraph above notwithstanding, there are researchers who claim that Shannon-type relations cannot capture all the dependencies inside multivariate random vectors [33]. Due to the novelty of that work, it is not clear how much the "standard" theory of Shannon measures would have to change to accommodate the objections raised to it in that respect. But this question seems to be off the mark for our purposes: the framework of channel balance equations and entropy triangles has not been developed to look into the question of dependency, but of *aggregate information transfer*, wherever that information comes from. It may be relevant to source balance equations and triangles [5]—which have a different purpose—but that still has to be researched into.

The normalizations involved in (6) and (26)—respectively, (8), (27) and (28)—are similar conceptually: to divide by the logarithm of the total size of the domains involved whether it is the size of $X \times Y$ or that of $\overline{X} \times \overline{Y}$. Notice, first, that this is the same as taking the logarithm base these sizes in the non-normalized equations. The resulting units would not be bits for the multivariate case proper, since the size of $\overline{X}$ or $\overline{Y}$ is at least $2 \times 2 = 4$. But since the entropy triangles represent compositions [12], which are inherently dimensionless, this allows us to represent many different, and otherwise incomparable systems, e.g., univariate and multivariate ones with the same kind of diagram. Second, this type of normalization allows for an interpretation of the extension of these measures to the continuous case as a limit in the process of equipartitioning a compact support, as done, for instance, for the Rényi entropy in ([34], Section 3) which is known to be a generalization of Shannon's. There are hopes, then for a continuous version of the balance equations for Renyi's entropy.

Finally, note that the application presented in Section 3.3 above, although principled in the framework presented here, is not conclusive on the quality of the analyzed transformations in general but only as applied to the particular dataset. For that, a wider selection of data transformation approaches, and many more datasets should be assessed. Furthermore, the feature selection process used the "filter" approach which for supervised tasks seems suboptimal. Future work will address this issue as well as how the technique developed here relates to the end-to-end assessment presented in [4] and the source characterization technique of [5].

4. Conclusions

In this paper, we have introduced a new way to assess quantitatively and visually the transfer of information from a multivariate source $\overline{X}$ to a multivariate sink of information $\overline{Y}$, using a heretofore unknown decomposition of the entropies around the joint distribution $P_{\overline{XY}}$. For that purpose, we have generalized a similar previous theory and visualization tools for bivariate sources, greatly extending the applicability of the results:

- We have been able to decompose the information of a random multivariate source into three components: (a) the non-transferable divergence from uniformity $\Delta H_{P_{\overline{XY}}}$, which is an entropy "missing" from $P_{\overline{XY}}$; (b) a transferable, but not transferred part, the variation of information $VI_{P_{\overline{XY}}}$; and (c) the transferable and transferred information $I_{P_{\overline{XY}}}$, which is a known, but never considered in this context, generalization of bivariate mutual information.
- Using the same principles as in previous developments, we have been able to obtain a new type of visualization diagram for this balance of information using de Finetti's ternary diagrams, which is actually an exploratory data analysis tool.

We have also shown how to apply these new theoretical developments and the visualization tools to the analysis of information transfer in unsupervised feature transformation and selection, a ubiquitous step in data analysis, and specifically, to apply it to the analysis of PCA and ICA. We believe this is a fruitful approach, e.g., for the assessment of learning systems, and foresee a bevy of applications to come. Further conclusions on this issue are left for a more thorough later investigation.

Author Contributions: Conceptualization, F.J.V.-A. and C.P.-M.; Formal analysis, F.J.V.-A. and C.P.-M.; Funding acquisition, C.P.-M.; Investigation, F.J.V.-A. and C.P.-M.; Methodology, F.J.V.-A. and C.P.-M.; Software, F.J.V.-A.; Supervision, C.P.-M.; Validation, F.J.V.-A. and C.P.-M.; Visualization, F.J.V.-A. and C.P.-M.; Writing—original draft, F.J.V.-A. and C.P.-M.; Writing—review & editing, F.J.V.-A. and C.P.-M.

Funding: This research was funded by he Spanish Government-MinECo projects TEC2014-53390-P and TEC2017-84395-P.

Conflicts of Interest: The authors declare no conflict of interest.

Abbreviations

The following abbreviations are used in this manuscript:

PCA   Principal Component Analysis
ICA   Independent Component Analysis
CMET Channel Multivariate Entropy Triangle
CBET   Channel Binary Entropy Triangle
SMET Source Multivariate Entropy Triangle

References

1. Goodfellow, I.; Bengio, Y.; Courville, A. *Deep Learning*; MIT Press: Cambridge, MA, USA, 2016.
2. Shwartz-Ziv, R.; Tishby, N. Opening the Black Box of Deep Neural Networks via Information. *arXiv* **2017**, arXiv:1703.00810v3.
3. Tishby, N.; Zaslavsky, N. Deep Learning and the Information Bottleneck Principle. In Proceedings of the IEEE 2015 Information Theory Workshop, San Diego, CA, USA, 1–6 February 2015.
4. Valverde-Albacete, F.J.; Peláez-Moreno, C. 100% classification accuracy considered harmful: The normalized information transfer factor explains the accuracy paradox. *PLOS ONE* **2014**, doi:10.1371/journal.pone.0084217. [CrossRef] [PubMed]
5. Valverde-Albacete, F.J.; Peláez-Moreno, C. The Evaluation of Data Sources using Multivariate Entropy Tools. *Expert Syst. Appl.* **2017**, *78*, 145–157. doi:10.1016/j.eswa.2017.02.010. [CrossRef]
6. Valverde-Albacete, F.J.; Peláez-Moreno, C. Two information-theoretic tools to assess the performance of multi-class classifiers. *Pattern Recognit. Lett.* **2010**, *31*, 1665–1671. [CrossRef]
7. Yeung, R. A new outlook on Shannon's information measures. *IEEE Trans. Inf. Theory* **1991**, *37*, 466–474. [CrossRef]
8. Reza, F.M. *An Introduction to Information Theory*; McGraw-Hill Electrical and Electronic Engineering Series; McGraw-Hill Book Co., Inc.: New York, NY, USA; Toronto, ON, Canada; London, UK, 1961.
9. MacKay, D.J.C. *Information Theory, Inference and Learning Algorithms*; Cambridge University Press: Cambridge, UK, 2003.
10. Shannon, C.E. A mathematical theory of communication. *Bell Syst. Tech. J.* **1948**, *XXVII*, 379–423, 623–656. [CrossRef]
11. Meila, M. Comparing clusterings—An information based distance. *J. Multivar. Anal.* **2007**, *28*, 875–893. [CrossRef]
12. Pawlowsky-Glahn, V.; Egozcue, J.J.; Tolosana-Delgado, R. *Modeling and Analysis of Compositional Data*; John Wiley & Sons: Chichester, UK, 2015.
13. Valverde-Albacete, F.J.; de Albornoz, J.C.; Peláez-Moreno, C. A Proposal for New Evaluation Metrics and Result Visualization Technique for Sentiment Analysis Tasks. In *CLEF 2013: Information Access Evaluation. Multilinguality, Multimodality and Visualization*; Forner, P., Müller, H., Paredes, R., Rosso, P., Stein, B., Eds.; Springer: Berlin/Heidelberg, Germany, 2013; Volume 8138, pp. 41–52.
14. Timme, N.; Alford, W.; Flecker, B.; Beggs, J.M. Synergy, redundancy, and multivariate information measures: An experimentalist's perspective. *J. Comput. Neurosci.* **2014**, *36*, 119–140. [CrossRef] [PubMed]
15. James, R.G.; Ellison, C.J.; Crutchfield, J.P. Anatomy of a bit: Information in a time series observation. *Chaos* **2011**, *21*, 037109. [CrossRef] [PubMed]
16. Watanabe, S. Information theoretical analysis of multivariate correlation. *J. Res. Dev.* **1960**, *4*, 66–82. [CrossRef]
17. Tononi, G.; Sporns, O.; Edelman, G.M. A measure for brain complexity: Relating functional segregation and integration in the nervous system. *Proc. Natl. Acad. Sci. USA* **1994**, *91*, 5033–5037. [CrossRef] [PubMed]
18. Studený, M.; Vejnarová, J. The Multiinformation Function as a Tool for Measuring Stochastic Dependence. In *Learning in Graphical Models*; Springer: Dordrecht, The Netherlands, 1998; pp. 261–297.
19. Han, T.S. Nonnegative entropy measures of multivariate symmetric correlations. *Inf. Control* **1978**, *36*, 133–156. [CrossRef]
20. Abdallah, S.A.; Plumbley, M.D. A measure of statistical complexity based on predictive information with application to finite spin systems. *Phys. Lett. A* **2012**, *376*, 275–281. [CrossRef]
21. Tononi, G. Complexity and coherency: Integrating information in the brain. *Trends Cognit. Sci.* **1998**, *2*, 474–484. [CrossRef]
22. McGill, W.J. Multivariate information transmission. *Psychometrika* **1954**, *19*, 97–116. [CrossRef]
23. Sun Han, T. Multiple mutual informations and multiple interactions in frequency data. *Inf. Control* **1980**, *46*, 26–45. [CrossRef]
24. Bell, A. The co-information lattice. In Proceedings of the Fifth International Workshop on Independent Component Analysis and Blind Signal Separation, Nara, Japan, 1–4 April 2003.
25. Abdallah, S.A.; Plumbley, M.D. *Predictive Information, Multiinformation and Binding Information*; Technical Report C4DM-TR10-10; Queen Mary, University of London: London, UK, 2010.

26. Valverde Albacete, F.J.; Peláez-Moreno, C. The Multivariate Entropy Triangle and Applications. In *Hybrid Artificial Intelligence Systems (HAIS 2016)*; Springer: Seville, Spain, 2016; pp. 647–658.

27. Witten, I.H.; Eibe, F.; Hall, M.A. *Data Mining. Practical Machine Learning Tools and Techniques*, 3rd ed.; Morgan Kaufmann: Burlington, MA, USA 2011.

28. Pearson, K. On Lines and Planes of Closest Fit to Systems of Points in Space. *Philos. Mag.* **1901**, 559–572. [CrossRef]

29. Bell, A.J.; Sejnowski, T.J. An Information-Maximization Approach to Blind Separation and Blind Deconvolution. *Neural Comput.* **1995**, *7*, 1129–1159. [CrossRef] [PubMed]

30. Hyvärinen, A.; Oja, E. Independent component analysis: Algorithms and applications. *IEEE Trans. Neural Netw.* **2000**, *13*, 411–430. [CrossRef]

31. Bache, K.; Lichman, M. *UCI Machine Learning Repository*; 2013.

32. Gu, G.; Fogla, P.; Dagon, D.; Lee, W.; Skorić, B. Measuring Intrusion Detection Capability: An Information-theoretic Approach. In Proceedings of the 2006 ACM Symposium on Information, Computer and Communications Security (ASIACCS '06), Taipei, Taiwan, 21–23 March 2006 ; ACM: New York, NY, USA, 2006; pp. 90–101. doi:10.1145/1128817.1128834. [CrossRef]

33. James, G.R.; Crutchfield, P.J. Multivariate Dependence beyond Shannon Information. *Entropy* **2017**, *19*, 531–545. [CrossRef]

34. Jizba, P.; Arimitsu, T. The world according to Rényi: Thermodynamics of multifractal systems. *Ann. Phys.* **2004**, *312*, 17–59. [CrossRef]

entropy

MDPI

Article

Cross-Sectoral Information Transfer in the Chinese Stock Market around Its Crash in 2015

Xudong Wang and Xiaofeng Hui *

School of Management, Harbin Institute of Technology, Harbin 150001, China; 13b310003@hit.edu.cn
* Correspondence: xfhui@hit.edu.cn; Tel.: +86-451-8640-2353

Received: 3 August 2018; Accepted: 24 August 2018; Published: 3 September 2018

Abstract: This paper applies effective transfer entropy to research the information transfer in the Chinese stock market around its crash in 2015. According to the market states, the entire period is divided into four sub-phases: the tranquil, bull, crash, and post-crash periods. Kernel density estimation is used to calculate the effective transfer entropy. Then, the information transfer network is constructed. Nodes' centralities and the directed maximum spanning trees of the networks are analyzed. The results show that, in the tranquil period, the information transfer is weak in the market. In the bull period, the strength and scope of the information transfer increases. The utility sector outputs a great deal of information and is the hub node for the information flow. In the crash period, the information transfer grows further. The market efficiency in this period is worse than that in the other three sub-periods. The information technology sector is the biggest information source, while the consumer staples sector receives the most information. The interactions of the sectors become more direct. In the post-crash period, information transfer declines but is still stronger than the tranquil time. The financial sector receives the largest amount of information and is the pivot node.

Keywords: information transfer; Chinese stock sectors; effective transfer entropy; market crash

1. Introduction

After decades of rapid growth, China has become the world's second largest economy. It plays an important role in global trade. However, its stock market has displayed poor performance since the US subprime crisis. Under the background of deepening economic reform, the Chinese stock market began to boom around July 2014 [1]. Tens of millions of new investors entered the market. The great majority of them were retail investors, which tended to exhibit herd behavior. Moreover, many of these novice investors engaged in leveraged trading through various channels, for example margin financing of brokerages, shadow banking, or grey-market (over-the-counter, OTC) margin lenders [2,3]. Huge amounts of borrowed money flooded into the market [3]. The Shanghai stock exchange composite index (SSECI) soared from 2050.38 on 1 July 2014, to a peak of 5166.35 on 12 June 2015. It increased about 152% in just one year. However, after the peak, the market plunged drastically. From late June to late August of 2015, the SSECI declined about 40% [4]. It was one of the biggest falls in global stock market history [2]. In order to stabilize the market, the Chinese government took a series of actions, including organizing state-backed financial firms collectively called the "national team" to buy stocks directly, banning short sales, stopping new initial public offerings, etc. [3]. Through these efforts, the market turbulence ended in February 2016 [5]. This crash brought heavy losses to Chinese investors and the economy. Market capitalization up to trillions of US dollars evaporated [6]. It also impacted the world markets.

Analyzing information transfer is one of the fundamental subjects for complex system studies. It characterizes the interactions between components and provides important insights into the structure and dynamics of the system. This issue attracts many researchers from different fields, for instance neuroscience [7,8], physics [9,10], climatology [11,12], and zoology [13,14], etc. In econophysics, Kwon and Yang [15] analyzed the strength and direction of the information flow in 25 stock indices. They found that the US market was the biggest information source, while most information receivers are located in the Asian Pacific region. Yang et al. [16] used the annual gross domestic product (GDP) data of 27 Chinese provinces and autonomous regions to study the information transmission before and after the reform and opening up policy in 1978. The results showed that the policy promoted regional economy development and changed the influence of different areas. Dimpfl and Peter [17] analyzed the information transfer between US and European stock markets. They discovered that there existed bidirectional information flow. The US subprime crisis enhanced this information exchange. Sandoval [18] investigated the transfer of information in 197 of the largest financial companies in the world. The bank and insurance companies were found to play important roles in the transmission of information. Sensoy et al. [19] employed nine developing countries' currency exchange rate and stock price data to research the exchange of information between them. The results suggested strong bidirectional information flow during the US subprime crisis. Bekios et al. [20] studied the information diffusion between commodity future and stock markets. The results indicated that finance, automobile, and energy stock sectors transmitted the most information to the commodity future market. Kim et al. [21] researched the information transfer in economy variables. It was discovered that Western countries had a strong influence in the world economic network. Japan's influence decreased after the Asian currency crisis in 1997.

The aim of this paper is to apply effective transfer entropy (ETE) to research the information transfer in Chinese stock market around the crash of 2015, and to reveal the impacts of this crash on the interactions between sectors. To the best of our knowledge, this question has not been studied systematically in the existing literature.

Transfer entropy (TE), introduced by Schreiber [22], is a very popular tool for measuring information transfer between time series. It has some remarkable properties. First, it is directional and can assess the direction of information. Second, it can be used in both linear and nonlinear environments. Third, it does not need specific model hypotheses. It is model-free and data-driven [23]. Because of these advantages, TE has been widely used in various domains [7–16]. In practice, the sample data is usually small and contains noise. To reduce these influences, Marschinski and Kantz combined a random shuffling procedure with TE and proposed ETE [24]. Lungarella et al. suggested that TE or ETE was the first choice when prior knowledge of the system is unknown [25]. In this paper, we consider the stock indices as continuous variables, avoiding information loss caused by data discretization. Since kernel density estimation (KDE) performs well in inferring probability density function [26,27], we adopt it to calculate ETE. The main contributions of this paper to the relevant literature are in three aspects: first, it analyzes the strength and scope of the information transfer in 10 Chinese stock sectors around the crash in 2015. Second, it applies node strength and betweenness centrality to assess sectors' influences in different sub-periods. Third, it uses Chu-Liu-Edmond's algorithm [28,29] to construct directed maximum spanning trees (MSTs) to research the backbones of the information transfer networks.

The rest of this paper is organized as follows. Section 2 introduces the methods. Section 3 describes the data and some preliminary analyses. Section 4 gives the results and some discussion. Section 5 concludes the paper.

2. Methodology

2.1. Transfer Entropy

Before introducing TE, we present the concept of Shannon entropy, which is fundamental for information theory. Let R^m denote the m-dimensional real space and $A \in R^m$. Its Shannon entropy $H(A)$ is defined as [30]:

$$H(A) = -\int_{R^m} p(A) \log p(A) dA \tag{1}$$

where $p(A)$ is the probability density function (PDF). Shannon entropy quantifies the amount of information that is needed to describe the variable, or the uncertainty of the variable. In this paper, the logarithm uses base 2; thus, the entropy is measured in bits.

Let another variable $B \in R^n$; the conditional entropy $H(A|B)$ is [31]:

$$H(A|B) = -\int_{R^{m+n}} p(A,B) \log p(A|B) dA dB \tag{2}$$

where $p(A,B)$ and $p(A|B)$ are the joint and conditional PDFs. They characterize the uncertainty of A given that B is known.

Given two stationary time series $X \in R^1$ and $Y \in R^1$, the TE from X to Y is defined as [32]:

$$\begin{aligned}
TE_{X \to Y} &= H(y_{t+1}|y_t^{(k)}) - H(y_{t+1}|y_t^{(k)}, x_t^{(l)}) \\
&= \int_{R^{k+l+1}} p(y_{t+1}, y_t^{(k)}, x_t^{(l)}) \log(\frac{p(y_{t+1}|y_t^{(k)}, x_t^{(l)})}{p(y_{t+1}|y_t^{(k)})}) dy_{t+1} dy_t^{(k)} dx_t^{(l)}
\end{aligned} \tag{3}$$

where $y_t^{(k)} = (y_t, y_{t-1}, \ldots, y_{t-k+1})$, $x_t^{(l)} = (x_t, x_{t-1}, \ldots, x_{t-l+1})$ are the past states; $p(y_{t+1}, y_t^{(k)}, x_t^{(l)})$, $p(y_{t+1}|y_t^{(k)}, x_t^{(l)})$, $p(y_{t+1}|y_t^{(k)})$ are the joint and conditional PDFs. $TE_{X \to Y}$ measures the uncertainty reduction or predictability improvement of y_{t+1}, which gains from $x_t^{(l)}$ that is not contained in $y_t^{(k)}$ itself [33]. In this way, it quantifies the predictive information transfer between variables [33,34]. By a simple transformation, Formula (3) can be rewritten as follows [31,35]:

$$TE_{X \to Y} = H(y_t^{(k)}, x_t^{(l)}) + H(y_{t+1}, y_t^{(k)}) - H(y_{t+1}, y_t^{(k)}, x_t^{(l)}) - H(y_t^{(k)}) \tag{4}$$

2.2. Effective Transfer Entropy

In practical applications, TE can produce spurious positive values for two independent time series because of limited data and random noise. ETE was proposed to reduce this bias. It is defined as [24,35]:

$$ETE_{X \to Y} = TE_{X \to Y} - \frac{1}{M} \sum TE_{Xshuffed \to Y} \tag{5}$$

where $X_{shuffed}$ is the random shuffled series of X. M is the number of shuffles and it is set to 1000. In this paper, we first applied $TE_{Xshuffed \to Y}$ to test the significance of $TE_{X \to Y}$. Referring to References [27,33], if $TE_{X \to Y}$ is larger than the 95th percentile of $TE_{Xshuffed \to Y}$, $TE_{X \to Y}$ is considered significant nonzero, and the ETE is calculated according to Formula (5). Otherwise, it is considered that there is no transmission of information, and ETE is 0.

2.3. Kernel Density Estimation

According to Formulas (4) and (5), ETE can be calculated through Shannon entropy. Because the KDE-based method has good performance [26,27], we can apply it to compute ETE in this study.

Let $u_1, u_2, \ldots, u_N$ be a sample of $U \in R^d$. Then, its PDF value $\hat{p}(u_j)$ estimated by KDE with a kernel function $K(\cdot)$ is [26]:

$$\hat{p}(u_j) = \frac{1}{Nh^d} \sum_{i=1}^{n} K(\frac{u_j - u_i}{h}) \tag{6}$$

where h is the bandwidth. In this paper we choose the Gaussian kernel, which is commonly used in practice. Therefore, Formula (6) can be written as [26,36]:

$$\hat{p}(u_j) = \frac{1}{Nh^d} \sum_{i=1}^{n} \frac{1}{\sqrt{(2\pi)^d \det(S)}} \exp\left(-\frac{(u_j - u_i)^T S^{-1}(u_j - u_i)}{2h^2}\right) \tag{7}$$

where S is the covariance matrix of the data; $\det(S)$ is the determinant of S.

The bandwidth is calculated by Formula (8), with reference to [26,36]:

$$h = \left(\frac{4}{d+2}\right)^{1/(d+4)} N^{-1/(d+4)} \tag{8}$$

After obtaining $\hat{p}(u_i)$, the Shannon entropy can be computed by Formula (9) [37,38]:

$$H(\boldsymbol{U}) = -\frac{1}{N}\sum_{t=1}^{N} \log \hat{p}(u_t) \tag{9}$$

where N is the length of the time series.

Thus, the TE in Formula (4) can be estimated by Formula (10):

$$TE_{X \to Y} = -\frac{1}{N}\left(\sum_{t=1}^{N} \log \hat{p}(y_t^{(k)}, x_t^{(l)}) + \sum_{t=1}^{N} \log \hat{p}(y_{t+1}, y_t^{(k)}) - \sum_{t=1}^{N} \log \hat{p}(y_{t+1}, y_t^{(k)}, x_t^{(l)}) - \sum_{t=1}^{N} \log \hat{p}(y_t^{(k)})\right) \tag{10}$$

Applying the above methodology, we calculate the ETE between two linear autoregressive processes [32]:

$$X_{i+1} = \alpha X_{i+1} + \eta_i^X; \quad Y_{i+1} = \beta Y_{i+1} + \gamma X_i + \eta_i^Y \tag{11}$$

where η^X and η^Y are random numbers that obey standard normal distributions; $\alpha = 0.5$ and $\beta = 0.6$. Let $k = l = 1$. Then, the analytical value of $TE_{X \to Y}$ for the two processes is [32]:

$$TE_{X \to Y} = \frac{1}{2} \log \frac{\det C(Y_i, X_j)\det C(Y_{i+1}, Y_i)}{\det C(Y_{i+1}, Y_i, X_i)\det C(Y_i)} \tag{12}$$

where $C(\cdot)$ is the theoretical covariance matrix; $\det(\cdot)$ denotes the determinant of a matrix.

For each γ, we generate 50 sample series of Formula (11) with a length of 200. This length is approximated to the sub-periods around the crash, which are divided in the later part of this paper. We then calculate the ETEs for these samples. The average values of these ETEs are displayed in Figure 1. It can be observed that the calculated ETEs match the theoretical TE well. The mean absolute error is just 0.0064. This supports the good performance of the methodology.

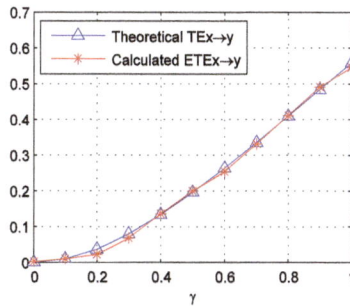

Figure 1. Theoretical $TE_{X \to Y}$ and calculated $ETE_{X \to Y}$ of the two linear autoregressive processes in Formula (11). γ increases from 0 to 1 with a step of 0.1.

3. Data

According to the CICS Industry Classification issued by China Securities Index (CSI) Co., Ltd, all mainland China listed companies are divided into 10 first-level sectors. This paper therefore uses the daily closing price of the 10 CSI sector indices for the study. Their numbers and names are listed in Table 1. All data are downloaded from the WIND database, which is a leading Chinese financial information provider.

Table 1. The numbers and names of the 10 China Securities Index (CSI) sector indices.

No.	Index Name	No.	Index Name
1	CSI Energy	6	CSI Health Care
2	CSI Materials	7	CSI Financials
3	CSI Industrials	8	CSI Information Technology
4	CSI Consumer Discretionary	9	CSI Telecommunication Services
5	CSI Consumer Staples	10	CSI Utilities

The time range of the data is from 1 July 2013 to 28 February 2017. According to the market states, we divide the time into four sub-periods: the tranquil, bull, crash, and post-crash periods. This can help to analyze the influence of market states on the information transfer between sectors. The tranquil period extends from 1 July 2013 to 30 June 2014—approximately one year. During this period, the market was quite calm [1]. The bull period extends from 1 July 2014 to 12 June 2015. The market soared in this stage [3,39]. It reached the peak on 12 June 2015. After that day, it plunged drastically [3,39]. So, we take this day as the end of the bull period. The crash period starts on 15 June 2015 and ends on 29 February 2016. Because the dates of 13 June 2015 and 14 June 2015 fall on weekends, the start of the crash period is considered to be the latest trading day, 15 June 2015. Zhai analyzed the structural breaks of the Chinese stock market, and found that the crash of 2015 ended in February 2016 [40]. We therefore take 29 February 2016 as the end of the crash. This time is also in agreement with the literature [5,41]. The post-crisis period is from 1 March 2016 to 28 February 2017. It is also about one year. During this period, the market became stable again. Figure 2 shows the SSECI and the four sub-periods.

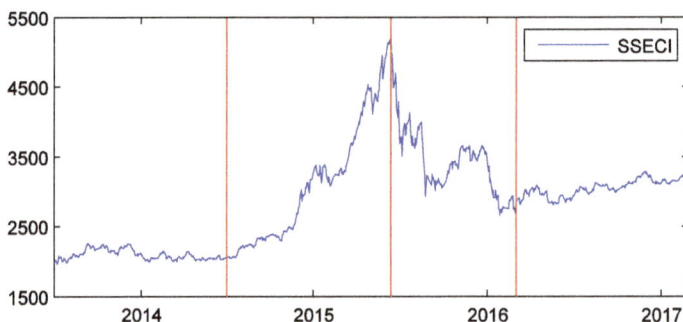

Figure 2. Shanghai stock exchange composite index (SSECI) and the four sub-periods around the time of the crash of 2015. The horizontal axis is the sample time ranging from 1 July 2013 to 28 February 2017. The red lines correspond to the dates of 1 July 2014, 12 June 2015, and 29 February 2016 from left to right.

We apply Formula (13) to calculate the daily logarithmic returns of the 10 sector indices:

$$R_t = \ln P_t - \ln P_{t-1} \tag{13}$$

where R_t represents the logarithmic return; P_t and P_{t-1} denotes the price on day t and $t-1$, respectively.

Since TE needs the time series to be stationary, we apply the augmented Dickey-Fuller (ADF) test to examine the stationarity of the whole return series. The lag length is selected by Schwarz Information Criterion. The maximum lags are defined as 20. We also use the Jarque-Bera test to examine whether the whole return series obey Gaussian distribution. Table 2 shows the results.

Table 2. Results of the augmented Dickey-Fuller (ADF) and Jarque-Bera tests for the whole return series of the 10 sectors.

No.	ADF Statistic	Jarque-Bera Statistic	No.	ADF Statistic	Jarque-Bera Statistic
1	−28.6154 ***	765.3783 ***	6	−28.4199 ***	902.8109 ***
2	−28.2453 ***	823.1612 ***	7	−28.7902 ***	953.3962 ***
3	−26.7147 ***	783.1250 ***	8	−27.1013 ***	351.2834 ***
4	−27.7279 ***	753.9022 ***	9	−27.6364 ***	548.0041 ***
5	−23.0576 ***	867.0495 ***	10	−27.9620 ***	970.4217 ***

Note: *** denotes statistical significance at the 1% level.

From Table 2, it can be concluded that all 10 return series are stationary. However, they do not obey Gaussian distribution as a whole. In addition, we conducted the ADF and Jarque-Bera tests on the four sub-periods of the return series. The results of the ADF test are all significant at the 1% level. This indicates that the series are also stationary in the four sub-periods. However, the results of Jarque-Bera test are mixed at the 1% significance level. Some are significant and some are not.

We use the autocorrelation function to determine the delay time k and l. The lag of its first zero-crossing, or the lag required for the function to decrease to $1/e$, can be selected as the delay time [42,43]. The results show that the first zero-crossings of all sectors' autocorrelation functions are between 1 and 2, and the functions' values of lag 1 are already below $1/e$. We therefore set $k = l = 1$ in this paper. This indicates the weak memory of the daily stock returns. This configuration is in accordance with the literature [15,18,20,44,45].

4. Results and Discussion

4.1. ETE between Sectors

We calculate the ETEs between 10 Chinese stock sectors during the four sub-periods. In order to visualize and compare them conveniently, we use colormaps to display them. All colorbars are adjusted to the same range. Figure 3 and Table 3 shows these colormaps and the mean ETE of each sub-period. The direction of the ETE is from the vertical axis to the horizontal axis. The numbers on the axes are the serial number of the sectors in Table 1. From Figure 3 and Table 3, it can be observed that the information transfer exists in only a few sectors in the tranquil period, and that its strength is weak. This indicates feeble interactions between sectors in this stage. In the bull period, the strength and scope of the information transfer increases, suggesting stronger interactions. In Figure 3b, there are two high ETE blocks with the coordinates (8, 3) and (8, 10). This means that the industry (No. 3) and the utility (No. 10) sector indices transmit much information to the information technology sector index (No. 8). One of the possible reasons for this is that during the booming market phase, informatization construction in the industry and the utility fields (for example, the state-prompted smart city, intelligent grid, "Made-in-China 2025", and "Internet-plus" plans [46–49]) provides huge demands and opportunities for information technology companies. This potential economical link may enhance the information transfer [50]. In the crash period, the strength and scope of the information flow increases further, and it reaches the maximum of the four sub-periods. This implies the strongest interactions between sectors. In the post-crash period, the market becomes stable again. The information transfer weakens, but it is still stronger than the tranquil period. The scope of the information transfer is also different from that in the tranquil stage.

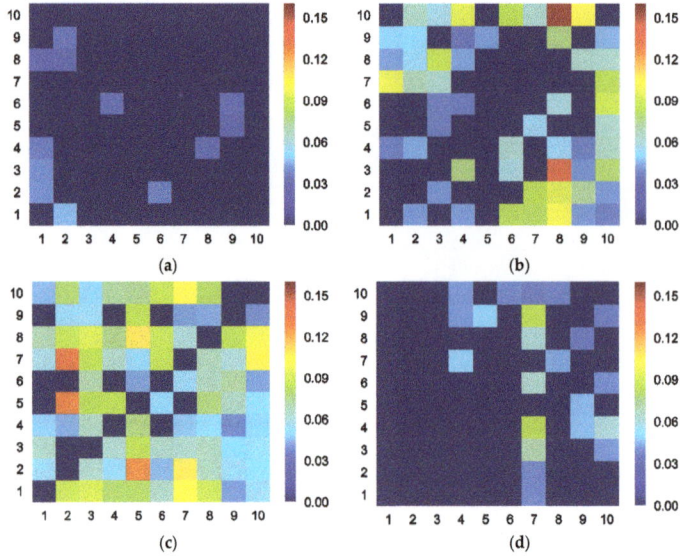

Figure 3. Colormaps of the effective transfer entropy (ETE) between the 10 sectors during the (**a**) tranquil, (**b**) bull, (**c**) crash, and (**d**) post-crash periods. The range of the colorbars is from 0 to 0.1533.

Table 3. Average ETE between sectors in the four sub-periods

	Tranquil	Bull	Crash	Post-crash
Average ETE	0.0049	0.0384	0.0619	0.0123

From the perspective of market microstructure, the movement of the stock price is determined by the arrival of new information, and by the process that absorbs the information into the price [51]. According to the Efficient Market Hypothesis, if the market is perfectly efficient, the price reflects all current information. Newly arrived information is incorporated instantaneously into the price. In this ideal condition, there is no predictability and information transfer between the stocks [52,53]. However, researchers have discovered that market frictions widely exist in capital markets; for instance, the limited attention of investors, asymmetric information, and noise traders, etc. [54]. They cause a difference in the speeds of the information absorption of prices, and result in the predictability and information transfer from the faster one to the slower one [50,54,55]. In practice, bidirectional information transfer can be seen. This is because different information may coexist in the market. Different stock may react to different information at different speeds. On the other hand, predictability can be an important indicator of market efficiency [56–58]. From this aspect, it can be inferred that the market efficiency in the tranquil period is relatively high. It deteriorates in the bull time, and it is the worst in the crash period. In the post-crash stage, it obtains some recovery. Contrasting with the drastic boom and crash of the stock market, the macroeconomic variables of China are steady. We conclude that their effect on the change of market statuses and ETE is weak in these periods. This conclusion is line with Song [2]. He applied multifactor models to examine the effect of macroeconomic variables and found that the bull market was not sensitive to the macroeconomic variables.

4.2. Centrality of Sectors

For the further study of the sectors' interactions, we construct the information transfer network. The sectors are considered as the nodes in the network. If there is a nonzero $ETE_{i \to j}$, a directed edge is added from sector i to sector j, with the weight of $ETE_{i \to j}$. Then we obtain the network. Node strength is a common centrality measure. For a weighted directed network, it can be divided into Out Node Strength NS_{out} and In Node Strength NS_{in}:

$$NS_{out}^i = \sum_j ETE_{ij}; \; NS_{in}^i = \sum_k ETE_{ki} \qquad (14)$$

Out Node Strength reflects the influence of a node on others. In Node Strength measures the influence of a node receiving from others [59]. Tables 4 and 5 display the Out and In Node Strengths of the sectors. The largest values are in bold. In the tranquil period, we can observe that the Out and In Node Strengths of the sectors are small. The energy (No. 1) and material (No. 2) sectors have relatively larger In Node Strengths than others. According to Cohen and Shahrur [50,60], the stock prices of downstream companies usually lead the upstream companies' stock prices. In supply chains, energy and material companies are usually upstream. It may therefore cause the two sectors to receive relatively more information. In the bull period, the Out and In Node Strengths all increase. The utility sector (No. 10) has the largest Out Node Strength. China has invested heavily in the infrastructure construction for years. There are great opportunities in this field. A booming market enhances investors' confidence. The utility sector, which is an important domain of infrastructures, may attract significant attention from investors. This could cause the utility sector to react fast to information and output more information than others in this period. The information technology (No. 8) sector receives the most information, indicating that it is affected greatly by other sectors. In the crash period, apart from the Out Node Strength of the utility sector, which exhibited a slight decrease, other sectors' values continue to increase. The information technology sector outputs the largest amount of information in this time. Since the herding activities in this sector are found to be stronger than others [61] and information technology companies usually have a high price-earnings (P/E) ratio, it may make this sector more sensitive to market turmoil. Meanwhile, the financial (No. 7) and energy sectors also have relatively strong Out Node Strengths. In this period, except for the utility sector, others' In Node Strengths all grow. The consumer staples (No. 5) sector receives the most information. Companies in this sector usually produce essential commodities in people's daily lives. This may make it less sensitive to market turbulence, and it results in receiving relatively more information from other sectors. In the post-crash period, the Out and In Node Strengths of the sectors all decline. The telecommunication service (No. 9) sector outputs the most information. The In Node Strength of the financial sector is much larger than others, indicating that it is heavily impacted by other sectors. This is because many large financial firms are part of the "national team", which aims to buy huge amount of stocks during the crash time in order to stabilize the market [3]. These financial firms therefore become the stakeholders and important money providers for many companies in different sectors. Other sectors' statuses therefore may influence the financial condition of these financial firms, as well as the investors' moods and strategies. This further impacts the price of these financial companies.

Table 4. Out Node Strengths of the 10 sectors in the four sub-periods.

No.	Sector Name	Pre-Bull	Bull	Crash	Post-Crash
1	Energy	0.0474	0.4301	0.7006	0.0396
2	Materials	**0.0811**	0.3366	0.6279	0.0392
3	Industrials	0.0407	0.3985	0.5149	0.1133
4	Consumer Discretionary	0.0714	0.3055	0.4283	0.1918
5	Consumer Staples	0.0316	0.1557	0.5451	0.0471
6	Health Care	0.0682	0.2251	0.3525	0.1048
7	Financials	0	0.3147	0.7130	0.0965
8	Information Technology	0.0628	0.3573	**0.7729**	0.0989
9	Telecommunication Services	0.0361	0.3045	0.3194	**0.2176**
10	Utilities	0	**0.6259**	0.5983	0.1565

Table 5. In Node Strengths of the 10 sectors in the four sub-periods.

No.	Sector Name	Pre-Bull	Bull	Crash	Post-Crash
1	Energy	**0.1509**	0.2314	0.3450	0
2	Materials	0.1156	0.3289	0.6136	0
3	Industrials	0	0.3278	0.6609	0
4	Consumer Discretionary	0.0345	0.3319	0.4932	0.1367
5	Consumer Staples	0	0.0435	**0.7335**	0.0521
6	Health Care	0.0393	0.2888	0.4924	0.0396
7	Financials	0	0.2835	0.5885	**0.4808**
8	Information Technology	0.0337	**0.6807**	0.6157	0.0832
9	Telecommunication Services	0.0653	0.3646	0.4743	0.1325
10	Utilities	0	0.5725	0.5558	0.1806

Betweenness is another centrality measure. It quantifies a node's underlying ability to control the information flow in the network [62]. Its definition is based on the number of shortest paths between nodes. For a weighted directed network, the shortest path d_{ij}^w between node i and j is [62]:

$$d_{ij}^w = \min\left(\frac{1}{w_{ih_0}} + \frac{1}{w_{h_0h_1}} + \cdots + \frac{1}{w_{h_kj}}\right) \tag{15}$$

where $v_{h_0}, v_{h_1}, \cdots, v_{h_k}$ are the intermediary nodes on the path from node v_i to v_j. The shortest path can be derived using Dijkstra's algorithm and the weighted betweenness centrality (WBC) is defined as [62]:

$$WBC(i) = \sum_{i \neq s, i \neq t, s \neq t} \frac{g_{st}^w(i)}{g_{st}^w} \tag{16}$$

where g_{st}^w is the number of the shortest paths from node v_s to v_t. $g_{st}^w(i)$ is the number of the shortest paths from v_s to v_t that also pass through v_i.

Figure 4 shows the WBCs of the 10 sectors in the four sub-periods. In the tranquil period, the material (No. 2) sector has the largest WBC. In the bull period, the utility (No. 10) sector has the largest WBC, while other sectors have much smaller values. This implies that the utility sector is the hub node, and it is very important for the information transmission in this stage. In the crash period, the WBCs are generally smaller. This is because in this phase, the information flow grows and sectors have more routes to connect with each other. The information technology (No. 8) sector has a relatively larger value. In the post-crash period, the financial sector (No. 7) has the largest WBC and is the pivot node of the information flow.

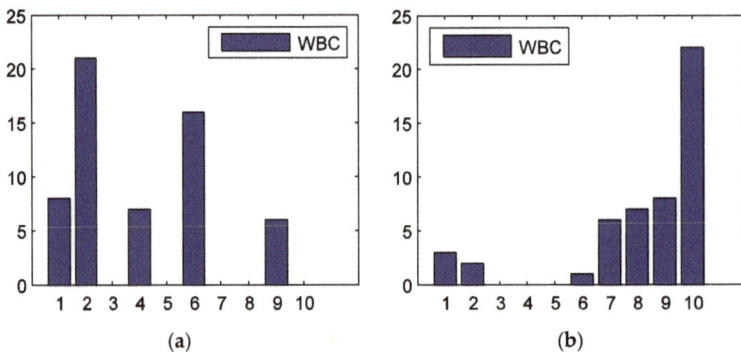

(a)   (b)

Figure 4. *Cont.*

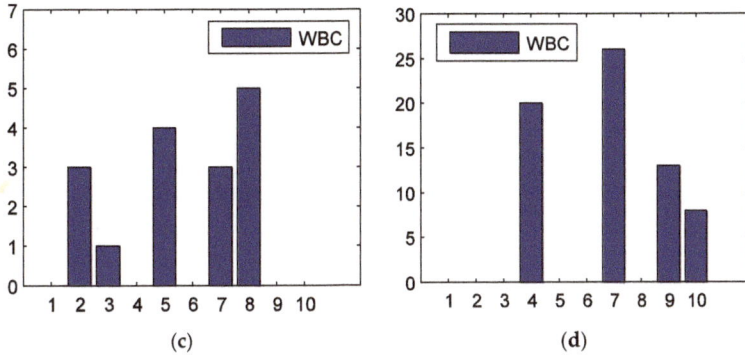

Figure 4. Weighted betweenness centrality (WBC) for the 10 stock sectors during the (**a**) tranquil, (**b**) bull, (**c**) crash, and (**d**) post-crash periods. The horizontal axis is the number of the sectors.

4.3. Directed Maximum Spanning Tree

The MST is one of the spanning trees of a network with the maximum total edge weights. It can help to disentangle the network and visualize the key structures [63]. For the undirected network, the MST can be built using the algorithms of Kruskal or Prim. However, in this paper, the information transfer network is a directed one. Thus, we adopt Chu-Liu-Edmond's algorithm [28,29] to build the directed MST. It is also called the maximum arborescence, and is the backbone of a network [29].

Figure 5 shows the directed MSTs in the four sub-phases. It can be seen that the structures of the trees are very different. This indicates that the market state has heavy impacts on the structure of the information transfer network. It also implies that information transfer can be a potential indicator for the market status. From Figure 5, we can observe that there are isolated sectors in both tranquil and post-crash periods. These isolated sectors have weak interactions with others, indicating that they could be references for portfolio diversification. In the bull and crash periods, the stronger information flow leads to the improvement of connectivity of the network. All sectors are contained in the trees. This means that the opportunity for asset diversification declines. Let the number of hops from the root node to itself be 0, so that the average hops of nodes to the root is 3.7 in the bull period. However, the value decreases to 1.9 in the crash time, suggesting that the tree has a more compact structure and the interactions between sectors are more direct in the crash stage.

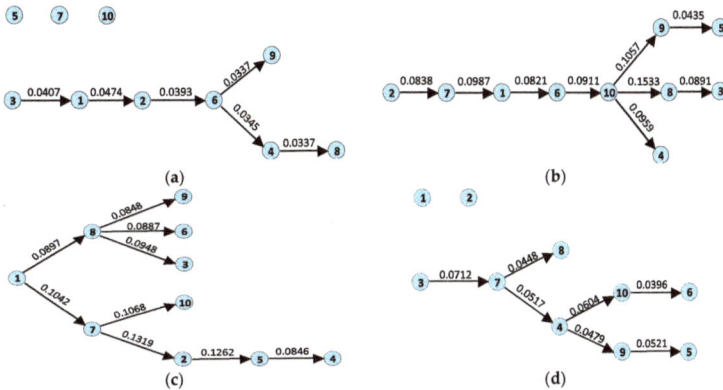

Figure 5. Directed maximum spanning trees (MSTs) of the 10 stock sectors during the (**a**) tranquil, (**b**) bull, (**c**) crash, and (**d**) post-crash periods. The values on the edges are their weights.

5. Conclusions

Using ETE and the 10 sectors' data from July 2013 to February 2017, this paper studies the information transfer in the Chinese stock market around its crash in 2015. According to the market states, the time range is divided into four sub-periods: the tranquil, bull, crash and post-crash periods. The stock data is considered as a continuous variable to avoid the subjectivity and information loss induced by data discretization. The KDE method is applied to compute the ETEs between sectors. Then, the information transfer network is constructed based on the ETEs. The influences of the sectors are analyzed by centrality measures. Lastly, a directed MST is used to disentangle the network.

The results display that the information transfer between sectors is weak in the tranquil period. The energy and material sectors are the main information receivers. In the bull period, the strength and scope of the information transmission both increase. The utility sector outputs much information, and it is the hub node. The information technology sector receives the most information in this stage. In the crash period, the information flow continues to grow, and it reaches the maximum of the four sub-periods, indicating the worst market efficiency. The information technology sector outputs the most information in this phase. The consumer staples sector is the biggest information receiver. The directed MST has a compact structure, suggesting more direct interactions between sectors. In the post-crash period, the strength and area of the information flow decreases, but it is still larger than that in the tranquil period. The telecommunication service sector emits the most information in this stage. The financial sector receives the largest amount of information, and it is the pivot node for the information transmission.

These findings can help us to understand the structure of the Chinese stock market. As information transfer reveals the predictability of the market, this study can also provide references to investors for the selection of investment strategies.

Author Contributions: W.X. and H.X. proposed the research framework together. W.X. collected the data, finished the computation, and wrote the documents. H.X. provided important guidance and advice during the process of this research.

Funding: This research was funded by the National Natural Science Foundation of China (Grant No. 71532004).

Conflicts of Interest: The authors declare no conflict of interest.

References

1. Qian, J. The 2015 Stock Panic of China: A Narrative. 2016. Available online: https://papers.ssrn.com/sol3/papers.cfm?abstract_id=2795543 (accessed on 21 September 2017).
2. Song, G. The Drivers of the Great Bull Stock Market of 2015 in China: Evidence and Policy Implications. 2015. Available online: https://papers.ssrn.com/sol3/papers.cfm?abstract_id=2643051 (accessed on 21 September 2017).
3. Zeng, F.; Huang, W.-C.; Hueng, J. On Chinese Government's Stock Market Rescue Efforts in 2015. *Mod. Econ.* **2016**, *07*, 411–418. [CrossRef]
4. Liu, D.; Gu, H.; Xing, T. The meltdown of the Chinese equity market in the summer of 2015. *Int. Rev. Econ. Financ.* **2016**, *45*, 504–517. [CrossRef]
5. Hou, Y.; Liu, F.; Gao, J.; Cheng, C.; Song, C. Characterizing Complexity Changes in Chinese Stock Markets by Permutation Entropy. *Entropy* **2017**, *19*, 514. [CrossRef]
6. Lu, L.; Lu, L. Unveiling China's Stock Market Bubble: Margin Financing, the Leveraged Bull and Governmental Responses. *J. Int. Bank. Law Rugul.* **2017**, *32*, 145–159.
7. Honey, C.J.; Kotter, R.; Breakspear, M.; Sporns, O. Network structure of cerebral cortex shapes functional connectivity on multiple time scales. *Proc. Natl. Acad. Sci. USA* **2007**, *104*, 10240–10245. [CrossRef] [PubMed]
8. Wibral, M.; Rahm, B.; Rieder, M.; Lindner, M.; Vicente, R.; Kaiser, J. Transfer entropy in magnetoencephalographic data: Quantifying information flow in cortical and cerebellar networks. *Prog. Biophys. Mol. Biophys.* **2011**, *105*, 80–97. [CrossRef] [PubMed]
9. Murari, A.; Peluso, E.; Gelfusa, M.; Garzotti, L.; Frigione, D.; Lungaroni, M.; Pisano, F.; Gaudio, P. Application of transfer entropy to causality detection and synchronization experiments in tokamaks. *Nucl. Fusion* **2016**, *56*, 026006. [CrossRef]

10. Van Milligen, B.P.; Hoefel, U.; Nicolau, J.H.; Hirsch, M.; García, L.; Carreras, B.A.; Hidalgo, C. Study of radial heat transport in W7-X using the transfer entropy. *Nucl. Fusion* **2018**, *58*, 076002. [CrossRef]

11. Bhaskar, A.; Ramesh, D.S.; Vichare, G.; Koganti, T.; Gurubaran, S. Quantitative assessment of drivers of recent global temperature variability: An information theoretic approach. *Clim. Dyn.* **2017**, *49*, 3877–3886. [CrossRef]

12. Oh, M.; Kim, S.; Lim, K.; Kim, S.Y. Time series analysis of the Antarctic Circumpolar Wave via symbolic transfer entropy. *Physica A* **2018**, *499*, 233–240. [CrossRef]

13. Hu, F.; Nie, L.-J.; Fu, S.-J. Information Dynamics in the Interaction between a Prey and a Predator Fish. *Entropy* **2015**, *17*, 7230–7241. [CrossRef]

14. Orange, N.; Abaid, N. A transfer entropy analysis of leader-follower interactions in flying bats. *Eur. Phys. J.-Spec. Top.* **2015**, *224*, 3279–3293. [CrossRef]

15. Kwon, O.; Yang, J.S. Information flow between stock indices. *EPL-Europhys. Lett.* **2008**, *82*, 68003. [CrossRef]

16. Yang, C.; Tang, M.; Cao, Y.; Chen, Y.; Deng, Q. The study on variation of influential regions in China from a perspective of asymmetry economic information flow. *Physica A* **2015**, *436*, 180–187. [CrossRef]

17. Dimpfl, T.; Peter, F.J. The impact of the financial crisis on transatlantic information flows: An intraday analysis. *J. Int. Financ. Mark. Inst. Money* **2014**, *31*, 1–13. [CrossRef]

18. Sandoval, L. Structure of a Global Network of Financial Companies Based on Transfer Entropy. *Entropy* **2014**, *16*, 4443–4482. [CrossRef]

19. Sensoy, A.; Sobaci, C.; Sensoy, S.; Alali, F. Effective transfer entropy approach to information flow between exchange rates and stock markets. *Chaos Soliton Fractals* **2014**, *68*, 180–185. [CrossRef]

20. Bekiros, S.; Nguyen, D.K.; Sandoval Junior, L.; Uddin, G.S. Information diffusion, cluster formation and entropy-based network dynamics in equity and commodity markets. *Eur. J. Oper. Res.* **2017**, *256*, 945–961. [CrossRef]

21. Kim, J.; Kim, G.; An, S.; Kwon, Y.K.; Yoon, S. Entropy-based analysis and bioinformatics-inspired integration of global economic information transfer. *PLoS ONE* **2013**, *8*, e51986. [CrossRef] [PubMed]

22. Schreiber, T. Measuring information transfer. *Phys. Rev. Lett.* **2000**, *85*, 461–464. [CrossRef] [PubMed]

23. Vicente, R.; Wibral, M.; Lindner, M.; Pipa, G. Transfer entropy—A model-free measure of effective connectivity for the neurosciences. *J. Comput. Neurosci.* **2011**, *30*, 45–67. [CrossRef] [PubMed]

24. Marschinski, R.; Kantz, H. Analysing the information flow between financial time series. *Eur. Phys. J. B* **2002**, *30*, 275–281. [CrossRef]

25. Lungarella, M.; Ishiguro, K.; Kuniyoshi, Y.; Otsu, N. Methods for quantifying the causal structure of bivariate time series. *Int. J. Bifurc. Chaos* **2007**, *17*, 903–921. [CrossRef]

26. Khan, S.; Bandyopadhyay, S.; Ganguly, A.R.; Saigal, S.; Erickson, D.J., 3rd; Protopopescu, V.; Ostrouchov, G. Relative performance of mutual information estimation methods for quantifying the dependence among short and noisy data. *Phys. Rev. E* **2007**, *76*, 026209. [CrossRef] [PubMed]

27. Lee, J.; Nemati, S.; Silva, I.; Edwards, B.A.; Butler, J.P.; Malhotra, A. Transfer entropy estimation and directional coupling change detection in biomedical time series. *Biomed. Eng. Online* **2012**, *11*, 19. [CrossRef] [PubMed]

28. Gibbons, A. Spanning-trees, branchings and connectivity. In *Algorithmic Graph Theory*; Cambridge University Press: London, UK, 1985; pp. 42–49. ISBN 0-521-24659-8.

29. Bellingeri, M.; Bodini, A. Food web's backbones and energy delivery in ecosystems. *Oikos* **2016**, *125*, 586–594. [CrossRef]

30. Chen, B.; Wang, J.; Zhao, H.; Principe, J. Insights into Entropy as a Measure of Multivariate Variability. *Entropy* **2016**, *18*, 196. [CrossRef]

31. Chávez, M.; Martinerie, J.; Le Van Quyen, M. Statistical assessment of nonlinear causality: Application to epileptic EEG signals. *J. Neurosci. Methods* **2003**, *124*, 113–128.

32. Kaiser, A.; Schreiber, T. Information transfer in continuous processes. *Physica D* **2002**, *166*, 43–62. [CrossRef]

33. Ruddell, B.L.; Kumar, P. Ecohydrologic process networks: 1. Identification. *Water Resour. Res.* **2009**, *45*, 1–22. [CrossRef]

34. Lizier, J.T.; Prokopenko, M. Differentiating information transfer and causal effect. *Eur. Phys. J. B* **2010**, *73*, 605–615. [CrossRef]

35. Qi, Y.; Im, W. Quantification of Drive-Response Relationships between Residues during Protein Folding. *J. Chem. Theory Comput.* **2013**, *9*, 3799–3805. [CrossRef] [PubMed]

36. Moon, Y.-I.; Rajagopalan, B.; Lall, U. Estimation of mutual information using kernel density estimators. *Phys. Rev. E* **1995**, *52*, 2318–2321. [CrossRef]

37. Ahmad, I.; Lin, P.E. A nonparametric estimation of the entropy for absolutely continuous distributions. *IEEE T. Inform. Theory* **1976**, *22*, 372–375. [CrossRef]

38. Le Caillec, J.M.; Itani, A.; Gueriot, D.; Rakotondratsimba, Y. Stock Picking by Probability–Possibility Approaches. *IEEE Trans. Fuzzy Syst.* **2017**, *25*, 333–349. [CrossRef]

39. Gao, Y.-C.; Tang, H.-L.; Cai, S.-M.; Gao, J.-J.; Stanley, H.E. The impact of margin trading on share price evolution: A cascading failure model investigation. *Physica A* **2018**, *505*, 69–76. [CrossRef]

40. Zhai, P.; Ma, R. An Analysis on the Structural Breaks in Dynamic Conditional Correlations among Equity Markets Based on the ICSS Algorithm: The Case from 2015–2016 Chinese Stock Market Turmoil. 2017. Available online: https://papers.ssrn.com/sol3/papers.cfm?abstract_id=2959830 (accessed on 11 October 2017).

41. Roni, B.; Abbas, G.; Wang, S. Return and Volatility Spillovers Effects: Study of Asian Emerging Stock Markets. *J. Syst. Sci. Inform.* **2018**, *6*, 97–119. [CrossRef]

42. Albano, A.M.; Muench, J.; Schwartz, C.; Mees, A.I.; Rapp, P.E. Singular-value decomposition and the Grassberger-Procaccia algorithm. *Phys. Rev. A* **1988**, *38*, 3017–3026. [CrossRef]

43. Shang, P.; Li, X.; Kamae, S. Chaotic analysis of traffic time series. *Chaos Soliton Fractals* **2005**, *25*, 121–128. [CrossRef]

44. Yook, S.-H.; Chae, H.; Kim, J.; Kim, Y. Finding modules and hierarchy in weighted financial network using transfer entropy. *Physica A* **2016**, *447*, 493–501. [CrossRef]

45. Daugherty, M.S.; Jithendranathan, T. A study of linkages between frontier markets and the U.S. equity markets using multivariate GARCH and transfer entropy. *J. Mult. Financ. Manag.* **2015**, *32–33*, 95–115. [CrossRef]

46. Wu, Y.; Zhang, W.; Shen, J.; Mo, Z.; Peng, Y. Smart city with Chinese characteristics against the background of big data: Idea, action and risk. *J. Clean. Prod.* **2018**, *173*, 60–66. [CrossRef]

47. Yuan, J.; Shen, J.; Pan, L.; Zhao, C.; Kang, J. Smart grids in China. *Renew. Sustain. Energy Rev.* **2014**, *37*, 896–906. [CrossRef]

48. Li, L. China's manufacturing locus in 2025: With a comparison of "Made-in-China 2025" and "Industry 4.0". *Technol. Forecast. Soc. Chang.* **2017**, in press. [CrossRef]

49. Xie, H.; Wang, S.; Chen, X.; Wu, J. Bibliometric analysis of "Internet-plus". *Inform. Learn. Sci.* **2017**, *118*, 583–595. [CrossRef]

50. Cohen, L.; Frazzini, A. Economic links and predictable returns. *J. Financ.* **2008**, *63*, 1977–2011. [CrossRef]

51. Andersen, T.G. Return volatility and trading volume: An information flow interpretation of stochastic volatility. *J. Financ.* **1996**, *51*, 169–204. [CrossRef]

52. Abhyankar, A.H. Return and volatility dynamics in the FT-SE 100 stock index and stock index futures markets. *J. Futures Mark.* **1995**, *15*, 457–488. [CrossRef]

53. Judge, A.; Reancharoen, T. An empirical examination of the lead–lag relationship between spot and futures markets: Evidence from Thailand. *Pac-Basin Financ. J.* **2014**, *29*, 335–358. [CrossRef]

54. Hou, K.; Moskowitz, T.J. Market Frictions, Price Delay, and the Cross-Section of Expected Returns. *Rev. Financ. Stud.* **2005**, *18*, 981–1020. [CrossRef]

55. Menzly, L.; Ozbas, O. Market segmentation and cross-predictability of returns. *J. Financ.* **2010**, *65*, 1555–1580. [CrossRef]

56. Chordia, T.; Roll, R.; Subrahmanyam, A. Liquidity and market efficiency. *J. Financ. Econ.* **2008**, *87*, 249–268. [CrossRef]

57. Zhang, Y.C. Toward a theory of marginally efficient markets. *Physica A* **1999**, *269*, 30–44. [CrossRef]

58. Zunino, L.; Zanin, M.; Tabak, B.M.; Pérez, D.G.; Rosso, O.A. Complexity-entropy causality plane: A useful approach to quantify the stock market inefficiency. *Physica A* **2010**, *389*, 1891–1901. [CrossRef]

59. Junior, L.S.; Mullokandov, A.; Kenett, D.Y. Dependency relations among international stock market indices. *J. Risk Financ. Manag.* **2015**, *8*, 227–265. [CrossRef]

60. Shahrur, H.; Becker, Y.L.; Rosenfeld, D. Return predictability along the supply chain: The international evidence. *Financ. Anal. J.* **2010**, *66*, 60–77. [CrossRef]

61. Zheng, D.; Li, H.; Chiang, T.C. Herding within industries: Evidence from Asian stock markets. *Int. Rev. Econ. Financ.* **2017**, *51*, 487–509. [CrossRef]

62. Lü, L.; Chen, D.; Ren, X.-L.; Zhang, Q.-M.; Zhang, Y.-C.; Zhou, T. Vital nodes identification in complex networks. *Phys. Rep.* **2016**, *650*, 1–63. [CrossRef]
63. Kwapien, J.; Oswiecimka, P.; Forczek, M.; Drozdz, S. Minimum spanning tree filtering of correlations for varying time scales and size of fluctuations. *Phys. Rev. E* **2017**, *95*, 052313. [CrossRef] [PubMed]

Article

Improving Entropy Estimates of Complex Network Topology for the Characterization of Coupling in Dynamical Systems

Teddy Craciunescu [1,*], **Andrea Murari** [2,3] **and Michela Gelfusa** [4]

[1]   National Institute for Laser, Plasma and Radiation Physics, RO-077125 Magurele-Bucharest, Romania
[2]   Consorzio RFX (CNR, ENEA, INFN, Universita' di Padova, Acciaierie Venete SpA), 35127 Padova, Italy; murari@igi.cnr.it.org
[3]   EUROfusion Consortium, JET, Culham Science Centre, Abingdon OX14 3DB, UK; Andrea.Murari@euro-fusion.org
[4]   Department of Industrial Engineering, University of Rome Tor Vergata, 00133 Rome, Italy; gelfusa@ing.uniroma2.it
*   Correspondence: teddy.craciunescu@gmail.com or teddy.craciunescu@inflpr.ro; Tel.: +40-766-326-625

Received: 24 October 2018; Accepted: 19 November 2018; Published: 20 November 2018

Abstract: A new measure for the characterization of interconnected dynamical systems coupling is proposed. The method is based on the representation of time series as weighted cross-visibility networks. The weights are introduced as the metric distance between connected nodes. The structure of the networks, depending on the coupling strength, is quantified via the entropy of the weighted adjacency matrix. The method has been tested on several coupled model systems with different individual properties. The results show that the proposed measure is able to distinguish the degree of coupling of the studied dynamical systems. The original use of the geodesic distance on Gaussian manifolds as a metric distance, which is able to take into account the noise inherently superimposed on the experimental data, provides significantly better results in the calculation of the entropy, improving the reliability of the coupling estimates. The application to the interaction between the El Niño Southern Oscillation (ENSO) and the Indian Ocean Dipole and to the influence of ENSO on influenza pandemic occurrence illustrates the potential of the method for real-life problems.

Keywords: system coupling; cross-visibility graphs; image entropy; geodesic distance

1. Introduction

The synchronization between systems connected through some form of coupling is a common phenomenon occurring in a wide variety of fields, like physics, engineering, biology, physiology, secure communication, environmental sciences, etc. Several types of synchronization have been identified during the last decades: complete synchronization, where the interaction between two identical systems is strong enough to lead to in step trajectories after a transient period [1]; generalized synchronization, which refers to completely different systems where the dynamic variables of one system (the response system) are determined by the other system (the drive system) [2]; phase synchronization, where the phase difference is asymptotically bounded while the amplitudes remain weakly correlated [3]; lag synchronization, which implies the existence of an asymptotic bound between the output of one system and the time-delayed output of a second one [4]; intermittent lag synchronization, which is equivalent to a lag synchronization interrupted by intervals of non-synchronous behavior; and almost synchronization [5], which implies the existence of an asymptotic finite difference between certain subsets of the variables in the two systems.

The first attempt to formulate a unified definition and a general formalism for synchronization has been proposed by Brown and Kocarev [6]. The systems X and Y are considered to be synchronized, with respect to the properties g_x and g_y, if there is a time independent mapping $h : \mathbb{R}^k \times \mathbb{R}^k \to \mathbb{R}^k$, t such that $\|h(g_x, g_y)\|$ approaches zero asymptotically when the time t goes to infinity. The choices of g_x, g_y, and h determine the type of synchronization. Boccaletti et al. [7] proposed a simplified approach, which is based on the notion of prediction and the use of a synchronization function such that a particular point in X is mapped, uniquely, to one point in Y. Therefore, synchronization means prediction of one system's values from another. However, a unifying framework for the study of synchronization of coupled dynamical systems is still an open problem.

From a practical point of view, measuring the degree of synchronization in coupled dynamical systems and the identification of possible causal relations is an important problem. A wide variety of methods has been proposed. A simple indicator of synchronization is based on the mutual information [8], which can be expressed by means of the Kullback–Leibler divergence (i.e., relative entropy [9] as the measure of the information gained by replacing the distribution $p_X \times p_Y$, corresponding to the independence between X and Y, with the joint probability distribution p_{XY}—where $p_Z(k)$ represents the probability of the k-th state in the Z time series. The transfer entropy [10,11] extends the concept of mutual information estimating the influence of the state of X on the transition probabilities in Y, taking into account the past history of Y. Therefore, the synchronization and the causal relationship of time series are investigated on the basis of predictability and information transfer. The synchronization likelihood [12] is closely related to the concept of generalized mutual information as introduced by Pawelzik et al. [13] and it is able to deal with non-stationary dynamics. A measure based on quantifying the correlation dimension [14] of the coupled system, in comparison with the constituent subsystems, was proposed in [15]. Various indices of phase synchronization were also proposed [16,17]. Hempel et al. [18] developed a permutation-based asymmetric association measure. The method is based on the observation that two interacting subsystems are monotonically increasing functions. Synchronization can be determined also by monitoring the quasi-simultaneous appearances of certain predefined events in the time series [19]. The definition of the events is dependent on the particular application of the method. Methods based on topological criteria have been first proposed for quantifying generalized synchronization [20]. In a more general approach [21], attempting to detect causality relations between time series, the cross-convergent maps method looks for the signature of X in Y by checking the correspondence between points in the attractor manifold built from Y and points in the X manifold, where the two manifolds are constructed from lagged coordinates of the time-series variables X and X, respectively. An analytical measure based on the time rate of information flowing from one series to the other was proposed in [22], with the advantage that the resulting formula involves only commonly used statistics. A detailed evaluation of several synchronization measures is reported in [23]. Most of these methods are based the assumption that a stronger coupling leads to a stronger synchronization. As noticed in [23], different systems may have different behaviors during their transition to synchronization and the Lyapunov exponent is not always strictly monotonic. However, this assumption remains a practical method to evaluate the effect of the coupling strength variations especially for real-life data.

In this paper, we are proposing a measure that is based on transforming the time series into graphs by means of the "cross-visibility networks" (CVN) method [24]. The quantification of the coupling is obtained by estimating the entropy of the network topology. To improve the reliability of the results, the uncertainties in the measurements are taken into account with the innovative approach of the geodesic distance on Gaussian manifolds (GD). With regard to the structure of the paper, the next section describes the proposed method. Section 3 is devoted to the evaluation of the method's performances by means of numerical tests, showing the advantages of adopting the GD in controlled conditions. For the demonstration of the approach potential to handle real-life data, an example from the field of atmospheric physics is presented in Section 4. Conclusions and indications of further developments are the subject of the last section of the paper.

2. Geodesic Distance to Improve Visibility Graphs for the Analysis of Synchronization Experiments

The transformation of time series into graphs was introduced to allow the study of time series dynamics by mean of the organization of networks. Zhang et al. [25] initially proposed to divide the time series into disjoint cycles and to consider each cycle as a node in a graph. Then the network representation can be obtained by connecting cycles for which the phase space distance is less than a predefined value. By using this method, the noisy periodic signals are mapped into random networks, while chaotic time series lead to complex networks exhibiting small-world and scale-free features [26].

In a very popular approach, Lacassa et al. [27] proposed to construct the mapping between the time domain and the network topology by considering a representation of time series using vertical bars; seeing this representation as a landscape, every bar in the time series is linked with those that can be seen from the top of the bar (Figure 1). Mathematically, two points, (t_i, y_i) and (t_j, y_j) in the time series, will be connected if the relation below is valid for any intermediate point (t_k, y_k):

$$y_k < y_j + (y_j - y_j)\frac{t_j - t_k}{t_j - t_i} \tag{1}$$

The resulting complex network, called "visibility graph" (VG), is connected as each node can be linked at least with its first order neighbors in an undirected way. As proven in [27], the visibility graph inherits certain properties of the original time series. For example, a periodic series is converted into a regular graph, a random series into a random graph, and a fractal series into a scale-free graph. VG encapsulates the same amount of information as the initial time series but it may make more visible certain properties that are difficult to capture when directly analyzing the time series.

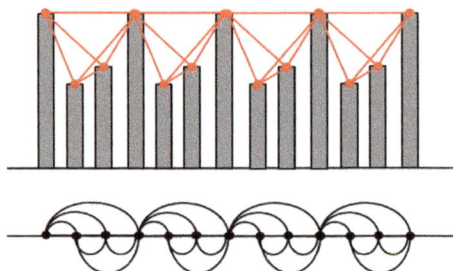

Figure 1. Illustration of the construction of a visibility graph from a time series.

While the visibility graph represents a novel view for analyzing time series, the recently introduced cross-visibility networks have been explicitly conceived to reveal the possible coupling between them [24]. Considering a pair of time series $\{x_i\}$ and $\{y_i\}$, first they should be normalized (to their mean and standard deviation in case of stationary sequences and to maximum values for the non-stationary case) in order to make them comparable. Then the network is constructed by mapping each component of $\{x_i\}$ in a node of the graph. The connections are constructed by the following rules:

$$y_k \leq y_i + \frac{x_j - x_i}{j - i}(k - i), \quad i < \forall k < j \tag{2}$$

or

$$y_k \geq y_i + \frac{x_j - x_i}{j - i}(k - i), \quad i < \forall k < j \tag{3}$$

Therefore, the node i is looking at the components of $\{x_i\}$ time series, through the obstacles of the shifted time series $\{y_k\} = \{y_k - y_i + x_i\}$. Equation (2) accounts for the visibility from the top view, while Equation (3) accounts for the visibility from the beneath view. Basically, the top view is determined by the reciprocal visibility of the peak values of the time series, whereas the beneath view by the reciprocal visibility of the valleys.

As emphasized in [24], the construction of the CVN is the result of local operations on the time series. This represents a different approach in comparison with methods like mutual information [8], Granger causality [28] and transfer entropy [10], which are investigating cross-correlations based on properties obtained by averaging over the whole times series.

The constructed CVN can be represented by the adjacency matrix, whose elements are given by the relation:

$$a_{ij} = \begin{cases} 1, & \text{if nodes } i \text{ and } j \text{ are connectd} \\ 0, & \text{otherwise} \end{cases} \tag{4}$$

Several studies have shown that more robust results can be obtained from complex networks by weighting the graph connections (see, e.g., [29–31]). As an evolution of this approach, we have modified the network adjacency matrix by weighting the connections with the metric distance between two connected values in the time series:

$$a_{ij}^{w} = \begin{cases} dist(y_i - y_j), & \text{if Equation (2) or Equation (3) is satisfied} \\ 0, & \text{otherwise} \end{cases} \tag{5}$$

where $dist(y_i - y_j)$ is a distance. Using the Euclidean distance for this metric is a very popular choice but it implicitly assumes that the data points are infinitely precise values. However, this assumption is rarely satisfied in practical applications. In many cases, measurements are affected by various noise sources which can be considered, from the statistical point of view, as independent random variables. This will lead to measurements with a global Gaussian distribution around the most probable value, which is the value of the actual measured quantity.

Therefore, a Gaussian probability density function, characterized by a specific mean μ and standard deviation σ, can be associated to each point in the time series data. In this view, the distance between two time series points is the distance between the corresponding Gaussian distributions, which can be calculated with the help of information geometry theory [32]. Various families of probability distribution functions (pdf) can be considered as lying on a Riemannian differential manifold. A point on this manifold corresponds to a specific pdf and the Fisher information constitutes a metric tensor (Fisher–Rao metric) on such manifold [33]. It can be demonstrated that the Fisher–Rao metric is unique, intrinsic and invariant under basic probabilistic transformations. For the case of two univariate Gaussian distributions $p_1(x|\mu_1, \sigma_1)$ and $p_2(x|\mu_2, \sigma_2)$, the geodesic distance (GD) on Gaussian manifolds is given by the relation:

$$GD(p_1||p_2) = \sqrt{2}\ln\frac{1+\delta}{1-\delta} = 2\sqrt{2}\tanh^{-1}\delta \tag{6}$$

where:

$$\delta = \left[\frac{(\mu_1 - \mu_2)^2 + 2(\sigma_1 - \sigma_2)^2}{(\mu_1 - \mu_2)^2 + 2(\sigma_1 + \sigma_2)^2}\right]^{\frac{1}{2}}$$

An illustrative example, showing the difference between the Euclidean and geodesic distance, is presented in Figure 2. Further details regarding the use of GD for the analysis of noisy time series can be found in [34].

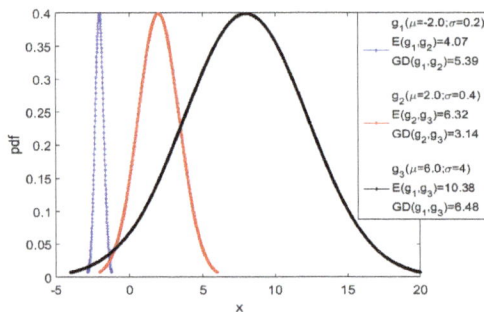

Figure 2. The geodesic distance GD between three Gaussian distributions g_1 (*blue*), g_2 (red), and g_3 (black) is compared to the Euclidean distance E between their modes. The Euclidean distance E(g1, g2) between g1 and g2 is less than half the distance E(g1, g3) between g1 and g3. On the other hand, since g3 has a standard deviation about an order of magnitudes higher than the other two, the geodesic distances GD(g1, g2) and GD(g1, g3) are almost the same [34].

The weighted adjacency matrix (WAM) given by Equation (5) can be used to monitor the changes of the CVN structure when the coupling between the two-time series is varied. As the strength of the coupling increases, the complexity of the network decreases. An appropriate measure of the network complexity should be used in order to follow quantitatively this evolution.

The network complexity has been evaluated traditionally by means of measures like, e.g., degree distribution, clustering coefficient, edge density. More recently methods derived from information theory have been used. Many of them are based on the use of the Shannon entropy [35] and the underlying idea is to quantify the information content of the network as a measure of its "typicality", as a tradeoff between random versus causal nature [36]. Implicitly, systems characterized by a low value of information, with low entropy, are considered to be "simple". The network entropy is defined with respect to a network invariant. Most popular choices are the graph degree distribution [37] and adjacency matrices [38]. The last one offers more robustness against changes in the network size and is correlated with its algebraic properties [39]. Certain caveats should also be considered: the network entropy is based on the selection of a network invariant but it is not itself a network invariant [40]. It can be argued also that entropy, measuring statistical randomness, is not aligned with intuitive human understanding of complexity [41]. However, the entropy based measure is computationally affordable and a number of successful applications like, e.g., the analysis of DNA sequences [42] or in molecular biology [43] have been reported.

The WAM adjacency matrix is expressed in real numbers instead of a binary representation and hence it can be represented as an image. Therefore, the image complexity evaluation is needed in this case in order to monitor the networks structure changes. Equivalently, image entropy will be used as a measure of complexity instead of network entropy.

Image entropy can be formulated starting from the Shannon's information theory [37] where entropy is used to measure the amount of information in a set of symbols such as an image. For a random variable X, the Shannon entropy is defined by:

$$H(X) = -\sum_i P_i log P_i \tag{7}$$

where P_i is the probability of the occurrence of $X = x_i$.

The entropy for a grayscale image is equivalent to the above equation in which p_i is calculated as follows:

$$p_i = \frac{H(i)}{\sum_i H(i)}, \quad i = 1, \ldots, N_H \tag{8}$$

where H is the histogram of pixel intensities in the image, $H(i)$ is the number of pixels with a certain intensity and N_H the number of intensity bins in the image.

A random noise image has the maximum value of entropy while the entropy of a uniform image equals zero. For usual images the areas characterized by a smooth changing of gray levels, the existence of blocks of uniform pixel values or the presence of repeating patterns of texture will have lower values than those for which the pixel values are changing rapidly or in a random way. Therefore, entropy reflects the non-uniformity and complexity of image texture [44,45].

The CVN structure evolves with the variation of the coupling strength. When the time series are not synchronized, the visibility of two points in the time series $\{y_k\}$ are more frequently interrupted by obstacles in the time series $\{x_k\}$. When the time series tends to synchronize due to increased coupling, the WAM evolves to a simpler structure, which translates into a relatively monotonic decrease of the image entropy (8). The entropy, as a measure of the degree of complexity, can therefore be used to define a measure of synchronization:

$$Q = -H(CVN) \tag{9}$$

where the minus sign has been introduced in order to have an increase of Q with the coupling strength, coherent with most synchronization measures.

The synchronization measure should exhibit a monotonic behavior with the increase of coupling, to allow distinguishing the effects on synchronization. The assessment of a such behavior has been evaluated by using the degree of monotonicity defined in [23]:

$$Mon(C) = \frac{2}{m(m-1)} \sum_{i=1}^{m-1} \sum_{j=i+1}^{m} sign(c_{i} - c_{j}) \tag{10}$$

where $C = \{c_1, c_2, \ldots, c_m\}$ is the sequence of monotonically increasing coupling strengths, $Mon(C) = 1$ for a strictly monotonical behavior.

3. Tests with Synthetic Data

The evaluation of the efficacy of the CVN entropy (ENT-CVN) measure in detecting different degrees of coupling has been tested first with synthetic data. We have considered three unidirectional coupled dynamical systems, based on strange attractors, that have been used also in [15,23]:

- the coupled Rössler system [46,47]:

$$\begin{aligned}
\dot{x}_2 &= 0.95x_1 + 0.15x_2 \\
\dot{x}_3 &= 0.2 + x_3(x_1 - 10) \\
\dot{y}_1 &= -1.05y_2 - y_3 + C(x_1 - y_1) \\
\dot{y}_2 &= 1.05y_1 + 0.15y_2 \\
\dot{y}_3 &= 0.2 + y_3(y_1 - 10)
\end{aligned} \tag{11}$$

- the coupled Hénon system [48,49]:

$$\begin{aligned}
x_1[n+1] &= 1.4 - x_1^2[n] + 0.3x_2[n] \\
x_2[n+1] &= x_1[n] \\
y_1[n+1] &= 1.4 - (Cx_1[n]y_1[n] + (1-C)y_1^2[n] \\
y_2[n+1] &= y_1[n]
\end{aligned} \tag{12}$$

- the coupled Lorenz system [50–52]:

$$
\begin{aligned}
\dot{x}_1 &= 10(x_2 - x_1) \\
\dot{x}_2 &= x_1(28 - x_3) - x_2 \\
\dot{x}_3 &= x_1 x_2 - \tfrac{8}{3} x_3 \\
\dot{y}_1 &= 10(y_2 - y_1) \\
\dot{y}_2 &= y_1(28.001 - y_3) - y_2 \\
\dot{y}_3 &= y_1 y_2 - \tfrac{8}{3} y_3 + C(x_3 - y_3)
\end{aligned}
\tag{13}
$$

The time series x_2 and y_2 were used in case of Rössler and Hénon systems while the time series x_1 and y_1 were used in case of the Lorenz system.

For the Hénon system, the synchronization between the driver and the responder system is reached for $C = 0.8$, as shown in [23], by following the plot of the responder attractor together with the plot of the driver versus responder components. For the Rössler and Lorenz systems, the evolution towards identity has been observed for $C = 2$. Therefore, the coupling strength C has been varied in the interval [0,0.8], in steps of 0.01, for the Hénon system, while for the Rössler and Lorenz systems C evolves in the interval [0, 2], in steps of 0.2.

The analysis has been carried out using relatively short time series, with 4000 samples, typical of real-life applications. These applications are characterized also by the presence of significant levels of noise in the experimental measurements. As already mentioned above, we assumed that the measurements are typically affected by a wide range of noise sources, which are often independent and additive. Therefore, the uncertainties in the measured values can be considered Gaussian distributions. The theoretical models (11–13) have been used to generate first clean data. For testing the robustness of the proposed coupling measure against noise, realistic experimental conditions have been simulated by adding noise with an amplitude of 10% *and* 20% of the standard deviation of the original synthetic data. The mean value over 10 different noise realizations and 10 different initial conditions of the coupled systems has been considered.

The evolution of the synchronization measure Q with the coupling strength C is presented in Figure 3, while the monotonicity values are listed in Table 1. In general, the evolution of the image entropy evolution is rather monotonic. An oscillatory behavior occurs in case of the Hénon system, but only for a very weak coupling $C < 0.15$. For the Lorenz system the image entropy decreases have a slow rate and it affected by certain oscillations. This behavior is similar to that observed in [23] for other synchronization measures and it can be justified by the fluctuations of the maximum Lyapunov exponent of the responder system up to intermediate values of the coupling strength. The superimposed Gaussian noise leads to a deterioration of the monotonicity, in range of [8–14%] for 10% added noise and in the range of [13–23%] for 20% added noise. The use of the geodesic distance on Gaussian manifold in the WAM Equation (5) allows much better counteracting the effects of noise, providing systematically improved estimates. The monotonicity values are improved of 8–12% for the Hénon and Lorenz systems.

Table 1. Evaluation of the monotonicity measure using clean data (CD), noise superimposed on data and Euclidian distance (N-ED) and noise superimposed on data and geodesic distance (N-GD).

	Rössler	Hénon	Lorenz
CD	0.92	0.95	0.89
10% noise, ED	0.84	0.83	0.75
10% noise, GD	0.88	0.91	0.83
20% noise, ED	0.79	0.74	0.71
20% noise, GD	0.81	0.86	0.82

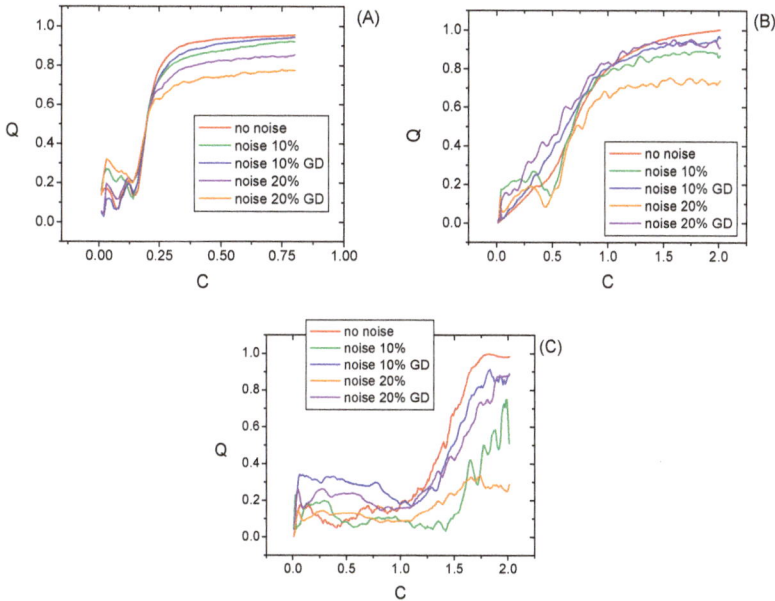

Figure 3. Dependence of the Q measure on the coupling strength for the Rössler (**A**), Hénon (**B**), and Lorenz (**C**). Q is normalized to its maximum value.

4. Real-World Applications

4.1. The Interaction between El Niño Southern Oscillation and the Indian Ocean Dipole

The identification of causal relations between time series has become an increasing focus of interest in climatology. In a pioneering approach reported in Marwan et al. [53], the influence of the El Niño Southern Oscillation (ENSO) irregular cyclicities, on the high variability of rainfall and river discharge in the Northwestern Argentine Andes, has been investigated by mean of cross-recurrence plots (CRP) [54]. Probably the most popular approach for identifying the coupling between time series in this field is based on the Granger causality technique [28] which exploits predictability to determine causation. It has been applied for exploring the causality between various phenomena like e.g. global average observed time series of carbon dioxide and temperature [55], changes in level of atmospheric CO_2 and the El Niño–Southern Oscillation [56], ENSO and rainfall-sensitive vegetation regions in Indonesia [57], atmosphere-ocean coupled circulation patterns and the global temperature variations [58], climate–vegetation dynamics [59], ENSO oscillation, and Indian summer monsoon (ISM) [60]. The coupling between ENSO and ISM has been investigated also by using statistical correlation tools [61]. The same kind of tools have been used also for investigating the causal influence of ENSO on the tropical plants reproduction and resource acquisition strategies [62]. The above list is far from comprehensive and it is intended only to give an idea about the increasing interest in this topic.

In order to validate the method proposed in this paper we will address the causal influence between ENSO and the Indian Ocean Dipole (IOD). This problem has been previously studied, using a similar methodology, but a different causation measure in [22].

ENSO is the most important coupled ocean-atmosphere phenomenon with profound consequences on the global climate and the ocean ecosystem on inter-annual time scales. It is believed to influence various phenomena, such as floods in South America and droughts in Southeast Asia and Southern Africa [63]. It has also been linked sea surface temperature (SST) anomalies over other ocean basins via the "atmospheric bridge" [64]. A typical example is IOD, which is also an air-sea coupled mode, which determines the SST periodic oscillations [65,66]. IOD influences the climate of Australia and of the countries surrounding the Indian Ocean Basin, determining high rainfall variability in this region. The general view assumes that IOD has a self-generating mechanism determined by the internal atmosphere–ocean coupling (see, e.g., [67]). However, increasing evidences on the existence of a link between ENSO and IOD have been reported: the occurrence of both El Niño/ La Niña and positive/negative events are described in [68]. ENSO events are able to influence the duration of IOD events. A reverse feedback may be induced by IOD on ENSO [69]. The ENSO–IOD interlink increased since 1970 together with the enhancement of the Walker circulation [70,71].

Several indices are used to monitor ENSO variations, all of them relying on sea surface temperatures (SST) anomalies averaged across a given region. The Niño-*n* (*n* = 1, 2, 3, 3.4, 4) indices correspond to regions crossed by different ships' tracks, which have enabled the historic records of ENSO. In this paper we used the Nino-4 index (5N–5S, 160E–150W) which captures SST anomalies in the central equatorial Pacific [72]. The IOD intensity is represented by mean of the Dipole Mode Index (DMI), which is the SST gradient between the western equatorial Indian Ocean (50E–70E and 10S–10N) and the south eastern equatorial Indian Ocean (90E–110E and 10S–0N) [73].

The existence of the linkage between ENSO and IOD is studied in this paper by analyzing the causal influence between the Nino-4 index and the Indian Ocean SST and also between the IOD index and the Pacific Ocean SST. The Nino-4 monthly index and the SST gridded data (latitude starting at −88.0 and increasing northward per 2 degrees up to +88.0 and longitude starting at 0.0 and increasing eastward per 2 degrees up to 358.0) are from the NOAA ESRL Physical Sciences Division [72], while monthly DMI series has been retrieved from the SST gradient by the Japan Agency for Marine-Earth Science and Technology (JAMSTEC) [73]. We used data in between years 1958 and 2010 because DMI data is available only since 1958 and also for a fair comparison with the results reported in [22].

The causal influence between the IOD index and the tropical Pacific SST is presented in Figure 4 which shows a clear causal influence characterized by an El Nino-like pattern. Figure 5 shows the causation between Nino4 and the Indian Ocean SST. The causation map is characterized by two poles. The map of the feedback from the Indian Ocean SST (Figure 5a, shows two positive poles, revealing the influence of the Indian Ocean on El Nino by mean of IOD. This confirms the findings first reported in Figure 5b of [22]. This influence is extremely important as it may lead to the amplifications of El Nino oscillations.

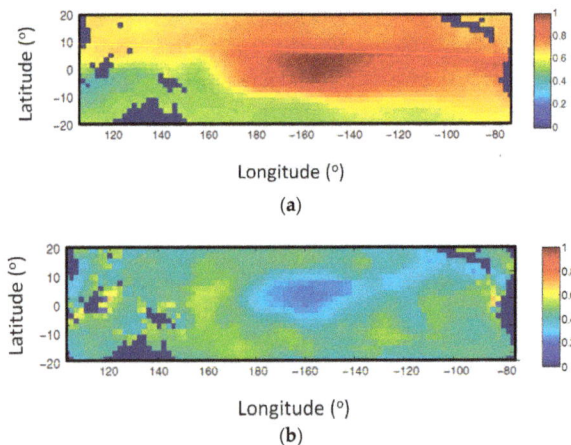

(a)

(b)

Figure 4. Map of the synchronization measure Q revealing the causation between the IOD index and the tropical Pacific SST (**a**), and vice versa (**b**). The Q map has been calculated with the spatial resolution of the SST gridded data (2 degrees for both latitude and longitude) for the time interval in between years 1958–2010.

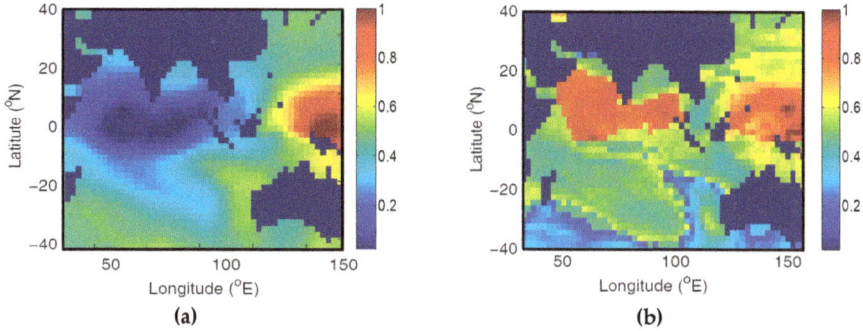

Figure 5. Map of the synchronization measure Q revealing the causation between Nino4 and the Indian Ocean SST (**a**) and vice-versa (**b**). The Q map has been calculated with the spatial resolution of the SST gridded data (2 degrees for both latitude and longitude) for the time interval in between years 1958–2010.

4.2. The Influence of the El Nino Southern Oscillation on Influenza Pandemic Occurrence

The analysis of the causal relation between climatologic phenomena and the outbreak of various diseases by means of time series analysis represents a relatively new topic. For example, the effect of ENSO on the leptospirosis outbreaks in New Caledonia has been investigated in [74]. It has been found that La Niña periods are associated with high rainfall and both these factors are temporarily associated with outbreaks of leptospirosis. Moreover, it was possible to forecast the outbreaks for the next few months based on the sea surface temperature. This opens the possibility of an effective preparation of the health authorities. The ENSO-driven climate variability connection with periodic major outbreaks of dengue (a mosquito-borne viral disease) in Venezuela has been studied in [75]. The most significant dengue events correspond to the warmer and dryer years of El Niño. It has been shown that the ENSO variations, on seasonal and inter-annual scales, drive the occurrence of dengue periodicity through local changes in temperature and rainfall. The relationship between the dengue incidence in 14 island nations of the South Pacific and ENSO has also been investigated in [76]. Positive correlations have been investigated in 10 cases. Propagation between neighboring islands proved to be related only to modulating factors (such as population density and travel) and independent of inter-annual climate variations. In [77] it has been shown that the cholera dynamics in Bangladesh is characterized by an inter-annual component, which corresponds to the dominant frequency of ENSO. For a review of the present understanding of ENSO health associations the reader is referred to [78]. The review advocates the idea that, as ENSO is a complex non-canonical phenomenon, simple correlations are not able to correctly describe the linkage with different health phenomena and that the analysis should use tools more sophisticated than purely statistical ones.

In this paper we analyze the causal relation between ENSO and the influenza pandemic occurrence. The incidence of the influenza epidemic is determined by the seasonal variation in virulence, transmission, and survival but also by the climatic factors. In particular ENSO induces the modulation the global precipitation (on time scales which extends from sub-annual to multi-decadal) [79]. A low precipitation rate increases the duration of suspension of aerosols in the air [80], leading to an increase influenza epidemic, as the aerosols represents the most effective mode of viruses' transmission.

This problem has been investigated previously in [81] by analyzing the Joint recurrence plot (JRP) constructed with the SOI and SST time series. It has been found that JRP revealed periodic, quasi-periodic, and chaotic regimes, dominated by chaos–chaos transitions. All peaks of multiple waves of the influenza pandemics can be related to high divergence of SOI and SST trajectories.

The problem has been analyzed in the present paper using a different tool. The coupling measure Q given by Equation (9) has been calculated for the SST and SOI time series, using a sliding temporal window of 18 months. Its evolution is presented in Figure 6 where the following historical records of onsets and peaks of influenza pandemic waves from 1876 to 2016 have been considered: December 1899, December 1900, March 1901, March 1918, July 1918, November 1919, January 1920, October 1957, February 1958, March 1969, December 1969, January 1970, June 2009, and October 2009 [81]. The pandemic waves coincide with a low correlation of the two processes. The results confirm, with a different and numerically independent method, the findings reported in [81].

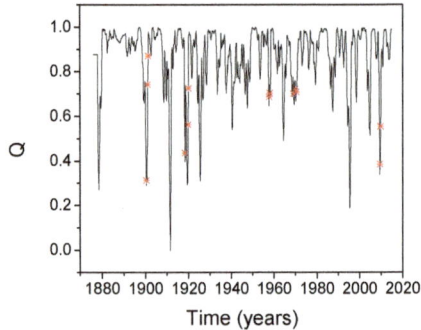

Figure 6. Synchronization measure Q calculated for the SOI and SST time series, using an 18-month sliding temporal window. The peaks of influenza pandemic waves are marked by red stars.

5. Conclusions

The new method of weighted cross-visibility networks has been applied to time series for the investigation of the coupling between dynamical systems. The strength of the coupling is derived from the topology of the CVN and is quantified by the Shannon entropy of the adjacency matrix. The method introduces a new approach to time series causality evaluation. Additionally, the original use of the geodesic distance on Gaussian manifolds in the calculation of the entropy allows taking into account the errors in the measurements. The reliability of the coupling estimates is significantly improved in all the numerical cases investigated, for relatively low, but realistic, levels of noise up to 20%. The application of the technique to the interaction between ENSO and the Indian Ocean Dipole and also to the influence ENSO on influenza pandemic occurrence proves the potential of the approach to handle actual measurements for the investigation of complex systems. As the proposed methodology is able to provide robust results when working with short and noisy time series it is, therefore, very promising for the study of other complex systems, such as thermonuclear plasmas [82,83], in which different complicated phenomena, ranging from the physics of the fast ions and the impurities to the pacing of various instabilities, need to be better understood [84,85].

Author Contributions: All the authors have contributed equally to conceptualization, methodology, software, validation and also for writing this paper.

Funding: In case of one of the authors (T.C.), this research was funded by National Institute of Lasers, Plasma, and Radiation Physics, Bucharest, Romania, grant number 1EU-4/2 4. In case of the other two authors (A.M. and M.G.), this research received no external funding.

Conflicts of Interest: The authors declare no conflict of interest. The funders had no role in the design of the study; in the collection, analyses, or interpretation of data; in the writing of the manuscript, or in the decision to publish the results.

References

1.   Pecora, L.M.; Carroll, T.L. Synchronization in Chaotic Systems. *Phys. Rev. Lett.* **1990**, *64*, 821–825. [CrossRef] [PubMed]

2.   Abarbanel, H.D.I.; Rulkov, N.F.; Sushchik, M.M.; Sushchik, M. Generalized synchronization of chaos: The auxiliary system approach. *Phys. Rev. E* **1996**, *53*, 4528–4535. [CrossRef]

3. Rosenblum, M.G.; Pikovsky, A.S.; Kurths, J. From phase to lag synchronization in coupled chaotic oscillators. *Phys. Rev. Lett.* **1997**, *78*, 4193–4196. [CrossRef]

4. Rosa, E.R.; Ott, E.; Hess, M.H. Transition to phase synchronization of chaos. *Phys. Rev. Lett.* **1998**, *80*, 1642–1645. [CrossRef]

5. Femat, R.; Solis-Perales, G. On the Chaos Synchronization Phenomena. *Phys. Lett. A* **1999**, *262*, 50–60. [CrossRef]

6. Brown, R.; Kocarev, L. A unifying definition of synchronization for dynamical systems. *Chaos* **2000**, *10*, 344–349. [CrossRef] [PubMed]

7. Boccaletti, S.; Pecora, L.M.; Pelaez, A. Unifying framework for synchronization of coupled dynamical systems. *Phys. Rev. E* **2001**, *63*, 066219. [CrossRef] [PubMed]

8. Cover, T.M.; Thomas, J.A. *Elements of Information Theory*; Wiley: New York, NY, USA, 1991.

9. Kullback, S.; Leibler, R.A. On information and sufficiency. *Ann. Math. Statist.* **1951**, *22*, 79–86. [CrossRef]

10. Schreiber, T. Measuring Information Transfer. *Phys Rev Lett* **2000**, *85*, 461–464. [CrossRef] [PubMed]

11. Hlaváčková-Schindlera, K.; Palub, M.; Vejmelkab, M.; Bhattacharya, J. Causality detection based on information-theoretic approaches in time series analysis. *Phys. Rep.* **2007**, *441*, 1–46.

12. Stama, C.J.; van Dijk, B.W. Synchronization likelihood: An unbiased measure of generalized synchronization in multivariate data sets. *Physica D* **2002**, *163*, 236–251. [CrossRef]

13. Pawelzik, K. *Nichtlineare Dynamik und Hirnaktivität: Charaktesierung Nichtlinearer Experimenteller Systeme durch Instabile Periodische Orbits, Vorhersagen und Informationsflüsse*; Verlag Harri Deutsch: Frankfurt, Germany, 1991. (In German)

14. Grassberger, P.; Procaccia, I. Measuring the Strangeness of Strange Attractors. *Physica D* **1983**, *9*, 189–208. [CrossRef]

15. Janjarasjittab, S.; Loparo, K.A. An approach for characterizing coupling in dynamical systems. *Physica D* **2008**, *237*, 2482–2486. [CrossRef]

16. Shabunin, A.; Demidov, V.; Astakhov, V.; Anishchenko, V. Information theoretic approach to quantify complete and phase synchronization of chaos. *Phys. Rev. E* **2002**, *65*, 056215. [CrossRef] [PubMed]

17. Palus, M.; Stefanovska, A. Direction of coupling from phases of interacting oscillators: An information-theoretic approach. *Phys. Rev. E* **2003**, *67*, 055201(R). [CrossRef] [PubMed]

18. Hempel, S.; Koseska, A.; Kurths, J.; Nikoloski, Z. Inner Composition Alignment for Inferring Directed Networks from Short Time Series. *PRL* **2011**, *107*, 054101. [CrossRef] [PubMed]

19. Quiroga, R.Q.; Kreuz, T.; Grassberger, P. Event synchronization: A simple and fast method to measure synchronicity and time delay patterns. *Phys. Rev. E* **2002**, *66*, 041904. [CrossRef] [PubMed]

20. Rulkov, N.F.; Sushchik, M.M.; Tsimring, L.S.; Abarbanel, H.D.I. Generalized synchronization of chaos in directionally coupled chaotic systems. *Phys. Rev. E* **1995**, *51*, 980–994. [CrossRef]

21. Sugihara, G.; May, R.; Ye, H.; Hsieh, C.-H.; Deyle, E.; Fogarty, M.; Munch, S. Detecting Causality in Complex Ecosystems. *Science* **2012**, *338*, 496–500. [CrossRef] [PubMed]

22. San Liang, X. Unraveling the cause-effect relation between time series. *Phys. Rev. E* **2014**, *90*, 052150. [CrossRef] [PubMed]

23. Kreuz, T.; Mormann, F.; Andrzejak, R.G.; Kraskov, A.; Lehnertz, K.; Grassberger, P. Measuring synchronization in coupled model systems: A comparison of different approaches. *Physica D* **2007**, *225*, 29–42. [CrossRef]

24. Mehraban, S.; Shirazi, A.H.; Zamani, M.; Jafari, G.R. Coupling between time series: A network view. *EPL* **2015**, *103*, 50011. [CrossRef]

25. Zhang, J.; Small, M. Complex Network from Pseudoperiodic Time Series: Topology versus Dynamics. *PRL* **2006**, *96*, 238701. [CrossRef] [PubMed]

26. Zhang, J.; Sun, J.; Luo, X.; Zhang, K.; Nakamura, T.; Small, M. Characterizing Pseudoperiodic Time Series through Complex Network Approach. *Physica D* **2008**, *237*, 2856–2865. [CrossRef]

27. Lacasa, L.; Luque, B.; Ballesteros, F.; Luque, J.; Nuño, J.C. From time series to complex networks: The visibility graph. *PNAS* **2008**, *105*, 4972–4975. [CrossRef] [PubMed]

28. Granger, C.W.J. Investigating Causal Relations by Econometric Models and Cross-spectral Methods. *Econometrica* **1969**, *37*, 424–438. [CrossRef]

29. Gonçalves, B.A.; Carpi, L.C.; Rosso, O.; Ravetti, M.G. Time series characterization via horizontal visibility graph and Information Theory. *Physica A* **2016**, *464*, 93–102. [CrossRef]

30. Zhu, G.; Li, Y.; Wen, P. Epileptic seizure detection in EEGs signals using a fast weighted horizontal visibility algorithm. *Comput. Methods Programs Biomed.* **2014**, *115*, 64–75. [CrossRef] [PubMed]

31. Supriya, S.; Siuly, S.; Wang, H.; Cao, J.; Zhang, Y. Weighted Visibility Graph with Complex Network Features in the Detection of Epilepsy. *IEEE Access* **2016**, *4*, 6554–6566. [CrossRef]

32. Amari, S.-I.; Nagaoka, H. *Methods of Information Geometry*; Oxford University Press and the American Mathematical Society: Oxford, UK, 2000.

33. Arwini, K.; Dodson, T.J. *Information Geometry: Near Randomness and Near Independence*; Springer: Berlin, Germany, 2008.

34. Craciunescu, T.; Murari, A. Geodesic distance on Gaussian Manifolds for the robust identification of chaotic systems. *Nonlinear Dyn.* **2016**, *86*, 677–693. [CrossRef]

35. Shannon, C.E. A Mathematical Theory of Communication. *Bell Syst. Tech. J.* **1948**, *27*, 379–423, 623–656. [CrossRef]

36. Bianconi, G. The entropy of randomized network ensembles. *EPL* **2007**, *81*, 28005. [CrossRef]

37. Korner, J.; Marton, K. Random access communication and graph entropy. *IEEE Trans. Inf. Theory* **1998**, *34*, 312–314. [CrossRef]

38. Orsini, C.; Mitrović Dankulov, M.; Jamakovic, A.; Mahadevan, P.; Colomer-de-Simón, P.; Vahdat, A.; Bassler, K.E.; Torokzcai, Z.; Boguñá, M.; Caldarelli, G.; et al. Quantifying randomness in real networks. *Nat. Commun.* **2015**, *6*, 8627. [CrossRef] [PubMed]

39. Estrada, E.; José, A.; Hatano, N. Walk entropies in graphs. *Linear Algebra Appl.* **2014**, *443*, 235–244. [CrossRef]

40. Zenil, H.; Kiani, N.A.; Tegnér, J. Low-algorithmic-complexity entropy-deceiving graphs. *Phys. Rev. E* **2017**, *96*, 012308. [CrossRef]

41. Morzy, M.; Kajdanowicz, T.; Kazienko, P. On Measuring the Complexity of Networks: Kolmogorov Complexity versus Entropy. *Complexity* **2017**, *2017*, 3250301. [CrossRef]

42. Sengupta, D.C.; Sengupta, J.D. Application of graph entropy in CRISPR and repeats detection in DNA sequences. *Comput. Mol. Biosci.* **2016**, *6*, 41–51. [CrossRef]

43. Mowshowitz, A.; Dehmer, M. Entropy and the complexity of graphs revisited. *Entropy* **2012**, *14*, 559–570. [CrossRef]

44. Albregtsen, F. Statistical Texture Measures Computed from Gray Level Coocurrence Matrices. Available online: https://www.uio.no/studier/emner/matnat/ifi/INF4300/h08/undervisningsmateriale/glcm.pdf (accessed on 7 November 2018).

45. Silva, L.E.V.; Senra Filho, A.C.S.; Fazan, V.P.S.; Felipe, J.C.; Murta Junior, L.O. Two-dimensional sample entropy: Assessing image texture through irregularity. *Biomed. Phys. Eng. Express* **2016**, *2*, 045002. [CrossRef]

46. Rössler, O.E. An Equation for Continuous Chaos. *Phys. Lett. A* **1976**, *57*, 397–398. [CrossRef]

47. Mormann, F.; Lehnertz, K.; David, P.; Elger, C.E. Mean phase coherence as a measure for phase synchronization and its application to the EEG of epilepsy patients. *Physica D* **2000**, *144*, 358–369. [CrossRef]

48. Henon, M. A two-dimensional mapping with a strange attractor. *Commun. Math. Phys.* **1976**, *50*, 69–77. [CrossRef]

49. Schiff, S.J.; So, P.; Chang, T.; Burke, R.E.; Sauer, T. Detecting dynamical interdependence and generalized synchrony through mutual prediction in a neural ensemble. *Phys. Rev. E* **1996**, *54*, 6708–6723. [CrossRef]

50. Lorenz, E.N. Deterministic Nonperiodic Flow. *J. Atmos. Sci.* **1963**, *20*, 130–141. [CrossRef]

51. Ma, H.-C.; Chen, C.-C.; Chen, B.-W. Dynamics and transitions of the coupled Lorenz system. *Phys. Rev.* **1997**, *56*, 1550–1555. [CrossRef]

52. Hahs, D.W.; Pethel, S.D. Distinguishing Anticipation from Causality: Anticipatory Bias in the Estimation of Information. Flow. *Phys. Rev. Lett.* **2011**, *107*, 128701. [CrossRef] [PubMed]

53. Marwan, N.; Romano, M.C.; Thiel, M.; Kurths, J. Recurrence plots for the analysis of complex systems. *Phys. Rep.* **2007**, *438*, 237–329. [CrossRef]

54. Marwan, N.; Trauth, M.H.; Vuille, M.; Kurths, J. Comparing modern and Pleistocene ENSO-like influences in NW Argentina using nonlinear time series analysis methods. *Clim. Dynam.* **2003**, *21*, 317–326. [CrossRef]

55. Kodra, E.; Chatterjee, S.; Ganguly, A.R. Exploring Granger causality between global average observed time series of carbon dioxide and temperature. *Theor. Appl. Climatol.* **2011**, *104*, 325–335. [CrossRef]

56. Leggett, L.M.W.; Ball, D.A. Granger causality from changes in level of atmospheric CO2 to global surface temperature and the El Niño–Southern Oscillation, and a candidate mechanism in global photosynthesis. *Atmos. Chem. Phys.* **2015**, *15*, 11571–11592. [CrossRef]

57. Arjasakusuma, S.; Yamaguchi, Y.; Hirano, Y.; Zhou, X. ENSO- and Rainfall-Sensitive Vegetation Regions in Indonesia as Identified from Multi-Sensor Remote Sensing Data. *ISPRS Int. J. Geo-Inf.* **2018**, *7*, 103. [CrossRef]

58. Attanasio, A.; Pasini, A.; Triacca, U. Has natural variability a lagged influence on global temperature? A multi-horizon Granger causality analysis. *Dyn. Stat. Climate Syst.* **2016**, *1*, dzw002. [CrossRef]

59. Papagiannopoulou, C.; Miralles, D.G.; Decubber, S.; Demuzere, M.; Verhoest, N.E.C.; Dorigo, W.A.; Waegeman, W. A non-linear Granger-causality framework to investigate climate–vegetation dynamics. *Geosci. Model Dev.* **2017**, *10*, 1945–1960. [CrossRef]

60. Mokhov, I.; Smirnov, D.A.; Nakonechny, P.I.; Kozlenko, S.S.; Seleznev, V.; Kurths, J. Alternating mutual influence of El-Niño/Southern Oscillation and Indian monsoon. *Geophys. Res. Lett* **2011**, *38*, L00F04. [CrossRef]

61. Berkelhammer, M.; Sinha, A.; Mudelsee, M.; Cheng, H.; Yoshimura, K.; Biswas, J. On the low-frequency component of the ENSO–Indian monsoon relationship: A paired proxy perspective. *Clim. Past.* **2014**, *10*, 733–744. [CrossRef]

62. Detto, M.; Wright, S.J.; Calderón, O.; Muller-Landau, H.C. Resource acquisition and reproductive strategies of tropical forest in response to the El Niño–Southern Oscillation. *Nat. Commun.* **2018**, *9*, 913. [CrossRef] [PubMed]

63. Cane, M.A. Oceanograhic events during El Nino. *Science* **1983**, *222*, 1189–1195. [CrossRef] [PubMed]

64. Klein, S.A.; Soden, B.J.; Lau, N.C. Remote Sea Surface Temperature Variations during ENSO: Evidence for a Tropical Atmospheric Bridge. *J. Clim.* **1999**, *12*, 917–932. [CrossRef]

65. Webster, P.J.; Moore, A.M.; Loschnigg, J.P.; Leben, R.R. Coupled ocean-atmosphere dynamics in the Indian Ocean during 1997-98. *Nature* **1999**, *401*, 356–360. [CrossRef] [PubMed]

66. Saji, N.H.; Goswami, B.N.; Vinayachandran, P.N.; Yamagata, T. A dipole mode in the tropical Indian Ocean. *Nature* **1999**, *401*, 360–363. [CrossRef] [PubMed]

67. Fischer, A.S.; Terray, P.; Guilyardi, E.; Gualdi, S.; Delecluse, P. Two independent triggers for the Indian Ocean dipole zonal mode in a coupled GCM. *J. Clim.* **2005**, *18*, 3428–3449. [CrossRef]

68. Annamalai, H.; Xie, S.-P.; McCreary, J.P.; Murtugudde, R. Impact of Indian Ocean sea surface temperature on developing El Niño. *J. Clim.* **2005**, *18*, 302–319. [CrossRef]

69. Yuan, Y.; Li, C. Decadal variability of the IOD-ENSO relationship. *Chin. Sci. Bull.* **2008**, *53*, 1745–1752. [CrossRef]

70. Fan, L.; LIU, Q.; Wang, C.; Guo, F. Indian Ocean Dipole Modes Associated with Different Types of ENSO. *Development. J. Clim.* **2017**, *30*, 2223–2249. [CrossRef]

71. Trenberth, K.E.; Dai, A.; Rasmussen, R.M.; Parsons, D.B. The Changing Character of Precipitation. *Bull. Am. Meteor. Soc.* **2003**, *84*, 1205–1218. [CrossRef]

72. NOAA ESRL Physical Sciences Division Data. Available online: https://www.esrl.noaa.gov/psd/data/gridded/rsshelp.html (accessed on 23 October 2018).

73. Japan Agency for Marine-Earth Science and Technology (JAMSTEC). Available online: http://www.jamstec.go.jp/e/ (accessed on 23 October 2018).

74. Weinberger, D.; Baroux, N.; Grangeon, J.-P.; Ko, A.I.; Goarant, C. El Niño Southern Oscillation and leptospirosis outbreaks in New Caledonia. *PLoS Negl. Trop. Dis.* **2014**, *8*, e2798. [CrossRef] [PubMed]

75. Vincenti-Gonzalez, M.F.; Tami, A.; Lizarazo, E.F.; Grillet, M.E. ENSO-driven climate variability promotes periodic major outbreaks of dengue in Venezuela. *Sci. Rep.* **2018**, *8*, 5727. [CrossRef] [PubMed]

76. Hales, S.; Weinstein, P.; Souares, Y.; Woodward, A. El Nino and the Dynamics of Vector-Borne Disease Transmission. *EHP* **1999**, *107*, 99–102. [PubMed]

77. Pascual, M.; Rodo, X.; Ellner, S.P.; Colwell, R.; Bouma, M.J. Cholera Dynamics and El Niño Southern Oscillation. *Science* **2000**, *289*, 1766–1769. [CrossRef] [PubMed]

78. McGregor, G.R.; Ebi, K. El Niño Southern Oscillation (ENSO) and Health: An Overview for Climate and Health Researchers. *Atmosphere* **2018**, *9*, 282. [CrossRef]

79. An, S.; Wang, B. Interdecadal change of the structure of the ENSO mode and its impact on the ENSO frequency. *J. Clim.* **2000**, *13*, 2044–2055. [CrossRef]

80. Lowen, A.C.; Mubareka, S.; Steel, J.; Palese, P. Influenza virus transmission is dependent on relative humidity and temperature. *PLoS Pathog.* **2007**, *3*, e151. [CrossRef] [PubMed]

81. Oluwole, O.S.A. Dynamic regimes of El Niño southern Oscillation and influenza Pandemic Timing. *Front Public Health.* **2017**, *5*, 301. [CrossRef] [PubMed]

82. Ongena, J. Towards the realization on JET of an integrated H-mode scenario for ITER. *Nucl. Fusion* **2004**, *44*, 124–133. [CrossRef]

83. Romanelli, F. Overview of JET results. *Nucl. Fusion* **2009**, *49*, 104006. [CrossRef]

84. Kiptily, V.G.; Perez von Thun, C.P.; Pinches, S.D.; Sharapov, S.E.; Borba, D.; Cecil, F.E.; Darrow, D.; Goloborod'ko, V.; Craciunescu, T.; Johnson, T.; et al. Recent progress in fast ion studies on JET. *Nucl. Fusion* **2009**, *49*, 065030. [CrossRef]

85. Puiatti, M.E.; Mattioli, M.; Telesca, G.; Valisa, M.; Coffey, I.; Dumortier, P.; Giroud, C.; Ingesson, L.C.; Lawson, K.D.; Maddison, G.; et al. Radiation pattern and impurity transport in argon seeded ELMy H-mode discharges in JET. *Plasma Phys. Control. Fusion* **2002**, *44*, 1863–1878. [CrossRef]

entropy

MDPI

Article

The Relationship between Postural Stability and Lower-Limb Muscle Activity Using an Entropy-Based Similarity Index

Chien-Chih Wang [1],*, Bernard C. Jiang [2] and Pei-Min Huang [3]

[1]   Department of Industrial Engineering and Management, Ming Chi University of Technology,
     New Taipei City 243, Taiwan
[2]   Department of Industrial Management, National Taiwan University of Science and Technology,
     Taipei City 106, Taiwan; bcjiang@mail.ntust.edu.tw
[3]   Department of Industrial Engineering and Management, Yuan Ze University, Chung-Li 320, Taiwan;
     e06hpm@gmail.com
*   Correspondence: ieccwang@mail.mcut.edu.tw; Tel.: +886-2-2908-9899

Received: 28 March 2018; Accepted: 21 April 2018; Published: 26 April 2018

Abstract: The aim of this study is to see if the centre of pressure (COP) measurements on the postural stability can be used to represent the electromyography (EMG) measurement on the activity data of lower limb muscles. If so, the cost-effective COP data measurements can be used to indicate the level of postural stability and lower limb muscle activity. The Hilbert–Huang Transform method was used to analyse the data from the experimental designed to examine the correlation between lower-limb muscles and postural stability. We randomly selected 24 university students to participate in eight scenarios and simultaneously measured their COP and EMG signals during the experiments. The Empirical Mode Decomposition was used to identify the intrinsic-mode functions (IMF) that can distinguish between the COP and EMG at different states. Subsequently, similarity indices and synchronization analyses were used to calculate the correlation between the lower-limb muscle strength and the postural stability. The IMF5 of the COP signals and the IMF6 of the EMG signals were not significantly different and the average frequency was 0.8 Hz, with a range of 0–2 Hz. When the postural stability was poor, the COP and EMG had a high synchronization with index values within the range of 0.010–0.015. With good postural stability, the synchronization indices were between 0.006 and 0.080 and both exhibited low synchronization. The COP signals and the low frequency EMG signals were highly correlated. In conclusion, we demonstrated that the COP may provide enough information on postural stability without the EMG data.

Keywords: experiment of design; empirical mode decomposition; signal analysis; similarity indices; synchronization analysis

1. Introduction

Postural stability is a complex process of coordination of the body that resists gravity and involves multiple coordination activities such as biomechanics, sensations, and mobility. Postural stability control is a key skill that affects motion performance. Static postural stability positively correlates with age from 2 years of age to 12 years of age. However, after reaching middle age, the standing postural stability becomes inversely correlated with age. Maintaining a good postural stability is a key factor for maintaining a good quality of life while ageing. When abnormalities occur in the vision, the vestibular or somatosensory systems, which controls postural stability and muscle endurance, can help for maintaining good postural stability [1]. Porter et al. [2] highlighted that the muscle endurance of the somatosensory system gradually decreases with age. Chodzko-Zajko et al. [3] indicated that resistance training can improve the postural stability and gait in elderly people, prevent the loss of muscle strength, and prevent the decline in cardiovascular circulation due to ageing. Recently, young people have been exercising less due to lifestyle changes. This has resulted in a poor muscle endurance and postural stability, which increases the risk of exercise injury or falls. In previous studies, mostly elderly subjects were used to show improvements in the postural stability and muscle strength. These studies did not clearly demonstrate the relationship between improved muscle strength and improved postural stability. Moreover, very few studies have used young subjects to simultaneously measure the centre of pressure (COP) and electromyography (EMG), and evaluate the correlation between postural stability and lower-limb muscles. Numerous noise sources may contaminate the EMG signal measurements and distort the signal, which can lead to interpretation errors of the EMG signal for investigating muscle activity. The variation in the COP measurements is relatively small. Therefore, entropy-based analyses for the application of COP may be enlarged if the relationship between the COP and EMG can be demonstrated.

The COP is the trajectory of the pressure centre acquired as a function of time as the body weight is transmitted to the ground when both feet are standing on the ground, which is important information for studying the postural stability and falls. There are some differences in the postural stability between young and elderly people. Hatton et al. [4] used conventional indicators of COP quantitation to examine patients who underwent cruciate-ligament reconstruction and found that there were no significant differences in the postural stability before and after surgery. Ozaki et al. [5] studied frail, elderly people and found that training with a newly-proposed, postural-stability-exercise-assist robot can improve postural stability; moreover, lower-limb muscle strength may improve faster using this technique compared to the conventional methods. Coelho et al. [6] studied the relationship between age and quality of standing balance in single and dual task conditions for the community-dwelling elderly. The results showed that the elderly were consistently associated with poor standing balance in single and dual task conditions. Bergamin et al. [7] studied the posture swaying in the various dual-task conditions of young people and old people. The results indicated that a simple verbal assignment was the secondary task which most influenced the postural balance and a dual-task condition seems to differently affect the balance variables, independently from age. The EMG measures weak potential changes produced during muscle contraction. The movement sequence of the agonist and antagonist muscles in a specific action can be revealed via EMG. Hägg et al. [8] highlighted that the EMG can be quantified to be a factor of the magnitude of force exerted and the level of fatigue. Muscles and bones play important roles in good postural stability. Müller et al. [9] studied the relationship between postural stability and lower-limb muscles in young and elderly people when standing using onset latency. The results demonstrated that the correlation coefficients between the tibialis anterior muscle and the anteroposterior direction in young and elderly people were 0.667 and 0.482, respectively. The correlation coefficients between the anteroposterior direction and the soleus and gastrocnemius muscles were less than 0.3. Borg et al. [10] studied the correlation between the postural stability and the gastrocnemius muscles in healthy people and patients with multiple sclerosis. He converted EMG into envelopes to calculate the correlation coefficient with anteroposterior COP. The results demonstrated that the correlation coefficient was 0.85, implying that gastrocnemius muscles are associated with the anteroposterior shaking of the body.

Previous studies utilised Fast Fourier Transform (FFT) for signal analysis. However, the FFT approach assumes a steady-state signal and linearity, which contradicts the signal characteristics of COP and EMG. Huang and Shen [11] proposed utilising the Hilbert–Huang Transform (HHT) for non-steady-state conditions and non-linear data, which can resolve the frequency range relevant for physical measurements. Amound et al. [12] adopted HHT for analysis and the results demonstrated that the frequency range from IMF1 to IMF5 can be used to identify the complex plane within a large area when the force plate vibrates. Andrade et al. [13] demonstrated that HHT can decrease the noise in EMG signals and that it performs better than wavelet transformation. Xie and Wang [14] calculated the mean EMG frequency during muscle fatigue, indicating that the HHT-estimated results have less variability and higher stability.

This study aims to see if the centre of pressure (COP) measurement on postural stability can be used to represent the electromyography (EMG) measurement of lower limb muscles activity data. If so, the cost-effective COP data measurement can be used to indicate the level of postural stability and lower limb muscles activity. Several experimental scenarios were designed for examining the effects of lower-limb muscles on postural stability in healthy, young people under dynamic and static exertion. An AMTI force plate and a NeXus-10 wireless physiological feedback system were used to collect the COP and EMG signals, respectively. HHT and Empirical Mode Decomposition (EMD) were used to identify the range of frequency characteristics under different states. Similarity indices and synchronization analyses were used to determine the correlation between lower-limb muscle strength and postural stability.

2. Methodology

The young people were used as experimental subjects, experiments were designed, and data analysis methods were proposed. Eight experimental scenarios were designed according to vision (open or closed eyes), posture (static or dynamic), and standing status (presence or absence of soft foam). An AMTI force plate and a NeXus-10 physiological feedback system were used to measure variations in postural stability and lower-limb muscles. Subsequently, EMD was utilised to decompose the COP and EMG signals into different IMF signals. Further, synchronization analyses were used to identify highly correlated COP and EMG data.

2.1. Test Subjects

The test subjects comprised of 12 healthy male and 12 healthy female undergraduate and graduate students with no mobility disorders. The eligibility criteria do not include the health or physical limitations that may affect the results of the study. The exclusion criteria included persons who had musculoskeletal or foot disorders or had undergone foot surgery or other invasive foot treatment procedures or mental illnesses. All participants provided their personal information before commencing the experiment. The average weight of males and females was 65 ± 7.53 kg and 45.7 ± 4.03 kg, respectively. Prior to the experiment, the subjects were informed about the experiment and experimental procedures in detail and the informed consent forms were signed.

2.2. Experimental Instrumentation

An AMTI force plate and a NeXus-10 physiological feedback system were used to collect the COP and EMG data, respectively (Figure 1a,b). The acquisition frequency of the force plate was 100 Hz. The acquisition frequency and measurement time of the wireless physiological feedback monitor were 1024 Hz and 60 s, respectively.

(a)   (b)

Figure 1. The experimental instruments. (a) AMTI force plate (AMTI®, Watertown, MA, USA) and (b) NeXus-10 physiological monitoring system (Mind Media®, BV, Herten, The Netherlands).

2.3. Experimental Procedures

A randomization method was used to arrange the experimental subjects. The rectus femoris (RF), vastus lateralis (VL), tibialis anterior (TA), and gastrocnemius medialis (GM) muscles were used as representatives of the lower-limb muscles. In the experimental procedure, each subject was swabbed with ethanol at the test muscle sites and electrode pads were attached (Figure 2). Subsequently, the maximum-voluntary-isometric contraction (MVC) was measured before the eight experimental scenarios were conducted in a random order with simultaneous COP and EMG measurements. In the study, a two-level experiment with three factors that includes vision (open or closed eyes), posture (static or dynamic) and standing status (presence or absence of soft foam) was designed with 2^3 scenarios to collect data. The static experimental scenario refers to the experiments where the subjects were standing still (Figure 3a). The dynamic experimental scenario refers to the process whereby two instructions were randomly given to the subject: (1) squat to a half-squat position whilst keeping their heels on the ground and then (2) automatically rise up (Figure 3b). The soft foam was placed on the force plate and the subjects were asked to stand on it (Figure 3c). Each experiment was conducted for 60 s with a 3 min rest interval between experiments in order to recover and to avoid muscle fatigue [15].

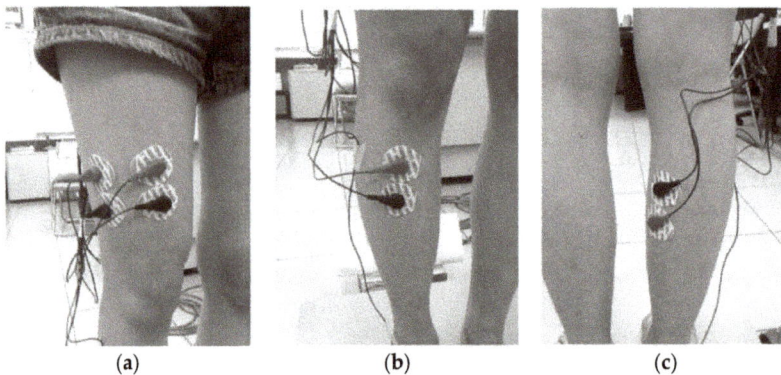

(a)   (b)   (c)

Figure 2. The electrode pads placed on different muscles: (a) rectus femoris and vastus lateralis; (b) tibialis anterior, and (c) gastrocnemius medialis.

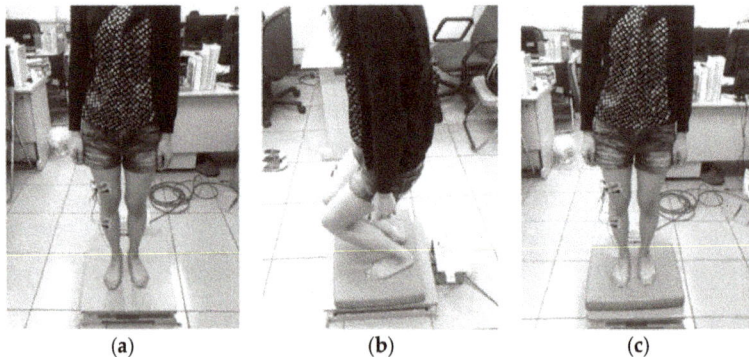

(a)   (b)   (c)

Figure 3. The experimental scenarios: (a) static experimental; (b) dynamic experimental; and (c) soft foam placed on the force plate.

The MVC measurements were used as a standard reference to decrease muscle differences between different individuals. During the MVC measurements of the rectus femoris and vastus lateralis muscles, the subject adopted a sitting posture on the table, with the calf closely pressed against the edge of the table and both hands grabbing the table edge to fix the position of the upper trunk. The kneecap extends to resist the external force from the ankle for 5 s continuously (Figure 4a). During the MVC measurements of tibialis anterior muscles, the subject adopted a single-leg standing position with both knees fully extended. The maximum dorsiflexion force was continuously exerted on the external force for 5 s using one leg as a support (Figure 4b). During MVC measurements of the medial-gastrocnemius muscles, the subjects held a two-leg standing position with both feet exerting upward forces to act as a plantar flexion force, which continuously resists the external forces from the shoulder for 5 s (Figure 4c).

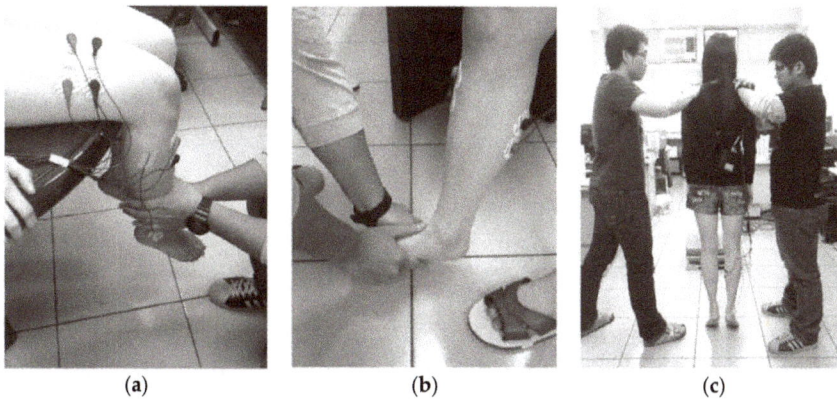

Figure 4. The measurement of maximum-voluntary-isometric contractions for (**a**) rectus femoris and vastus lateralis; (**b**) tibialis anterior; and (**c**) gastrocnemius medialis.

2.4. Analysis Algorithms

A two-stage procedure was used to analyse the COP and EMG signals. Firstly, the EMD was used to decompose the COP and EMG signals into several IMFs and two IMFs, and similar signal frequencies with COP and EMG were identified. During the second stage, the IMFs obtained from COP and EMG was used for correlation analyses to determine the relationship between postural stability and changes in the lower-limb muscles.

2.4.1. Empirical Mode Decomposition

The greatest difference between HHT and wavelet transformations is the absence of basic functions. The HHT uses decomposition rules to replace basis functions and EMD is used to decompose signals into summations of IMFs. Subsequently, the Hilbert Transform method is used to obtain meaningful instantaneous frequencies. This is advantageous since it can analyse non-linear and non-steady-state signals and present the original physical characteristics of the signals. EMD and Hilbert Transform are the main procedures for HHT [11]. Figure 5 depicts the HHT analyses procedures.

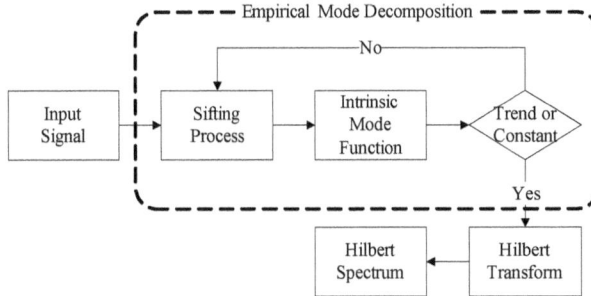

Figure 5. A flowchart of the Hilbert–Huang Transform analysis.

EMD utilises the characteristic time scales in signals to define the vibrational mode and provide good vibrational mode analyses. A non-zero mean signal can also be used, and the decomposition procedure is termed sifting process [16–18], which is as follows:

Step 1. Identify the local maxima and local minima of the signal $X(t)$ and use cubic splines to define the local maxima as the upper envelope and the local minima as the lower envelope. The mean of the upper and lower envelopes is taken as the mean envelope $m_1(t)$.

Step 2. The mean envelope $m_1(t)$ is subtracted from the signal $X(t)$ to obtain the component $h_1(t)$.

$$h_1(t) = X(t) - m_1(t)$$

Step 3. Validate whether $h_1(t)$ fulfils the IMF conditions. The stopping criterion is that the difference between the number of extreme points and the number of cross-zero points must be less than or equal to one, and the mean envelope must be zero. If the stopping criterion is not fulfilled, return to Step 1 to use $h_1(t)$ as $X(t)$ for the second sifting process to obtain $h_{11}(t)$.

$$h_{11}(t) = h_1(t) - m_{11}(t)$$

Step 4. The excessive screening will result in the loss of original physical characteristics from the results. Therefore, the sifting process must obey the convergence criteria to ensure that the physical characteristics of IMF are maintained. The convergence criteria are that the difference between the number of local extreme points and cross-zero points must be zero and the standard deviation should lie between 0.2 and 0.3. The formula is given as

$$SD = \sum_{t=0}^{T} \left[\frac{\left(h_{1(k-1)}(t) - h_{1k}(t) \right)}{h_{1(k-1)}} \right]^2$$

Step 5. After the k sifting processes, if the IMF stopping or convergence criteria are met, then component $h_{1k}(t)$ can be taken to be the first IMF and is represented by $c_1(t)$. $c_1(t)$ as the component with the shortest cycle in $X(t)$.

$$h_{1k}(t) = h_{1(k-1)}(t) - m_{1k}(t)$$

$$c_1(t) = h_{1k}(t)$$

Step 6. $c_1(t)$ is subtracted from the signal $X(t)$ to obtain the residual function $r_1(t)$. If $r_1(t)$ contains other components with longer cycles, then Steps 1–5 are repeated for $r_1(t)$ until n number of IMFs are selected.

$$X(t) - c_1(t) = r_1(t)$$

$$r_1(t) - c_2(t) = r_2(t)$$

$$\vdots$$

$$r_{n-1}(t) - c_n(t) = r_n(t)$$

Step 7. When IMFs with physical characteristics cannot be obtained from the residual function $r_n(t)$, the sifting process is stopped. The last $r_n(t)$ obtained is used as the mean trend. The summation of every IMF and the mean trend can be used to reconstruct the initial signal $X(t)$.

$$X(t) = \sum_{k=1}^{n} c_k(t) + r_n(t)$$

After COP and EMG signals were decomposed by EMD into several IMFs, Hilbert Transform was used to calculate the instantaneous frequency of every IMF, using the following formula:

$$Y(t) = \frac{1}{\pi} PV \int_{-\infty}^{\infty} \frac{X(\tau)}{t - \tau} d\tau$$

where Hilbert transform $Y(t)$ can be obtained for any time series $X(t)$ and PV represents Cauchy Principal Value.

Every $X(t)$ that undergoes Hilbert transform becomes as follows

$$X(t) = \sum_{j=1}^{n} a_j(t) \exp\left(i \int w_j(t) dt \right)$$

2.4.2. Similarity Index and Synchronization Analysis

This study uses the similarity index and synchronization analyses to investigate the correlation between the COP and EMG [19–21]. The similarity index is used to evaluate the similarity between two sets of signal fluctuation models and is obtained as follows:

Step 1. Decimal signals are converted to binary signals using the following formula:

$$I_n = \begin{cases} 0, & if\ x_n \leq x_{n-1} \\ 1, & if\ x_n > x_{n-1} \end{cases}$$

In Figure 6, the first data point is greater than the second data point and the interval between the two is coded to equal zero. The same applies to the second interval. The third data point is smaller than the fourth data point, and the third interval is coded to equal one, and so on and so forth.

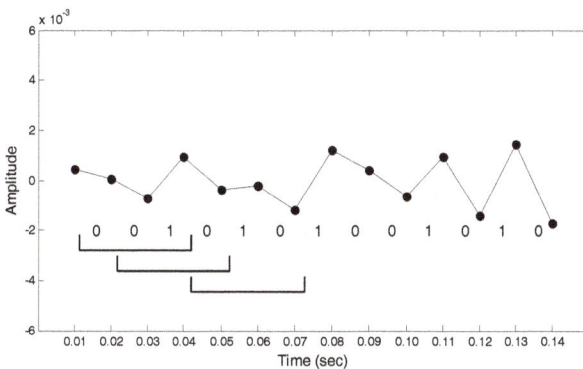

Figure 6. The original signal coding diagram for Steps 1 and 2.

Step 2. The parameter *m* is set as the size of the moving window for use as the basis for the binary conversion to decimal. When *m* is set as 4, using the first moving window in Figure 6 as an example, the binary time-series of the window was 0010 and converted to decimal as 2. The binary time series of the second moving window was 0101 and converted to decimal as 5, and so on and so forth, to convert the binary time-series in the entire set into decimals.

Step 3. As *m* was set to be 4, the numbers 0 to 15 will appear in the binary to decimal conversion. The decimal time-series after conversion are tallied, the probabilities for the appearance of the numbers 0 to 15 are calculated, these are plotted as a histogram and the probabilities are ranked from largest to smallest. Figure 7 depicts the COP-ML (centre of pressure-medial lateral) signal results and the probability ranking of the numeral 0 was 2.

Step 4. Steps 1–3 are repeated for another set of signals. Figure 8 depicts the results for the EMG-RF (electromyography-rectus femoris) muscles.

Step 5. Scatterplots are generated for the numeral ranking of two sets of signals that are to be compared. Figure 9 illustrates the results of COP-ML and EMG for the RF muscles. If the points in the scatterplot are more concentrated at the 45-degree diagonal dotted line, the similarity in the fluctuation methods of both sets of signals is high.

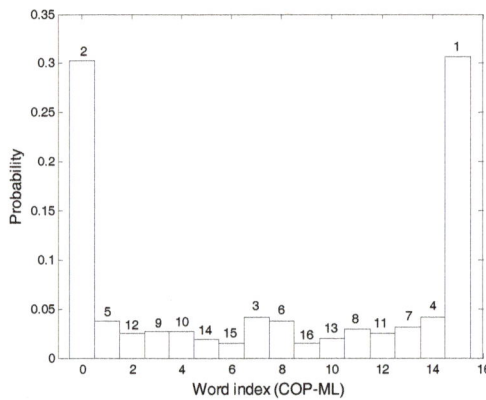

Figure 7. The example COP-ML (centre of pressure-medial lateral) signal results for Step 3.

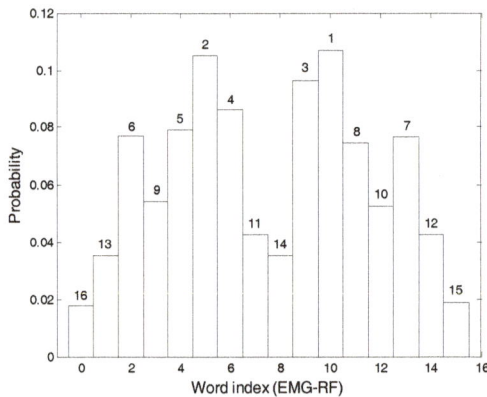

Figure 8. The example EMG-RF (electromyography-rectus femoris) signal results for Step 4.

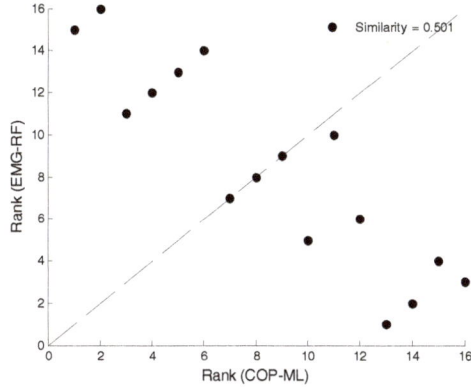

Figure 9. The scatterplot between Rank (EMG-RF) and Rank (COP-ML).

Step 6. The similarity index $D_m(\text{COP}, \text{EMG})$ of COP and EMG is

$$D_m(\text{COP}, \text{EMG}) = 1 - \frac{\sum_{k=1}^{2^m} |R_1(w_k) - R_2(w_k)| F(w_k)}{2^m - 1}$$

$$F(w_k) = \frac{(-p_1(w_k) \log p_1(w_k) - p_2(w_k) \log p_2(w_k))}{\sum_{k=1}^{2^m} (-p_1(w_k) \log p_1(w_k) - p_2(w_k) \log p_2(w_k))}$$

If $D_m(\text{COP}, \text{EMG})$ approaches one, the two sets of signals are similar. $R_1(w_k)$ represents the ranking of the k-th numeral of the first signal set and $p_1(w_k)$ is the probability of the k-th numeral appearing in the first signal set.

Synchronization indices are used to evaluate the degree of phase coordination between two sets of signals [22,23]. In this study, synchronization indices were used to evaluate the correlation of COP and EMG. The calculation of synchronization indices is based on the instantaneous phase and Shannon entropy and the process is as follows:

Step 1. The $Y(t)$ obtained from the Hilbert Transform is expressed as complex numbers, $Y(t) = a(t) + ib(t)$. The instantaneous phase $\theta(t)$ is calculated from the real part $a(t)$ and the imaginary part $b(t)$.

$$\theta(t) = \tan^{-1}\left(\frac{b(t)}{a(t)}\right)$$

Step 2. The instantaneous phases obtained from the two sets of signals are subtracted from each other to obtain the phase difference.

$$\text{phase difference} = \theta_{signal1}(t) - \theta_{signal2}(t)$$

Step 3. The absolute value of the phase difference is obtained to plot the histogram of N number of intervals.

$$\text{phase difference}(t) = |\text{phase difference}(t)|$$

$N = e^{0.626 + 0.4 \times \ln(n-1)}$, n is the total length of the data.

Step 4. The Shannon entropy of the histogram is calculated.

$$\text{Shannon entropy} = \sum_{i=1}^{N} p(i) \times \ln(p(i))$$

p is the probability of the occurrence for that interval.

Step 5. The synchronization indices are calculated.

$$\text{Synchronization index} = 1 - \frac{\text{Shannon entropy}}{\ln(N)}$$

3. Experimental Results

Firstly, the sampling was set to 100 Hz and experiments were designed to simultaneously measure the COP and EMG data. Subsequently, the EMD was used to decompose the COP and EMG signals into several IMFs and the average frequency of each IMF was calculated. The IMFs with similar COP and EMG signal-frequency-vibration modes were identified for synchronization analyses. Using data from closed eyes, standing still, and presence of soft foam as examples, Table 1 depicts the mean frequencies of each intrinsic-mode function in the left-right direction and anterior-posterior direction COP and the EMG of the RF, VL, TA, and GM. Statistical tests found that the IMF5 frequency of the COP signal and the IMF6 frequency of the EMG signal were not significantly different. Figure 10 depicts the frequency distribution of COP_{IMF5} and EMG_{IMF6}. The frequency distribution of the two sets of signals were similar; the average frequency was 0.8 Hz with a range of 0–2 Hz.

Table 1. The statistical test results for each IMF of the left-right direction and anterior-posterior direction COP, and the EMG of the RF, VL, TA and GM.

	COP			EMG				p-Value			
	ML		RF	VL	TA	GM	ML-RF	ML-VL	ML-TA	ML-GM
IMF1	31.59 ± 0.25	IMF1	32.45 ± 1.00	32.73 ± 1.02	32.47 ± 0.77	32.75 ± 0.62	0.00	0.00	0.00	0.00
IMF2	15.02 ± 0.65	IMF2	15.69 ± 0.38	15.91 ± 0.44	15.75 ± 0.34	15.74 ± 0.32	0.00	0.00	0.00	0.00
IMF3	6.59 ± 0.59	IMF3	7.90 ± 0.32	7.99 ± 0.36	7.99 ± 0.26	8.08 ± 0.21	0.00	0.00	0.00	0.00
IMF4	2.28 ± 0.39	IMF5	1.82 ± 0.14	1.88 ± 0.16	1.88 ± 0.17	1.92 ± 0.23	0.00	0.00	0.00	0.00
IMF5	0.81 ± 0.12	IMF6	0.80 ± 0.12	0.80 ± 0.13	0.81 ± 0.16	0.88 ± 0.14	0.77	0.87	0.97	0.07
IMF6	0.38 ± 0.05	IMF7	0.34 ± 0.07	0.32 ± 0.07	0.34 ± 0.08	0.38 ± 0.09	0.02	0.00	0.05	0.96
	AP		RF	VL	TA	GM	AP-RF	AP-VL	AP-TA	AP-GM
IMF1	31.46 ± 0.31	IMF1	32.45 ± 1.00	32.73 ± 1.02	32.47 ± 0.77	32.75 ± 0.62	0.00	0.00	0.00	0.00
IMF2	15.24 ± 0.56	IMF2	15.69 ± 0.38	15.91 ± 0.44	15.75 ± 0.34	15.74 ± 0.32	0.00	0.00	0.00	0.00
IMF3	6.12 ± 1.09	IMF3	7.90 ± 0.32	7.99 ± 0.36	7.99 ± 0.26	8.08 ± 0.21	0.00	0.00	0.00	0.00
IMF4	2.26 ± 0.33	IMF5	1.82 ± 0.14	1.88 ± 0.16	1.88 ± 0.17	1.92 ± 0.23	0.00	0.00	0.00	0.00
IMF5	0.87 ± 0.14	IMF6	0.80 ± 0.12	0.80 ± 0.13	0.81 ± 0.16	0.88 ± 0.14	0.07	0.09	0.19	0.75
IMF6	0.40 ± 0.04	IMF7	0.34 ± 0.07	0.32 ± 0.07	0.34 ± 0.08	0.38 ± 0.09	0.00	0.00	0.00	0.29

Note: The bold text indicates statistical significance (p-value < 0.05), where ML-RF indicates ML and RF were used to perform statistical hypothesis tests. IMF: intrinsic-mode functions; COP: centre of pressure; EMG: electromyography; RF: rectus femoris; VL: vastus lateralis; TA: tibialis anterior; GM: gastrocnemius medialis; ML: medial-lateral; AP: anterior-posterior.

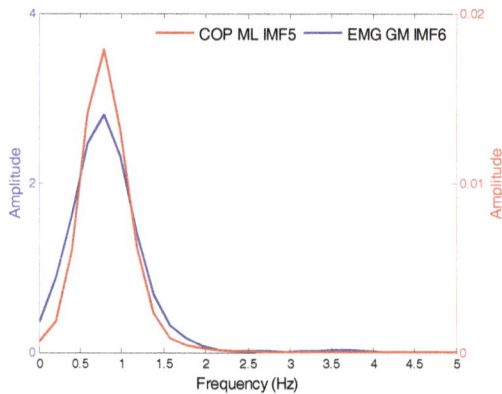

Figure 10. The COP and EMG frequency distribution for the experimental scenarios with eyes closed, standing still, and with soft foam. IMF: intrinsic-mode functions; EMG: Electromyography.

The similarity index was then used to compare the differences between the COP_{IMF5} and EMG_{IMF6} signals, and the original COP and EMG signals. Under the conditions of closed eyes, standing still, and presence of soft foam, Figure 11a,b depicts the raw-signal graph of the anterior-posterior COP the EMG of the medial-gastrocnemius muscles and the similarity index, which was 0.494. Figure 12a,b depicts that the signal graphs of IMF5 of anterior-posterior COP, the IMF6 of the EMG of medial-gastrocnemius muscles and the similarity index, which was 0.956. The fluctuation modes of the two sets of signals are extremely similar and the similarity index was higher relative to the raw-signal (0.462).

For the four muscles, the similarity between the raw signals and decomposed signals of the left-right direction and anterior-posterior direction, respectively, the COP and the EMG data were compared. The black histogram in Figure 13 depicts the similarity between the two sets of signals, and the white histogram depicts the similarity between the two sets of signals after decomposition. ML-RF is the similarity between the left-right COP and EMG of rectus femoris muscles. The * symbol indicates that there are statistically-significant differences between the two signals. The mean similarity index of the raw signals was 0.5–0.6. The similarity index between the IMF5 of the COP signal and the IMF6 of the EMG signal was increased to greater than 0.95. The COP_{IMF5} and EMG_{IMF6} obtained from the statistical tests and the similarity index was used to replace the raw signals for subsequent analysis.

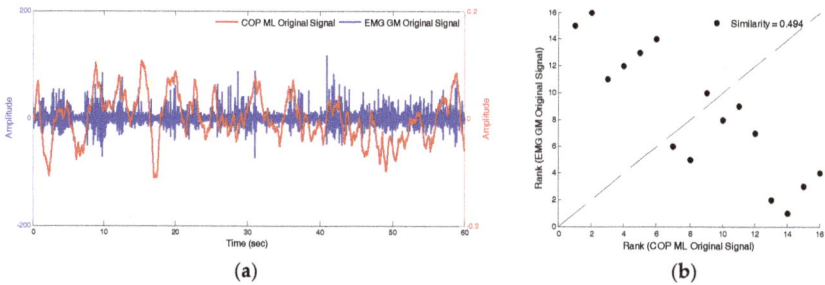

Figure 11. The raw-signal of the anterior-posterior COP and the EMG of MG muscles and the similarity index.

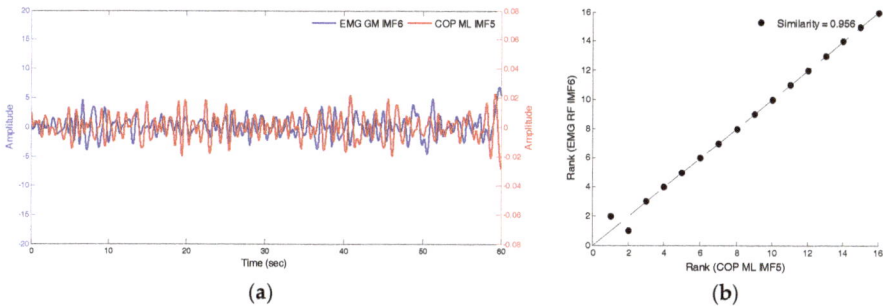

Figure 12. The signal functions of IMF5 of anterior-posterior COP and IMF6 of the EMG of MG muscles and the similarity index.

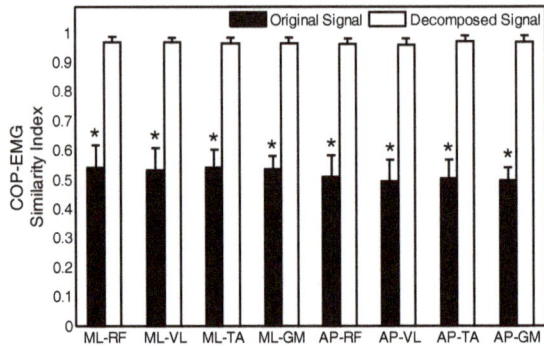

Figure 13. The comparative analyses of the similarity index between the original and decomposed signals. RF: rectus femoris; VL: vastus lateralis; TA: tibialis anterior; GM: gastrocnemius medialis; ML: medial-lateral; AP: anterior-posterior.

4. Discussion and Conclusions

In this study, different scenarios were designed to simultaneously collect COP and lower-limb-muscle EMG data to investigate the correlation between these two sets of parameters. The Empirical Mode Decomposition was used to identify the intrinsic-mode functions (IMF) that can distinguish between COP and EMG at different states. Alickovic et al. [18] studied seizure onset detection and seizure onset prediction using EEG signals by the EMD, DWT (discrete wavelet transform), and WPD (wavelet packet decomposition). The results indicated that DWT or WPD is better than EMD for feature extraction. This study applied the empirical mode decomposition that has been demonstrated that COP may provide enough information on postural stability without EMG data. Therefore, there is no comparison with other methods.

The experimental analyses found that the IMFs that are similar for the COP and EMG signals are IMF5 and IMF6, respectively, and the average frequency was 0.8 Hz. The similarity index between the two was 0.95 to 1. This is higher than the similarity between the raw signals, which ranged from 0.5 to 0.6, and the difference was statistically significant. This is consistent with the results of Berg who highlighted that EMGs with frequencies less than 1 Hz were correlated with COP signals [23]. By analysing the synchronization indices, it was found that when postural stability is poor, EMG and COP were more synchronised with synchronisation indices approximately ranging from 0.010 to 0.015. With better postural stability, synchronization indices ranged from approximately 0.006 to 0.08 and both EMG and COP exhibited lower synchronization. Postural stability is relatively unstable when there is a greater synchronization between lower-limb muscle and postural stability. Conversely, when synchronization between lower-limb muscle and postural stability is low, postural stability is in a relatively stable state.

The results of this study and those from Manor et al. [24] are consistent. Manor et al. [24] remarked that there is an inverse correlation between the synchronization of respiratory flow and postural stability and postural stability. Conversely, lower synchronization leads to improved postural stability. Manor et al. [24] also noted that the synchronization index between respiratory flow and COP in normal people and stroke patients were approximately 0.11 and 0.25, respectively. This difference is greater than the synchronization index between EMG and COP, which ranged from 0.006 to 0.015. This is mainly because for muscles that affect postural stability, the same muscle does not continuously affect postural stability to maintain postural stability. Therefore, postural stability will possibly be affected by RF in the initial 10 s, and then the TA on the right leg will affect postural stability in the subsequent 10 s. This is different from breathing, which continuously affects postural stability. Therefore, the values from synchronization analyses in this study are less than or equal to respiratory flow and COP.

No fatigue was observed in the experimental data, which was used for joint analyses of the EMG spectrum and application, for any lower-limb muscle when subjects were standing for 1 min. In future research, increasing the duration that the subjects stand will be used to investigate how lower-limb-muscle fatigue affects the maintenance of postural stability. It is expected that the synchronization indices will increase when muscles are fatigued.

Author Contributions: B.C.J., P.-M.H. and C.-C.W. designed the study. P.-M.H. was responsible for data collection and analysis. P.-M.H., C.-C.W. and B.C.J. reviewed relevant literature and interpreted the acquired data. C.-C.W. drafted the manuscript. All authors have read and approved the final manuscript.

Funding: This research was funded by [the Ministry of Science and Technology, Taiwan] grant number [104-2221-E-131 -005 -MY3].

Conflicts of Interest: The authors declare no conflict of interest.

References

1. Lepers, R.; Bigard, A.X.; Diard, J.P.; Gouteyron, J.F.; Guezennec, C.Y. Posture control after prolonged exercise. *Eur. J. Appl. Physiol. Occup. Physiol.* **1997**, *76*, 55–61. [CrossRef] [PubMed]
2. Porter, M.M.; Vandervoort, A.A.; Lexell, J. Aging of human muscle: Structure, function and adaptability. *Scand. J. Med. Sci. Sports* **1995**, *5*, 129–142. [CrossRef] [PubMed]
3. Chodzko-Zajko, W.J.; Proctor, D.N.; Singh, M.A.F.; Minson, C.T.; Nigg, C.R.; Salem, G.J.; Skinner, J.S. Exercise and physical activity for older adults. *Med. Sci. Sports Exerc.* **2009**, *41*, 1510–1530. [CrossRef] [PubMed]
4. Hatton, A.L.; Crossley, K.M.; Clark, R.A.; Whitehead, T.S.; Morris, H.G.; Culvenor, A.G. Between-leg differences in challenging single-limb balance performance one year following anterior cruciate ligament reconstruction. *Gait Posture* **2017**, *52*, 22–25. [CrossRef] [PubMed]
5. Ozaki, K.; Kondo, I.; Hirano, S.; Kagaya, H.; Saitoh, E.; Osawa, A.; Fujinori, Y. Training with a balance exercise assist robot is more effective than conventional training for frail older adults. *Geriatr. Gerontol. Int.* **2017**, *17*, 1982–1990. [CrossRef] [PubMed]
6. Coelho, T.; Fernandes, Â.; Santos, R.; Paúl, C.; Fernandes, L. Quality of standing balance in community-dwelling elderly: Age-related differences in single and dual task conditions. *Arch. Gerontol. Geriatr.* **2016**, *67*, 34–39. [CrossRef] [PubMed]
7. Bergamin, M.; Gobbo, S.; Zanotto, T.; Sieverdes, J.C.; Alberton, C.L.; Zaccaria, M.; Ermolao, A. Influence of age on postural sway during different dual-task conditions. *Front. Aging Neurosci.* **2014**, *6*, 271. [CrossRef] [PubMed]
8. Hägg, G.M.; Luttmann, A.; Jäger, M. Methodologies for evaluating electromyographic field data in ergonomics. *J. Electromyogr. Kinesiol.* **2000**, *10*, 301–312. [CrossRef]
9. Müller, M.L.; Redfern, M.S. Correlation between EMG and COP onset latency in response to a horizontal platform translation. *J. Biomech.* **2004**, *37*, 1573–1581. [CrossRef] [PubMed]
10. Borg, F.; Finell, M.; Hakala, I.; Herrala, M. Analyzing gastrocnemius EMG-activity and sway data from quiet and perturbed standing. *J. Electromyogr. Kinesiol.* **2007**, *17*, 622–634. [CrossRef] [PubMed]
11. Huang, N.E.; Shen, S.S. *Hilbert-Huang Transform and Its Applications*; World Scientific: Singapore, 2014.
12. Amoud, H.; Snoussi, H.; Hewson, D.J.; Duchêne, J. Hilbert-Huang transformation: Application to postural stability analysis. In Proceedings of the Conference of the IEEE Engineering Medicine Biology Society, Lyon, France, 22–26 August 2007; pp. 1562–1565.
13. Andrade, A.O.; Nasuto, S.; Kyberd, P.; Sweeney-Reed, C.M.; Van Kanijn, F.R. EMG signal filtering based on empirical mode decomposition. *Biomed. Signal Process. Control* **2006**, *1*, 44–55. [CrossRef]
14. Xie, H.; Wang, Z. Mean frequency derived via Hilbert-Huang transform with application to fatigue EMG signal analysis. *Comput. Methods Programs Biomed.* **2006**, *82*, 114–120. [CrossRef] [PubMed]
15. Wang, C.C.; Jiang, B.C.; Lin, W.C. Evaluation of effects of balance training from using wobble board-based exergaming system by MSE and MMSE techniques. *J. Ambient Intell. Humaniz. Comput.* **2017**, 1–10. [CrossRef]
16. Pachori, R.B.; Avinash, P.; Shashank, K.; Sharma, R.; Acharya, U.R. Application of empirical mode decomposition for analysis of normal and diabetic RR-interval signals. *Expert Syst. Appl.* **2015**, *42*, 4567–4581. [CrossRef]

Entropy **2018**, *20*, 320

17. Safi, K.; Mohammed, S.; Albertsen, I.M.; Delechelle, E.; Amirat, Y.; Khalil, M.; Gracies, J.M.; Hutin, E. Automatic analysis of human posture equilibrium using empirical mode decomposition. *Signal Image Video Process.* **2017**, *11*, 1081–1088. [CrossRef]

18. Alickovic, E.; Kevric, J.; Subasi, A. Performance evaluation of empirical mode decomposition, discrete wavelet transform, and wavelet packed decomposition for automated epileptic seizure detection and prediction. *Biomed. Signal Process. Control* **2018**, *39*, 94–102. [CrossRef]

19. Cui, X.; Chang, E.; Yang, W.H.; Jiang, B.C.; Yang, A.C.; Peng, C.K. Automated Detection of Paroxysmal Atrial Fibrillation Using an Information-Based Similarity Approach. *Entropy* **2017**, *19*, 677. [CrossRef]

20. Yang, A.C.; Goldberger, A.L.; Peng, C.K. Genomic classification using an information-based similarity index: Application to the SARS coronavirus. *J. Comput. Biol.* **2005**, *12*, 1103–1116. [PubMed]

21. Peng, C.K.; Yang, A.C.C.; Goldberger, A.L. Statistical physics approach to categorize biologic signals: From heart rate dynamics to DNA sequences. *Chaos* **2007**, *17*, 015115. [CrossRef] [PubMed]

22. Holmes, M.L.; Manor, B.; Hsieh, W.H.; Hu, K.; Lipsitz, L.A.; Li, L. Tai Chi training reduced coupling between respiration and postural control. *Neurosci. Lett.* **2016**, *610*, 60–65. [CrossRef] [PubMed]

23. Berg, K.O.; Maki, B.E.; Williams, J.I.; Holliday, P.J.; Wood-Dauphinee, S.L. Clinical and laboratory measures of postural balance in an elderly population. *Arch. Phys. Med. Rehabilit.* **1992**, *73*, 1073–1080.

24. Manor, B.D.; Hu, K.; Peng, C.K.; Lipsitz, L.A.; Novak, V. Posturo-respiratory synchronization: Effects of aging and stroke. *Gait Posture* **2012**, *36*, 254–259. [CrossRef] [PubMed]

entropy

MDPI

Article

Vague Entropy Measure for Complex Vague Soft Sets

Ganeshsree Selvachandran [1], Harish Garg [2,*] and Shio Gai Quek [3]

[1] Department of Actuarial Science and Applied Statistics, Faculty of Business & Information Science,
    UCSI University, Jalan Menara Gading, Cheras 56000, Kuala Lumpur, Malaysia;
    Ganeshsree@ucsiuniversity.edu.my
[2] School of Mathematics, Thapar Institute of Engineering & Technology (Deemed University), Patiala 147004,
    Punjab, India
[3] A-Level Academy, UCSI College KL Campus, Lot 12734, Jalan Choo Lip Kung, Taman Taynton View,
    Cheras 56000, Kuala Lumpur, Malaysia; queksg@ucsicollege.edu.my
*   Correspondence: harishg58iitr@gmail.com; Tel.: +91-86-9903-1147

Received: 12 March 2018; Accepted: 17 May 2018; Published: 24 May 2018

Abstract: The complex vague soft set (CVSS) model is a hybrid of complex fuzzy sets and soft sets that have the ability to accurately represent and model two-dimensional information for real-life phenomena that are periodic in nature. In the existing studies of fuzzy and its extensions, the uncertainties which are present in the data are handled with the help of membership degree which is the subset of real numbers. However, in the present work, this condition has been relaxed with the degrees whose ranges are a subset of the complex subset with unit disc and hence handle the information in a better way. Under this environment, we developed some entropy measures of the CVSS model induced by the axiomatic definition of distance measure. Some desirable relations between them are also investigated. A numerical example related to detection of an image by the robot is given to illustrate the proposed entropy measure.

Keywords: vague entropy; distance induced vague entropy; distance; complex fuzzy set; complex vague soft set

1. Introduction

Classical information measures deal with information which is precise in nature, while information theory is one of the trusted ways to measure the degree of uncertainty in data. In our day-to-day life, uncertainty plays a dominant role in any decision-making process. In other words, due to an increase of the system day-by-day, decision makers may have to give their judgments in an imprecise, vague and uncertain environment. To deal with such information, Zadeh [1] introduced the theory of fuzzy sets (FSs) for handling the uncertainties in the data by defining a membership function with values between 0 and 1. In this environment, Deluca and Termini [2] proposed a set of axioms for fuzzy entropy. Liu [3] and Fan and Xie [4] both studied information measures related to entropy, distance, and similarity for fuzzy sets. With the growing complexities, researchers are engaged in extensions such as intuitionistic fuzzy set (IFS) [5], vague set (VS) [6], interval-valued IFS [7] to deal with the uncertainties. Under these extensions, Szmidt and Kacprzyk [8] extended the axioms of Deluca and Termini [2] to the IFS environment. Later on, corresponding to Deluca and Termini's [2] fuzzy entropy measure, Vlachos and Sergiadis [9] extended their measure in the IFS environment. Burillo and Bustince [10] introduced the entropy of intuitionistic fuzzy sets (IFSs), as a tool to measure the degree of intuitionism associated with an IFS. Garg et al. [11] presented a generalized intuitionistic fuzzy entropy measure of order α and degree β to solve decision-making problems. In addition to the mentioned examples, other authors have also addressed the problem of decision-making by using the different information measures [12–25].

All the above-defined work is successfully applied to the various disciplines without considering the parameterization factor during the analysis. Therefore, under some certain cases, these existing theories may be unable to classify the object. To cope with such situations, many researchers are paying more attention to soft set (SS) theory [26]. After its discovery, researchers are engaged in its extensions. For instance, Maji et al. [27,28] combined the theory of SSs with FSs and IFSs and came up with a new concept of the fuzzy soft set (FSS) and intuitionistic fuzzy soft set (IFSS). Further, the concept of the hybridization of the SSs with the others, such as generalized fuzzy soft set [29,30], generalized intuitionistic fuzzy soft set [31,32], distance measures [33–36], and fuzzy number intuitionistic fuzzy soft sets [37] plays a dominant role during the decision making process. IFSS plays a dominant role in handling the uncertainties in the data by incorporating the idea of the expert as well as the parametric factors. In that environment, Arora and Garg [38,39] presented some aggregation operators for intuitionistic fuzzy soft numbers. Garg and Arora [40] presented some non-linear methodology for solving decision-making problems in an IFSS environment. In terms of the information measures, Garg and Arora [34] developed various distance and similarity measures for dual hesitant FSS. Recently, Garg and Arora [41] presented Bonferroni mean aggregation operators for an IFSS environment. Apart from these, vague soft set [42] is an alternative theory which is the hybridization of the vague set [6] and soft set [26]. In this field, Chen [43] developed some similarity measures for vague sets. Wang and Qu [44] developed some entropy, similarity and distance measures for vague sets. Selvachandran et al. [45] introduced distance induced entropy measures for generalized intuitionistic fuzzy soft sets.

The above theories using FSs, IFSs, IFSSs, VSs, FSSs are widely employed by researchers but they are able to handle only the uncertainty in the data. On the other hand, none of these models will be able to handle the fluctuations of the data at a given phase of time during their execution, but in today's life, the uncertainty and vagueness of the data changes periodicity with the passage of time and hence the existing theories are unable to consider this information. To overcome this deficiency, Ramot et al. [46] presented a complex fuzzy set (CFS) in which the range of membership function is extended from real numbers to complex numbers with the unit disc. Ramot et al. [47] generalized traditional fuzzy logic to complex fuzzy logic in which the sets used in the reasoning process are complex fuzzy sets, characterized by complex valued membership functions. Later on, Greenfield et al. [48] extended the concept of CFS by taking the grade of the membership function as an interval-number rather than single numbers. Yazdanbakhsh and Dick [49] conducted a systematic review of CFSs and logic and discussed their applications. Later on, Alkouri and Salleh [50] extended the concepts of CFS to complex intuitionistic fuzzy (CIF) sets (CIFSs) by adding the degree of complex non-membership functions and studied their basic operations. Alkouri and Salleh [51] introduced the concepts of CIF relation, composition, projections and proposed a distance measure between the two CIFSs. Rani and Garg [52] presented some series of distance measures for a CIFS environment. Kumar and Bajaj [53] proposed some distance and entropy measures for CIF soft sets. In these theories, a two-dimensional information (amplitude and phase terms) are represented as a single set. The function of the phase term is to model the periodicity and/seasonality of the elements. For instance, when dealing with an economics-related situation, the phase term represents the time taken for the change in an economic variable to impact the economy. On the other hand, in robotics, the phase term can represent direction, whereas in image processing, the phase term can represent the non-physical attributes of the image.

As an alternative to these theories, the concept of the complex vague soft set (CVSS) [54] handles the two-dimensional information by combining the properties of CFSs [46], soft sets [26] and vague sets [6]. The CVSSs differs from the existing sets with the features that they contain: (1) an interval-based membership structure that provides users with the means of recording their hesitancy in the process of assigning membership values for the elements; (2) the ability to handle the partial ignorance of the data; (3) adequate parameterization abilities that allow for a more comprehensive representation of the parameters. Selvachandran et al. [54,55] investigated complex vague soft sets

(CVSSs). Selvachandran et al. [56] presented similarity measures for CVSSs and their applications to pattern recognition problems.

Thus, motivated from the concept of CVSS, the focus of this work is to explore the structural characteristics of CVSSs and to present some information measures for handling the uncertainties in the data. Per our knowledge, in the aforementioned studies, the information measures cannot be utilized to handle the CVSS information. Thus, in order to achieve this, we develop the axiomatic definition of the distance and entropy measures between CVSSs and hence propose some new entropy measures. Some of the algebraic properties of these measures and the relations between them are also verified. The proposed measures have the following characteristics: (1) they serve as a complement to the CVSS model and its relations in representing and modeling time-periodic phenomena; (2) they have elegant properties that increase their reach and applicability; (3) they have important applications in many real-world problems in the areas of image detection, pattern recognition, image processing; (4) they add to the existing collection of methodologies and techniques in artificial intelligence and soft computing, where it is often necessary to determine the degree of vagueness of the data, in order to make optimal decisions. This provides support of the increasingly widespread trend in the use of mathematical tools to complement scientific theories and existing procedures, in the handling and solving of real-life problems that involve vague, unreliable and uncertain two-dimensional information. Furthermore, an effort has been put forth to solve the classification problem in multi-dimensional complex data sets. To elaborate the proposed method, we will be focusing on the representation and recognition of digital images defined by multi-dimensional complex data sets using the properties of CVSSs and a new distance and entropy measure for this model.

The rest of the manuscript is organized as follows: in Section 2, we briefly review the basic concepts of SSs and CVSSs. In Section 3, we define the axiomatic definition of the distance and entropy measures for CVSSs. In Section 4, some basic relationships between the distance and entropy measures are defined. In Section 5, the utility of the CVSS model and its entropy measure is illustrated by applying it in a classification of the digital image with multi-dimensional data. Finally, conclusions and future work are stated in Section 6.

2. Preliminaries

In this section, we briefly reviewed some basic concepts related to the VSs, SSs, CVSSs defined over the universal set U.

Definition 1 [6]. *A vague set (VS) V in U is characterized by the truth and falsity membership functions t_V, f_V: $U \rightarrow [0.1]$ with $t_V(x) + f_V(x) \leq 1$ for any $x \in U$. The values assigned corresponding to $t_V(x)$ and $f_V(x)$ are the real numbers of $[0, 1]$. The grade of membership for x can be located in $[t_V(x), 1 - f_V(x)]$ and the uncertainty of x is defined as $(1 - f_V(x)) - t_V(x)$.*

It is clearly seen from the definition that VSs are the generalization of the fuzzy sets. If we assign $1 - f_V(x)$ to be $1 - t_V(x)$ then VS reduces to FS. However, if we set $1 - t_V(x)$ to be $v_A(x)$ (called the non-membership degree) then VS reduces to IFS. On the other hand, if we set $t_V(x) = \mu_V^L(x)$ and $1 - f_V(x) = \mu_V^U(x)$ then VS reduces to interval-valued FS. Thus, we conclude that VSs are the generalization of the FSs, IFSs and interval-valued FSs.

Definition 2 [6]. *Let $A = \{< x, [t_A(x), 1 - f_A(x)] >: x \in U\}$ and $B = \{< x, [t_B(x), 1 - f_B(x)] >: x \in U\}$ be two VSs defined on U then the basic operational laws between them are defined as follows:*

(i) $A \subseteq B$ *if* $t_A(x) \leq t_B(x)$ *and* $1 - f_A(x) \leq 1 - f_B(x)$ *for all* x.

(ii)   *Complement:* $A^C = \{<x, [f_A(x), 1 - t_A(x)]>: x \in U\}$.

(iii) *Union:* $A \cup B = \{<x, [\max(t_A(x), t_B(x)), \max(1 - f_A(x), 1 - f_B(x))]>: x \in U\}$

(iv) *Intersection:* $A \cap B = \{<x, [\min(t_A(x), t_B(x)), \min(1 - f_A(x), 1 - f_B(x))]>: x \in U\}$

Definition 3 [26]. *Let $P(U)$ denote the power set of U. A pair (F, A) is called a soft set (SS) over V where F is a mapping given by $F : A \to P(U)$.*

Definition 4 [42]. *Let $V(U)$ be the power set of VSs over U. A pair $(\hat{F}, A)$ is called a vague soft set (VSS) over U, where $\hat{F}$ is a mapping given by $\hat{F} : A \to V(U)$. Mathematically, VSS can be defined as follows:*

$$(\hat{F}, A) = \left\{ \langle x, \left[t_{\hat{F}_{(e)}}(x), 1 - f_{\hat{F}_{(e)}}(x) \right] \rangle : x \in U, e \in A \right\}$$

It is clearly seen that this set is the hybridization of the SSs and VSs.

Definition 5 [57]. *A complex vague set (CVS) is defined as an ordered pair defined as*

$$A = \{ \langle x, [t_A(x), 1 - f_A(x)] \rangle : x \in U \}$$

where $t_A : U \to \{a : a \in C, |a| \leq 1\}$, $f_A : U \to \{a : a \in X, |a| \leq 1\}$ are the truth and falsity membership functions with unit disc and are defined as $t_A(x) = r_{t_A}(x).e^{iw^r_{t_A}(x)}$ and $1 - f_A(x) = \left(1 - k_{f_A}(x)\right).e^{i(2\pi - w^k_{f_A}(x))}$ where $i = \sqrt{-1}$.

Definition 6 [54]. *Let $P(U)$ denote the complex vague power set of U and E be the set of parameters. For any $A \subset E$, a pair (F, A) is called a complex vague soft set (CVSS) over U, where $F : A \to P(U)$, defined as:*

$$F(x_j) = \left\{ \left(x_j, \left[r_{t_{F_a}}(x_j), 1 - k_{t_{F_a}}(x_j) \right].e^{i \left[w^r_{t_{F_a}}(x_j), 2\pi - w^k_{f_{F_a}}(x_j) \right]} \right) : x_j \in U \right\}$$

where $j = 1, 2, 3, \ldots$ is the number of parameters, $\left[r_{t_{F_a}}(x), 1 - k_{t_{F_a}}(x) \right]$ are real-valued $\in [0, 1]$, the phase terms $\left[w^r_{t_{F_a}}(x), 2\pi - w^k_{f_{F_a}}(x) \right]$ are real-valued in the interval $(0, 2\pi]$, $0 \leq r_{t_{F_a}}(x) + k_{F_a}(x) \leq 1$ and $i = \sqrt{-1}$.

The major advantages of the CVSS are that it represents two-dimensional information in a single set and each object is characterized in terms of its magnitude as well as its phase term. Further, the soft set component in CVSS provides an adequate parameterization tool to represent the information.

Definition 7 [54]. *Let two CVSSs (F, A) and (G, B) over U, the basic operations between them are defined as*

(i) $(F, A) \subset (G, B)$ *if and only if the following conditions are satisfied for all $x \in U$:*

(a) $r_{t_{F_a}}(x) \leq r_{t_{G_b}}(x)$ *and* $k_{f_{G_b}}(x) \leq k_{f_{F_a}}(x)$;

(b) $w^r_{t_{F_a}}(x) \leq w^r_{t_{G_b}}(x)$ *and* $w^k_{f_{G_b}}(x) \leq w^k_{f_{F_a}}(x)$.

(ii) *Null CVSS: $(F, A) = \phi$ if $r_{t_{F_a}}(x) = 0, k_{f_{F_a}}(x) = 1$ and $w^r_{t_{F_a}}(x) = 0\pi, w^k_{f_{F_a}}(x) = 2\pi$ for all $x \in U$.*

(iii) *Absolute CVSS: $(F, A) = 1$ if $r_{t_{F_a}}(x) = 1, k_{f_{F_a}}(x) = 0$ and $w^r_{t_{F_a}}(x) = 2\pi, w^k_{f_{F_a}}(x) = 0\pi$ for all $x \in U$.*

3. Axiomatic Definition of Distance Measure and Vague Entropy

Let E be a set of parameters and U be the universe of discourse. In this section, we present some information measures namely distance and entropy for the collections of CVSSs, which are denoted by CVSS(U).

Definition 8. *Let (F, A), (G, B), $(H, C) \in CVSS(U)$. A complex-value function $d : CVSS(U) \times CVSS(U) \to \{a, a \in U, |a| \leq 1\}$ is called a distance measure between CVSSs if it satisfies the following axioms:*

(D1) $d((F, A), (G, B)) = d((G, B), (F, A))$

(D2) $d((F, A), (G, B)) = 0 \Longleftrightarrow (F, A) = (G, B)$

(D3) $d((F, A), (G, B)) = 1 \Longleftrightarrow \forall e \in E, x \in U$, *both* (F, A) *and* (G, B) *are crisp sets in* U, i.e.,

$$(F, A) = \left\{\left(x, [0, 0]e^{i[0\pi, 0\pi]}\right)\right\} \text{ and } (G, B) = \left\{\left(x, [1, 1]e^{i[2\pi, 2\pi]}\right)\right\},$$
$$\text{or } (F, A) = \left\{\left(x, [0, 0]e^{i[2\pi, 2\pi]}\right)\right\} \text{ and } (G, B) = \left\{\left(x, [1, 1]e^{i[0\pi, 0\pi]}\right)\right\},$$
$$\text{or } (F, A) = \left\{\left(x, [1, 1]e^{i[2\pi, 2\pi]}\right)\right\} \text{ and } (G, B) = \left\{\left(x, [0, 0]e^{i[0\pi, 0\pi]}\right)\right\},$$
$$\text{or } (F, A) = \left\{\left(x, [1, 1]e^{i[0\pi, 0\pi]}\right)\right\} \text{ and } (G, B) = \left\{\left(x, [0, 0]e^{i[2\pi, 2\pi]}\right)\right\}.$$

(D4) *If* $(F, A) \subseteq (G, B) \subseteq (H, C)$, *then* $d((F, A), (H, C)) \geq max(d((F, A), (G, B)), d((G, B), (H, C)))$.

Next, we define the axiomatic definition for the vague entropy for a CVSS.

Definition 9. *A complex-valued function* $M : CVSS(U) \rightarrow \{a : a \in \mathbb{C}, |a| \leq 1\}$ *is called vague entropy of CVSSs, if it satisfies the following axioms for any* $(F, A), (G, B) \in CVSS(U)$.

(M1) $0 \leq |M(F, A)| \leq 1$.

(M2) $M(F, A) = 0 \Longleftrightarrow (F, A)$ *is a crisp set on* U *for all* $a \in A$ *and* $x \in U$, i.e., $r_{t_{F_a}}(x) = 1$, $k_{f_{F_a}}(x) = 0$
$\quad$ *and* $w^r_{t_{F_a}}(x) = 2\pi$, $w^k_{f_{F_a}}(x) = 0\pi$ *or* $r_{t_{F_a}}(x) = 1$, $k_{f_{F_a}}(x) = 0$ *and* $w^r_{t_{F_a}}(x) = 0\pi$, $w^k_{f_{F_a}}(x) = 2\pi$
$\quad$ *or* $r_{t_{F_a}}(x) = 0$, $k_{f_{F_a}}(x) = 1$ *and* $w^r_{t_{F_a}}(x) = 0\pi$, $w^k_{f_{F_a}}(x) = 2\pi$ *or* $r_{t_{F_a}}(x) = 0$, $k_{f_{F_a}}(x) = 1$ *and*
$\quad$ $w^r_{t_{F_a}}(x) = 2\pi$, $w^k_{f_{F_a}}(x) = 0\pi$.

(M3) $M(F, A) = 1 \Longleftrightarrow \forall a \in A$ *and* $x \in U$, (F, A) *is completely vague*

$\quad$ i.e., $r_{t_{F_a}}(x) = k_{f_{F_a}}(x)$ *and* $w^r_{t_{F_a}}(x) = w^k_{f_{F_a}}(x)$.

(M4) $M(F, A) = M((F, A)^c)$

(M5) *If the following two cases holds for all* $a \in A$ *and* $x \in U$,

$$\text{Case 1} : r_{t_{F_a}}(x) \leq r_{t_{G_b}}(x), \ k_{f_{F_a}}(x) \geq k_{f_{G_b}}(x) \text{whenever } r_{t_{G_b}}(x) \leq k_{f_{G_b}}(x);$$

$$\text{and } w^r_{t_{F_a}}(x) \leq w^r_{t_{G_b}}(x), \ w^k_{f_{F_a}}(x) \geq w^k_{f_{G_b}}(x) \text{ whenever } w^r_{t_{G_b}}(x) \leq w^k_{f_{G_b}}(x);$$

$$\text{Case 2} : r_{t_{F_a}}(x) \geq r_{t_{G_b}}(x), \ k_{f_{F_a}}(x) \leq k_{f_{G_b}}(x) \text{whenever } r_{t_{G_b}}(x) \geq k_{f_{G_b}}(x)$$

$$\text{and } w^r_{t_{F_a}}(x) \geq w^r_{t_{G_b}}(x), w^k_{f_{F_a}}(x) \leq w^k_{f_{G_b}}(x) \text{whenever } w^r_{t_{G_b}}(x) \geq w^k_{f_{G_b}}(x);$$

then $M(F, A) \leq M(G, B)$.

Based on this definition, it is clear that a value close to 0 indicates that the CVSS has a very low degree of vagueness whereas a value close to the 1 implies that the CVSS is highly vague. For all $x \in U$, the nearer $r_{t_{F_a}}(x)$ is to $k_{f_{F_a}}(x)$, the larger the vague entropy measure and it reaches a maximum when $r_{t_{F_a}}(x) = k_{f_{F_a}}(x)$. Condition M5 on the other hand, is slightly different as it is constructed using the sharpened version of a vague soft set as explained in Hu et al. [58], instead of the usual condition of $(F, A) \subset (G, B)$ implies that the entropy of (F, A) is higher than the entropy of (G, B). In [58], Hu et al. proved that this condition is inaccurate and provided several counter-examples to disprove this condition. Subsequently, they replaced this flawed condition with two new cases. We generalized these two cases to derive condition (M5) in this paper, in a bid to increase the accuracy of our proposed vague entropy. We refer the readers to [58] for further information on these revised conditions.

4. Relations between the Proposed Distance Measure and Vague Entropy

In the following, let U be universal and ϕ be empty over CVSSs. Then based on the above definition, we define some of the relationship between them as follows:

Theorem 1. *Let (F, A) be CVSS and d is the distance measure between CVSSs, then the equations M_1, M_2 and M_3 defined as below*

(i) $M_1(F, A) = 1 - d((F, A), (F, A)^c)$
(ii) $M_2(F, A) = d((F, A) \cup (F, A)^c, U)$
(iii) $M_3(F, A) = 1 - d((F, A) \cup (F, A)^c, (F, A) \cap (F, A)^c)$

are the valid vague entropies of CVSSs.

Proof. Here, we shall prove only the part (i), while others can be proved similarly.

It is clearly seen from the definition of vague entropies that M_1 satisfies conditions (M1) to (M4). So we need to prove only (M5). For it, consider the two cases stated in Definition 9. We only prove that the condition (M5) is satisfied for Case 1; the proof for Case 2 is similar and is thus omitted.

From the conditions given in Case 1 of (M5), we obtain the following relationship:

$$r_{t_{F_a}}(x) \leq r_{t_{G_b}}(x) \leq k_{f_{G_b}}(x) \leq k_{f_{F_a}}(x)$$

$$\text{and } w^r_{t_{F_a}}(x) \leq w^r_{t_{G_b}}(x) \leq w^k_{f_{G_b}}(x) \leq w^k_{f_{F_a}}(x).$$

Therefore, we have:

$$\phi \subset (F, A) \subset (G, B) \subset (G, B)^c \subset (F, A)^c \subset U.$$

Hence, it follows that:

$$((F, A), (F, A)^c) \geq d((G, B), (G, B)^c).$$

Now, by definition of M_1, we have:

$$M_1(F, A) = 1 - d((F, A), (F, A)^c)$$

$$\leq 1 - d((G, B), (G, B)^c)$$

$$= M_1(G, B).$$

This completes the proof. $\square$

Theorem 2. *If d is the distance measure between CVSSs, then:*

$$M_4(F, A) = \frac{d((F, A) \cup (F, A)^c, U)}{d((F, A) \cap (F, A)^c, U)}$$

is a vague entropy of CVSSs.

Proof. For two CVSSs (F, A) & (G, B), clearly seen that M_4 satisfies conditions (M1)–(M4). So, it is enough to prove that M_4 satisfy the condition (M5).

Consider the case:

$$r_{t_{F_a}}(x) \leq r_{t_{G_b}}(x), \; k_{f_{F_a}}(x) \geq k_{f_{G_b}}(x) \text{ whenever } r_{t_{G_b}}(x) \leq k_{f_{G_b}}(x)$$

$$\text{and } w^r_{t_{F_a}}(x) \leq w^r_{t_{G_b}}(x), \; w^k_{f_{F_a}}(x) \geq w^k_{f_{G_b}}(x) \text{ whenever } w^r_{t_{G_b}}(x) \leq w^k_{f_{G_b}}(x)$$

which implies that:

$$r_{t_{F_a}}(x) \leq r_{t_{G_b}}(x) \leq k_{f_{G_b}}(x) \leq k_{f_{F_a}}(x)$$

$$\text{and } w^r_{t_{F_a}}(x) \le w^r_{t_{G_b}}(x) \le w^k_{f_{G_b}}(x) \le w^k_{f_{F_a}}(x).$$

Thus, we obtain:

$$\phi \subset (F, A) \cap (F, A)^c \subset (G, B) \cap (G, B)^c \subset (G, B) \cup (G, B)^c \subset (F, A) \cup (F, A)^c \subset U.$$

Therefore, we have:

$$d((F, A) \cup (F, A)^c, U) \le d((G, B) \cup (G, B)^c U)$$

$$\text{and } d((F, A) \cap (F, A)^c, U) \le d((G, B) \cap (G, B)^c U).$$

Hence, by definition of M_4, we have:

$$M_4(F, A) = \frac{d((F, A) \cup (F, A)^c, U)}{d((F, A) \cap (F, A)^c, U)}$$

$$\le \frac{d((G, B) \cup (G, B)^c U)}{d((G, B) \cap (G, B)^c U)}$$

$$= M_4(G, B)$$

Similarly, we can obtain for other case i.e., when $r_{t_{F_a}}(x) \ge r_{t_{G_b}}(x)$, $k_{f_{F_a}}(x) \le k_{f_{G_b}}(x)$ whenever $r_{t_{G_b}}(x) \ge k_{f_{G_b}}(x)$ and $w^r_{t_{F_a}}(x) \ge w^r_{t_{G_b}}(x)$, $w^k_{f_{F_a}}(x) \le w^k_{f_{G_b}}(x)$ whenever $w^r_{t_{G_b}}(x) \ge w^k_{f_{G_b}}(x)$, we have $M_4(F, A) \le M_4(G, B)$. Hence (M5) satisfied.

Therefore, M_4 is a valid entropy measure. $\square$

Theorem 3. *For CVSS* (F, A) *and if d is the distance measure between CVSSs, then:*

$$M_5(F, A) = \frac{d((F, A) \cap (F, A)^c, \phi)}{d((F, A) \cup (F, A)^c, \phi)}$$

is a vague entropy of CVSSs.

Proof. It can be obtained as similar to Theorem 2, so we omit here. $\square$

Theorem 4. *For two CVSSs* (F, A) *and* (G, B). *If d is a distance measure between CVSSs such that:*

$$d((F, A), (G, B)) = d((F, A)^c, (G, B)^c),$$

then the entropies M_4 and M_5 satisfies the equation $M_4 = M_5$.

Proof. By definition of M_4 and M_5, we have:

$$M_4(F, A) = \frac{d((F, A) \cup (F, A)^c, U)}{d((F, A) \cap (F, A)^c, U)}$$

$$= \frac{d(((F, A) \cup (F, A)^c)^c, U^c)}{d(((F, A) \cap (F, A)^c)^c, U^c)}$$

$$= \frac{d((F, A) \cap (F, A)^c, \phi)}{d((F, A) \cup (F, A)^c, \phi)}$$

$$= M_5(F, A).$$

Theorem 5. *For a CVSS* (F, A), *if* d *is the distance measure between CVSSs and satisfies:*

$$d((F, A), U) = d((F, A), \phi),$$

Then:

$$M_6(F, A) = \frac{d((F, A) \cup (F, A)^c, U)}{d((F, A) \cup (F, A)^c, \phi)}$$

is a vague entropy of CVSSs.

Theorem 6. *If* d *is the distance measure between CVSSs and satisfies* $d((F, A), U) = d((F, A), \phi)$, *then:*

$$M_7(F, A) = \frac{d((F, A) \cap (F, A)^c, \phi)}{d((F, A) \cap (F, A)^c, U)}$$

is a vague entropy of CVSSs.

Theorem 7. *If* d *is a distance measure between CVSSs that satisfies:*

$$d((F, A), (G, B)) = d((F, A)^c, (G, B)^c),$$

then $M_6 = M_7$.

Proof. The proof of the Theorems 5–7 can be obtained as similar to above, so we omit here. $\square$

Theorem 8. *If* d *is a distance measure between CVSSs, then:*

$$M_8(F, A) = 1 - d((F, A) \cap (F, A)^c, U) + d((F, A) \cup (F, A)^c, U)$$

is a vague entropy of CVSSs.

Theorem 9. *If* d *is a distance measure between CVSSs, then:*

$$M_9(F, A) = 1 - d((F, A) \cup (F, A)^c, \phi) + d((F, A) \cap (F, A)^c, \phi)$$

is a vague entropy of CVSSs.

Theorem 10. *If* d *is a distance measure between CVSSs* (F, A) *and* (G, B) *such that:*

$$d((F, A), (G, B)) = d((F, A)^c, (G, B)^c),$$

then $M_8 = M_9$.

Theorem 11. *If* d *is a distance measure between CVSSs, then:*

$$M_{10}(F, A) = 1 - d((F, A) \cup (F, A)^c, \phi) + d((F, A) \cup (F, A)^c, U)$$

is a vague entropy of CVSSs.

Theorem 12. *If* d *is a distance measure between CVSSs, then:*

$$M_{11}(F, A) = 1 - d((F, A) \cap (F, A)^c, U) + d((F, A) \cap (F, A)^c, \phi)$$

is a vague entropy of CVSSs.

Theorem 13. *If d is a distance measure between CVSSs (F,A) and (G,B) such that:*

$$d((F, A), (G, B)) = d((F, A)^c, (G, B)^c),$$

then $M_{10} = M_{11}$.

Proof. The proof of these Theorems can be obtained as similar to above, so we omit here. □

5. Illustrative Example

In this section, we present a scenario which necessitates the use of CVSSs. Subsequently, we present an application of the entropy measures proposed in Section 4 to an image detection problem to illustrate the validity and effectiveness of our proposed entropy formula.

Firstly, we shall define the distance between any two CVSSs as follows:

Definition 10. *Let (F, A) and (G, B) be two CVSSs over U. The distance between (F, A) and (G, B) is as given below:*

$$d((F, A), (G, B)) = \frac{1}{4mn} \sum_{j=1}^{n} \sum_{i=1}^{m} \left[\begin{array}{c} \max\left\{ \left| r_{t_{F_{(a_i)}}}(x_j) - r_{t_{G_{(b_i)}}}(x_j) \right|, \left| k_{f_{G_{(b_i)}}}(x_j) - k_{f_{F_{(a_i)}}}(x_j) \right| \right\} \\ + \frac{1}{2\pi}\left(\max\left\{ \left| w^r_{t_{F_{(a_i)}}}(x_j) - w^r_{t_{G_{(b_i)}}}(x_j) \right|, \left| w^k_{f_{G_{(b_i)}}}(x_j) - w^k_{f_{F_{(a_i)}}}(x_j) \right| \right\}\right) \end{array} \right]$$

In order to demonstrate the utility of the above proposed entropy measures $M_i (i = 1, 2, \ldots, 11)$, we demonstrate it with a numerical example. For it, consider a CVSS (F, A) whose data sets are defined over the parameters $e_1, e_2 \in E$ and $x_1, x_2, x_3 \in U$ as follows:

$$(F, A) = \begin{array}{c} x_1 \\ x_2 \end{array} \left[\begin{array}{ccc} [0.2, 0.8]e^{i[0.1(2\pi), 0.2(2\pi)]} & [0.3, 0.5]e^{i[0.2(2\pi), 0.4(2\pi)]} & [0.3, 0.6]e^{i[0.4(2\pi), 0.5(2\pi)]} \\ [0.3, 0.6]e^{i[0.4(2\pi), 0.5(2\pi)]} & [0.2, 0.3]e^{i[0.2(2\pi), 0.4(2\pi)]} & [0.7, 0.9]e^{i[0.4(2\pi), 0.5(2\pi)]} \end{array} \right]$$

and hence the complement of CVSS is:

$$(F, A)^c = \begin{array}{c} x_1 \\ x_2 \end{array} \left[\begin{array}{ccc} [0.2, 0.8]e^{i[0.8(2\pi), 0.9(2\pi)]} & [0.5, 0.7]e^{i[0.6(2\pi), 0.8(2\pi)]} & [0.4, 0.7]e^{i[0.5(2\pi), 0.6(2\pi)]} \\ [0.2, 0.5]e^{i[0.8(2\pi), 0.9(2\pi)]} & [0.7, 0.8]e^{i[0.6(2\pi), 0.8(2\pi)]} & [0.1, 0.3]e^{i[0.5(2\pi), 0.6(2\pi)]} \end{array} \right]$$

Then, the distance measure based on the Definition 10, we get $((F, A), (F, A)^c) = 0.1708$, $d((F, A) \cup (F, A)^c, U) = 0.2167$, $d((F, A) \cup (F, A)^c, (F, A) \cap (F, A)^c) = 0.1708$, $d((F, A) \cap (F, A)^c, U) = 0.3875$, $d((F, A) \cup (F, A)^c, \phi) = 0.3875$, $d((F, A) \cap (F, A)^c, \phi) = 0.2167$. Therefore, the values of the entropy measures defined on the Theorem 1 to Theorem 11 are computed as:

(i) $M_1(F, A) = 1 - d((F, A), (F, A)^c) = 1 - 0.1708 = 0.8292$.

(ii) $M_2(F, A) = d((F, A) \cup (F, A)^c, U) = 0.2167$.

(iii) $M_3(F, A) = 1 - d((F, A) \cup (F, A)^c, (F, A) \cap (F, A)^c) = 1 - 0.1708 = 0.8292$.

(iv) $M_4(F, A) = \frac{d((F, A) \cup (F, A)^c, U)}{d((F, A) \cap (F, A)^c, U)} = \frac{0.2167}{0.3875} = 0.5592$.

(v) $M_5(F, A) = \frac{d((F, A) \cap (F, A)^c, \phi)}{d((F, A) \cup (F, A)^c, \phi)} = \frac{0.2167}{0.3875} = 0.5592$.

(vi) $M_6(F, A) = \frac{d((F, A) \cup (F, A)^c, U)}{d((F, A) \cup (F, A)^c, \phi)} = \frac{0.2167}{0.3875} = 0.5592$.

(vii) $M_7(F, A) = \frac{d((F, A) \cap (F, A)^c, \phi)}{d((F, A) \cap (F, A)^c, U)} = \frac{0.2167}{0.3875} = 0.5592$.

(viii) $M_8(F, A) = 1 - d((F, A) \cap (F, A)^c, U) + d((F, A) \cup (F, A)^c, U) = 1 - 0.3875 + 0.2167 = 0.5592$

(ix) $M_9(F, A) = 1 - d((F, A) \cup (F, A)^c, \phi) + d((F, A) \cap (F, A)^c, \phi) = 1 - 0.3875 + 0.2167 = 0.5592$

(x) $M_{10}(F, A) = 1 - d((F, A) \cup (F, A)^c, \phi) + d((F, A) \cup (F, A)^c, U) = 1 - 0.3875 + 0.2167 = 0.5592$

(xi) $M_{11}(F, A) = 1 - d((F, A) \cap (F, A)^c, U) + d((F, A) \cap (F, A)^c, \phi) = 1 - 0.3875 + 0.2167 = 0.5592$

Next, we give an illustrative example from the field of pattern recognition which are stated and demonstrated as below.

5.1. The Scenario

A type of robot has a single eye capable of capturing (and hence memorizing) things it sees as an 850×640, 24 bit bitmap image. The robot was shown an object (a pillow with a smiley), and the image that was captured by the robot's eye at that instant is shown in Figure 1. This image was saved as pic001.bmp in the memory of the robot.

Figure 1. The image of the object captured by the robot.

The robot was then given a way (in this example, it is done by human input) to recognize the object, whenever the robot encounters the object again, by retrieving the colors at certain coordinates of its field of vision, and then comparing this with the same coordinates from image pic001.bmp stored in its memory. In order to distinguish noises, the coordinates are chosen in clusters of four, as shown in Figure 2. The coordinates of the clusters of the images are summarized in Table 1.

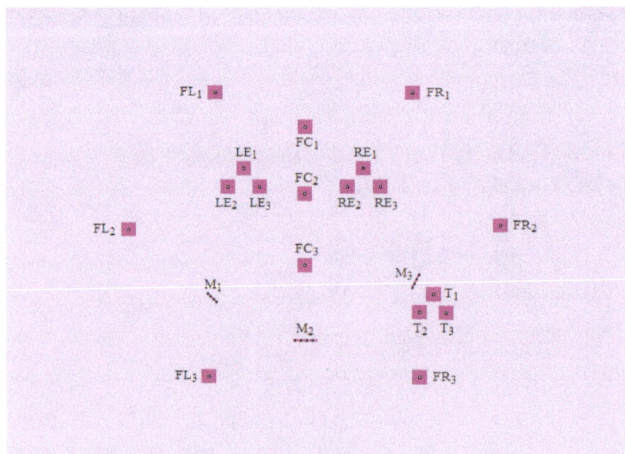

Figure 2. The clusters of the image pic001.bmp for recognition purposes.

Table 1. The coordinates of the clusters for image pic001.bmp from Figure 2.

	1st Position ($n = 1$)	2nd Position ($n = 2$)	3rd Position ($n = 3$)
"Left Eye" (LE_n)	(323, 226), (324, 226), (323, 227), (324, 227),	(301, 252), (302, 252), (301, 253), (302, 253),	(345, 252), (346, 252), (345, 253), (346, 253),
"Right Eye" (RE_n)	(486, 226), (487, 226), (486, 227), (487, 227),	(464, 252), (465, 252), (464, 253), (465, 253),	(509, 252), (510, 252), (509, 253), (510, 253),
"Left side of Face" (LF_n)	(284, 119), (285, 119), (284, 120), (285, 120),	(167, 312), (168, 312), (167, 313), (168, 313),	(275, 519), (276, 519), (275, 520), (276, 520),
"Centre of Face" (CF_n)	(407, 168), (408, 168), (407, 169), (408, 169),	(406, 262), (407, 262), (406, 263), (407, 263),	(406, 363), (407, 363), (406, 364), (407, 364),
"Right side of Face" (RF_n)	(553, 120), (554, 120), (553, 121), (554, 121),	(671, 307), (672, 307), (671, 308), (672, 308),	(562, 521), (563, 521), (562, 522), (563, 522),
"Tongue" (T_n)	(581, 404), (582, 404), (581, 405), (582, 405),	(562, 429), (563, 429), (562, 430), (563, 430),	(598, 430), (599, 430), (598, 431), (599, 431),
"Mouth" (M_n)	(274, 403), (278, 407), (282, 411), (286, 415),	(393, 469), (401, 469), (409, 469), (417, 469),	(553, 395), (556, 389), (559, 383), (562, 377),

Remark: For a computer image, the top-leftmost pixel is labeled (0, 0).

We now have three images, namely image A, image B and image C. The robot needs to recognize if the objects shown in image A, B and C is the same as the image shown in image pic001.bmp stored in the robot's memory. Images A, B and C are shown in Figures 3–5, respectively. For comparison purposes, pic001.bmp is shown alongside all the three images.

Figure 3. Image A (**left**) and the original image pic001.bmp (**right**).

Figure 4. Image B (**left**) and the original image pic001.bmp (**right**).

Figure 5. Image C (**left**) and the original image pic001.bmp (**right**).

From a human perspective, it is clear that the object shown in image A will be recognized as the same object shown in image pic001.bmp, and it will be concluded that the object is shown in image B (a red airplane) is not the image pic001.bmp stored in the memory of the robot. No conclusion can be deduced from image C as it is made up of only noise, and therefore we are unable to deduce the exact object behind the noise. By retrieving the coordinates from Table 1, we now obtain the following sets of colors which are given in Table 2.

Table 2. The sets of colors for image A, B, C and image pic001.bmp.

	pic001.bmp (Memory)			Image A			Image B			Image C		
	$n = 1$	$n = 2$	$n = 3$	$n = 1$	$n = 2$	$n = 3$	$n = 1$	$n = 2$	$n = 3$	$n = 1$	$n = 2$	$n = 3$
LE												
RE												
LF												
CF												
RF												
T												
M												

The luminosity and hue of the pixels are obtained using a picture editing program, and these are given in Tables 3 and 4, respectively.

Luminosity, $\mathfrak{L}_{m,k,n,b}$, where $k \in \{LE, RE, LF, CF, RF, T, M\}$, $b \in \{0, 1, 2, 3\}$ (cluster of four pixels).

Hue, $\mathfrak{H}_{m,k,n,b}$, where $k \in \{LE, RE, LF, CF, RF, T, M\}$, $b \in \{0, 1, 2, 3\}$ (cluster of four pixels).

Table 3. The values of the luminosity for image A, B, C and image pic001.bmp.

	pic001.bmp (Memory), $m=0$			Image A, $m=1$			Image B, $m=2$			Image C, $m=3$		
	$n=1$	$n=2$	$n=3$	$n=1$	$n=2$	$n=3$	$n=1$	$n=2$	$n=3$	$n=1$	$n=2$	$n=3$
LE	23	23	24	24	34	32	5	97	72	57	54	180
	24	24	24	25	38	31	6	116	80	78	107	120
RE	24	21	23	43	41	47	7	8	7	48	0	2
	23	22	24	45	41	48	6	8	7	0	67	0
LF	101	104	90	78	55	96	3	162	122	24	31	62
	96	106	91	78	55	95	3	163	125	14	61	40
CF	85	80	78	97	103	102	6	79	24	0	22	109
	88	81	78	97	103	99	5	81	33	3	139	27
RF	83	64	60	119	136	139	6	16	62	93	94	56
	84	65	59	120	135	138	6	16	61	20	30	45
T	60	59	58	143	144	152	116	27	127	0	75	81
	59	61	60	144	145	150	117	24	125	104	3	0
M	26	15	9	21	34	33	81	175	36	112	15	14
	25	13	8	20	43	36	87	181	79	121	115	31

Entropy 2018, 20, 403

Table 4. The values of the hue of the pixels for image A, B, C and image pic001.bmp.

	pic001.bmp (Memory), $m = 0$						**Image A** $m = 1$						**Image B** $m = 2$						**Image C** $m = 3$					
	$n = 1$		$n = 2$		$n = 3$		$n = 1$		$n = 2$		$n = 3$		$n = 1$		$n = 2$		$n = 3$		$n = 1$		$n = 2$		$n = 3$	
LE	12	12	10	10	12	12	17	17	15	15	18	16	187	160	8	8	6	7	55	80	23	160	226	27
	13	12	9	10	13	12	17	17	15	15	18	16	160	160	8	8	6	7	214	173	127	70	168	66
RE	13	13	13	13	12	11	19	19	18	18	20	20	160	160	160	160	160	160	112	64	227	160	220	119
	13	13	13	12	12	11	19	19	18	18	19	20	160	160	160	160	160	160	152	150	160	177	160	77
LF	31	31	31	31	31	30	28	28	27	27	31	31	187	187	173	167	7	7	12	167	169	67	57	211
	31	30	31	31	31	30	29	28	27	27	31	30	187	187	171	167	9	7	23	237	18	76	186	199
CF	29	29	29	28	29	29	30	29	30	30	31	30	160	160	8	8	224	230	155	86	160	214	177	0
	29	29	29	29	29	29	30	29	30	30	30	30	180	180	8	8	213	220	42	51	200	107	97	165
RF	29	30	29	29	31	31	30	30	32	32	33	33	160	160	160	160	139	139	154	5	238	68	94	36
	30	30	29	29	31	31	30	30	32	32	32	33	160	160	160	153	137	141	68	160	192	76	131	158
T	11	11	11	12	10	10	12	12	12	12	13	12	13	9	168	165	7	7	119	212	160	45	160	130
	11	11	9	9	9	10	12	12	11	12	13	12	9	13	164	160	8	10	29	205	31	67	160	200
M	11	10	13	13	18	14	16	15	18	20	17	16	10	14	19	18	208	183	145	118	195	26	124	160
	11	12	13	13	14	18	15	16	21	19	16	17	13	14	17	15	184	5	42	29	49	181	115	220

5.2. Formation of CVSS and Calculation of Entropies

Let $U = \{x_1, x_2, x_3\}$, and $A = \{$LE, RE, LF, CF, RF, T, M$\}$. We now form three CVSSs $(\mathcal{F}_\delta, A)$, $\delta \in \{1, 2, 3\}$, which denote image A, B and C, respectively using the formula given below:

$$\mathcal{F}_{\delta_{(k)}}(x_n) = \left[e^{-\left(\frac{(y_{\delta,(k,n)})^2}{\rho}\right)}, e^{-\left(\frac{(\gamma_{\delta,(k,n)})^2}{\rho}\right)} \right] e^{2\pi i \left[e^{-\left(\frac{(\theta_{\delta,(k,n)})^2}{\varrho}\right)}, e^{-\left(\frac{(\vartheta_{\delta,(k,n)})^2}{\varrho}\right)} \right]},$$

where:

$$y_{\delta,(k,n)} = \min\left\{ \left| \mathfrak{L}_{\delta,k,n,p} - \mathfrak{L}_{0,k,n,q} \right| : p,q \in \{0,1,2,3\} \right\}, \gamma_{\delta,(k,n)} = \max\left\{ \left| \mathfrak{L}_{\delta,k,n,p} - \mathfrak{L}_{0,k,n,q} \right| : p,q \in \{0,1,2,3\} \right\},$$

$$\theta_{\delta,(k,n)} = \min\left\{ \left| \mathfrak{H}_{\delta,k,n,p} - \mathfrak{H}_{0,k,n,q} \right| : p,q \in \{0,1,2,3\} \right\}, \vartheta_{\delta,(k,n)} = \max\left\{ \left| \mathfrak{H}_{\delta,k,n,p} - \mathfrak{H}_{0,k,n,q} \right| : p,q \in \{0,1,2,3\} \right\}.$$

We choose $\rho = 6400$ and $\rho = 6400$ for this scenario. The CVSSs that were formed for this scenario are as given in Tables 5–7.

Table 5. Tabular representation of $(\mathcal{F}_1, A)$.

			n	
		1	2	3
	LE	$[0.996, 1.000]e^{2\pi i[0.996, 0.997]}$	$[0.960, 0.987]e^{2\pi i[0.994, 0.996]}$	$[0.987, 0.994]e^{2\pi i[0.994, 0.998]}$
	RE	$[0.920, 0.945]e^{2\pi i[0.994, 0.994]}$	$[0.927, 0.955]e^{2\pi i[0.994, 0.996]}$	$[0.899, 0.927]e^{2\pi i[0.987, 0.992]}$
	LF	$[0.920, 0.955]e^{2\pi i[0.998, 0.999]}$	$[0.666, 0.708]e^{2\pi i[0.997, 0.997]}$	$[0.990, 0.997]e^{2\pi i[0.999, 1.000]}$
k	CF	$[0.955, 0.990]e^{2\pi i[0.999, 1.000]}$	$[0.913, 0.939]e^{2\pi i[0.999, 0.999]}$	$[0.913, 0.939]e^{2\pi i[0.999, 0.999]}$
	RF	$[0.798, 0.825]e^{2\pi i[0.999, 1.000]}$	$[0.434, 0.475]e^{2\pi i[0.998, 0.998]}$	$[0.349, 0.386]e^{2\pi i[0.999, 0.999]}$
	T	$[0.323, 0.358]e^{2\pi i[0.999, 0.999]}$	$[0.314, 0.349]e^{2\pi i[0.998, 1.000]}$	$[0.251, 0.298]e^{2\pi i[0.997, 0.999]}$
	M	$[0.992, 0.999]e^{2\pi i[0.994, 0.998]}$	$[0.868, 0.945]e^{2\pi i[0.990, 0.996]}$	$[0.884, 0.913]e^{2\pi i[0.998, 0.999]}$

Table 6. Tabular representation of $(\mathcal{F}_2, A)$.

			N	
		1	2	3
	LE	$[0.945, 0.955]e^{2\pi i[0.008, 0.034]}$	$[0.258, 0.444]e^{2\pi i[0.999, 0.999]}$	$[0.591, 0.708]e^{2\pi i[0.992, 0.996]}$
	RE	$[0.950, 0.960]e^{2\pi i[0.034, 0.034]}$	$[0.955, 0.973]e^{2\pi i[0.032, 0.034]}$	$[0.955, 0.969]e^{2\pi i[0.031, 0.032]}$
	LF	$[0.222, 0.258]e^{2\pi i[0.021, 0.022]}$	$[0.580, 0.612]e^{2\pi i[0.042, 0.055]}$	$[0.807, 0.876]e^{2\pi i[0.913, 0.933]}$
k	CF	$[0.332, 0.377]e^{2\pi i[0.028, 0.068]}$	$[0.994, 1.000]e^{2\pi i[0.933, 0.939]}$	$[0.623, 0.728]e^{2\pi i[0.001, 0.005]}$
	RF	$[0.386, 0.405]e^{2\pi i[0.068, 0.071]}$	$[0.666, 0.708]e^{2\pi i[0.068, 0.090]}$	$[0.996, 0.999]e^{2\pi i[0.150, 0.172]}$
	T	$[0.559, 0.623]e^{2\pi i[0.999, 0.999]}$	$[0.798, 0.852]e^{2\pi i[0.019, 0.032]}$	$[0.475, 0.516]e^{2\pi i[0.998, 1.000]}$
	M	$[0.465, 0.623]e^{2\pi i[0.997, 1.000]}$	$[0.007, 0.071]e^{2\pi i[0.994, 0.999]}$	$[0.454, 0.892]e^{2\pi i[0.002, 0.987]}$

Table 7. Tabular representation of $(\mathcal{F}_3, A)$.

			N	
		1	2	3
	LE	$[0.178, 0.999]e^{2\pi i[0.001, 0.759]}$	$[0.332, 0.868]e^{2\pi i[0.028, 0.973]}$	$[0.021, 0.999]e^{2\pi i[0.000, 0.969]}$
	RE	$[0.229, 0.997]e^{2\pi i[0.048, 0.666]}$	$[0.718, 0.933]e^{2\pi i[0.000, 0.034]}$	$[0.222, 0.939]e^{2\pi i[0.001, 0.516]}$
	LF	$[0.749, 0.876]e^{2\pi i[0.001, 0.992]}$	$[0.266, 0.749]e^{2\pi i[0.051, 0.973]}$	$[0.495, 0.996]e^{2\pi i[0.005, 0.899]}$
k	CF	$[0.559, 0.997]e^{2\pi i[0.083, 0.973]}$	$[0.340, 0.612]e^{2\pi i[0.004, 0.386]}$	$[0.655, 0.955]e^{2\pi i[0.032, 0.876]}$
	RF	$[0.332, 0.969]e^{2\pi i[0.068, 0.913]}$	$[0.728, 0.884]e^{2\pi i[0.001, 0.788]}$	$[0.738, 0.998]e^{2\pi i[0.080, 0.996]}$
	T	$[0.559, 0.973]e^{2\pi i[0.001, 0.950]}$	$[0.548, 0.973]e^{2\pi i[0.028, 0.945]}$	$[0.046, 0.965]e^{2\pi i[0.003, 0.105]}$
	M	$[0.282, 0.999]e^{2\pi i[0.057, 0.955]}$	$[0.161, 1.000]e^{2\pi i[0.005, 0.973]}$	$[0.906, 0.996]e^{2\pi i[0.001, 0.229]}$

By using Definition 10, the entropy values for images A, B and C are as summarized in Table 8.

Table 8. Summary of the entropy values for image A, B and C.

Entropy Measure	Image A $(\mathcal{F}_1, A)$	Image B $(\mathcal{F}_2, A)$	Image C $(\mathcal{F}_3, A)$
$M_1(\mathcal{F}_i, A)$	0.571	0.647	0.847
$M_2(\mathcal{F}_i, A)$	0.039	0.089	0.328
$M_3(\mathcal{F}_i, A)$	0.571	0.647	0.847
$M_4(\mathcal{F}_i, A)$	0.084	0.202	0.682
$M_5(\mathcal{F}_i, A)$	0.084	0.202	0.682
$M_6(\mathcal{F}_i, A)$	0.084	0.202	0.682
$M_7(\mathcal{F}_i, A)$	0.084	0.202	0.682
$M_8(\mathcal{F}_i, A)$	0.571	0.647	0.847
$M_9(\mathcal{F}_i, A)$	0.571	0.647	0.847
$M_{10}(\mathcal{F}_i, A)$	0.571	0.647	0.847
$M_{11}(\mathcal{F}_i, A)$	0.571	0.647	0.847

From these values, it can be clearly seen that $M_i(\mathcal{F}_3, A) > M_i(\mathcal{F}_2, A) > M_i(\mathcal{F}_1, A)$ for all $i = 1, 2, \ldots, 11$. Hence it can be concluded that Image A is the image that is closest to the original image pic001.bmp that is stored in the memory of the robot, whereas Image C is the image that is the least similar to the original pic001.bmp image. The high entropy value for $(\mathcal{F}_3, A)$ is also an indication of the abnormality of Image C compared to Images A and B. These entropy values and the results obtained for this scenario prove the effectiveness of our proposed entropy formula. The entropy values obtained in Table 8 further verifies the validity of the relationships between the 11 formulas that was proposed in Section 4.

6. Conclusions

The objective of this work is to introduce some entropy measures for the complex vague soft set environment to measure the degree of the vagueness between sets. For this, we define firstly the axiomatic definition of the distance and entropy measures for two CVSSs and them some desirable relations between the distance and entropy are proposed. The advantages of the proposed measures are that they are defined over the set where the membership and non-membership degrees are defined as a complex number rather than real numbers. All of the information measures proposed here complement the CVSS model in representing and modeling time-periodic phenomena. The proposed measures are illustrated with a numerical example related to the problem of image detection by a robot. Furthermore, the use of CVSSs enables efficient modeling of the periodicity and/or the non-physical attributes in signal processing, image detection, and multi-dimensional pattern recognition, all of which contain multi-dimensional data. The work presented in this paper can be used as a foundation to further extend the study of the information measures for complex fuzzy sets or its generalizations. On our part, we are currently working on studying the inclusion measures and developing clustering algorithms for CVSSs. In the future, the result of this paper can be extended to some other uncertain and fuzzy environment [59–68].

Author Contributions: Conceptualization, Methodology, Validation, Writing-Original Draft Preparation, G.S.; Writing-Review & Editing, H.G.; Investigation and Visualization, S.G.Q.; Funding Acquisition, G.S.

Funding: This research was funded by the Ministry of Higher Education, Malaysia, grant number FRGS/1/2017/STG06/UCSI/03/1 and UCSI University, Malaysia, grant number Proj-In-FOBIS-014.

Acknowledgments: The authors are thankful to the editor and anonymous reviewers for their constructive comments and suggestions that helped us in improving the paper significantly. The authors would like to gratefully acknowledge the financial assistance received from the Ministry of Education, Malaysia under grant no. FRGS/1/2017/STG06/UCSI/03/1 and UCSI University, Malaysia under grant no. Proj-In-FOBIS-014.

Conflicts of Interest: The authors declare no conflict of interest.

References

1. Zadeh, L.A. Fuzzy sets. *Inf. Control* **1965**, *8*, 338–353. [CrossRef]
2. Deluca, A.; Termini, S. A definition of non-probabilistic entropy in setting of fuzzy set theory. *Inf. Control* **1971**, *20*, 301–312. [CrossRef]
3. Liu, X. Entropy, distance measure and similarity measure of fuzzy sets and their relations. *Fuzzy Sets Syst.* **1992**, *52*, 305–318.
4. Fan, J.; Xie, W. Distance measures and induced fuzzy entropy. *Fuzzy Sets Syst.* **1999**, *104*, 305–314. [CrossRef]
5. Atanassov, K.T. Intuitionistic fuzzy sets. *Fuzzy Sets Syst.* **1986**, *20*, 87–96. [CrossRef]
6. Gau, W.L.; Buehrer, D.J. Vague sets. *IEEE Trans. Syst. Man Cybern.* **1993**, *23*, 610–613. [CrossRef]
7. Atanassov, K.; Gargov, G. Interval-valued intuitionistic fuzzy sets. *Fuzzy Sets Syst.* **1989**, *31*, 343–349. [CrossRef]
8. Szmidt, E.; Kacprzyk, J. Entropy for intuitionistic fuzzy sets. *Fuzzy Sets Syst.* **2001**, *118*, 467–477. [CrossRef]
9. Vlachos, I.K.; Sergiadis, G.D. Intuitionistic fuzzy information—Application to pattern recognition. *Pattern Recognit. Lett.* **2007**, *28*, 197–206. [CrossRef]
10. Burillo, P.; Bustince, H. Entropy on intuitionistic fuzzy sets and on interval-valued fuzzy sets. *Fuzzy Sets Syst.* **1996**, *78*, 305–316. [CrossRef]
11. Garg, H.; Agarwal, N.; Tripathi, A. Generalized intuitionistic fuzzy entropy measure of order α and degree β and its applications to multi-criteria decision making problem. *Int. J. Fuzzy Syst. Appl.* **2017**, *6*, 86–107. [CrossRef]
12. Liao, H.; Xu, Z.; Zeng, X. Distance and similarity measures for hesitant fuzzy linguistic term sets and their application in multi-criteria decision making. *Inf. Sci.* **2014**, *271*, 125–142. [CrossRef]
13. Garg, H.; Agarwal, N.; Tripathi, A. A novel generalized parametric directed divergence measure of intuitionistic fuzzy sets with its application. *Ann. Fuzzy Math. Inform.* **2017**, *13*, 703–727.
14. Gou, X.; Xu, Z.; Liao, H. Hesitant fuzzy linguistic entropy and cross-entropy measures and alternative queuing method for multiple criteria decision making. *Inf. Sci.* **2017**, *388–389*, 225–246. [CrossRef]
15. Liu, W.; Liao, H. A bibliometric analysis of fuzzy decision research during 1970–2015. *Int. J. Fuzzy Syst.* **2017**, *19*, 1–14. [CrossRef]
16. Garg, H. Hesitant pythagorean fuzzy sets and their aggregation operators in multiple attribute decision making. *Int. J. Uncertain. Quantif.* **2018**, *8*, 267–289. [CrossRef]
17. Garg, H. Distance and similarity measure for intuitionistic multiplicative preference relation and its application. *Int. J. Uncertain. Quantif.* **2017**, *7*, 117–133. [CrossRef]
18. Dimuro, G.P.; Bedregal, B.; Bustince, H.; Jurio, A.; Baczynski, M.; Mis, K. QL-operations and QL-implication functions constructed from tuples (O, G, N) and the generation of fuzzy subsethood and entropy measures. *Int. J. Approx. Reason.* **2017**, *82*, 170–192. [CrossRef]
19. Liao, H.; Xu, Z.; Viedma, E.H.; Herrera, F. Hesitant fuzzy linguistic term set and its application in decision making: A state-of-the art survey. *Int. J. Fuzzy Syst.* **2017**, 1–27. [CrossRef]
20. Garg, H.; Agarwal, N.; Tripathi, A. Choquet integral-based information aggregation operators under the interval-valued intuitionistic fuzzy set and its applications to decision-making process. *Int. J. Uncertain. Quantif.* **2017**, *7*, 249–269. [CrossRef]
21. Garg, H. Linguistic Pythagorean fuzzy sets and its applications in multiattribute decision-making process. *Int. J. Intell. Syst.* **2018**, *33*, 1234–1263. [CrossRef]
22. Garg, H. A robust ranking method for intuitionistic multiplicative sets under crisp, interval environments and its applications. *IEEE Trans. Emerg. Top. Comput. Intell.* **2017**, *1*, 366–374. [CrossRef]
23. Yu, D.; Liao, H. Visualization and quantitative research on intuitionistic fuzzy studies. *J. Intell. Fuzzy Syst.* **2016**, *30*, 3653–3663. [CrossRef]
24. Garg, H. Generalized intuitionistic fuzzy entropy-based approach for solving multi-attribute decision-making problems with unknown attribute weights. *Proc. Natl. Acad. Sci. India Sect. A Phys. Sci.* **2017**, 1–11. [CrossRef]
25. Garg, H. A new generalized improved score function of interval-valued intuitionistic fuzzy sets and applications in expert systems. *Appl. Soft Comput.* **2016**, *38*, 988–999. [CrossRef]
26. Molodtsov, D. Soft set theory—First results. *Comput. Math. Appl.* **1999**, *27*, 19–31. [CrossRef]
27. Maji, P.K.; Biswas, R.; Roy, A. Intuitionistic fuzzy soft sets. *J. Fuzzy Math.* **2001**, *9*, 677–692.

28. Maji, P.K.; Biswas, R.; Roy, A.R. Fuzzy soft sets. *J. Fuzzy Math.* **2001**, *9*, 589–602.

29. Yang, H.L. Notes on generalized fuzzy soft sets. *J. Math. Res. Expos.* **2011**, *31*, 567–570.

30. Majumdar, P.; Samanta, S.K. Generalized fuzzy soft sets. *Comput. Math. Appl.* **2010**, *59*, 1425–1432. [CrossRef]

31. Agarwal, M.; Biswas, K.K.; Hanmandlu, M. Generalized intuitionistic fuzzy soft sets with applications in decision-making. *Appl. Soft Comput.* **2013**, *13*, 3552–3566. [CrossRef]

32. Garg, H.; Arora, R. Generalized and group-based generalized intuitionistic fuzzy soft sets with applications in decision-making. *Appl. Intell.* **2018**, *48*, 343–356. [CrossRef]

33. Majumdar, P.; Samanta, S. Similarity measure of soft sets. *New Math. Nat. Comput.* **2008**, *4*, 1–12. [CrossRef]

34. Garg, H.; Arora, R. Distance and similarity measures for dual hesitant fuzzy soft sets and their applications in multi criteria decision-making problem. *Int. J. Uncertain. Quantif.* **2017**, *7*, 229–248. [CrossRef]

35. Kharal, A. Distance and similarity measures for soft sets. *New Math. Nat. Comput.* **2010**, *6*, 321–334. [CrossRef]

36. Jiang, Y.; Tang, Y.; Liu, H.; Chen, Z. Entropy on intuitionistic fuzzy soft sets and on interval-valued fuzzy soft sets. *Inf. Sci.* **2013**, *240*, 95–114. [CrossRef]

37. Garg, H.; Agarwal, N.; Tripathi, A. Fuzzy number intuitionistic fuzzy soft sets and its properties. *J. Fuzzy Set Valued Anal.* **2016**, *2016*, 196–213. [CrossRef]

38. Arora, R.; Garg, H. Prioritized averaging/geometric aggregation operators under the intuitionistic fuzzy soft set environment. *Sci. Iran. E* **2018**, *25*, 466–482. [CrossRef]

39. Arora, R.; Garg, H. Robust aggregation operators for multi-criteria decision making with intuitionistic fuzzy soft set environment. *Sci. Iran. E* **2018**, *25*, 931–942. [CrossRef]

40. Garg, H.; Arora, R. A nonlinear-programming methodology for multi-attribute decision-making problem with interval-valued intuitionistic fuzzy soft sets information. *Appl. Intell.* **2017**, 1–16. [CrossRef]

41. Garg, H.; Arora, R. Bonferroni mean aggregation operators under intuitionistic fuzzy soft set environment and their applications to decision-making. *J. Oper. Res. Soc.* **2018**, 1–14. [CrossRef]

42. Xu, W.; Ma, J.; Wang, S.; Hao, G. Vague soft sets and their properties. *Comput. Math. Appl.* **2010**, *59*, 787–794. [CrossRef]

43. Chen, S.M. Measures of similarity between vague sets. *Fuzzy Sets Syst.* **1995**, *74*, 217–223. [CrossRef]

44. Wang, C.; Qu, A. Entropy, similarity measure and distance measure of vague soft sets and their relations. *Inf. Sci.* **2013**, *244*, 92–106. [CrossRef]

45. Selvachandran, G.; Maji, P.; Faisal, R.Q.; Salleh, A.R. Distance and distance induced intuitionistic entropy of generalized intuitionistic fuzzy soft sets. *Appl. Intell.* **2017**, 1–16. [CrossRef]

46. Ramot, D.; Milo, R.; Fiedman, M.; Kandel, A. Complex fuzzy sets. *IEEE Trans. Fuzzy Syst.* **2002**, *10*, 171–186. [CrossRef]

47. Ramot, D.; Friedman, M.; Langholz, G.; Kandel, A. Complex fuzzy logic. *IEEE Trans. Fuzzy Syst.* **2003**, *11*, 450–461. [CrossRef]

48. Greenfield, S.; Chiclana, F.; Dick, S. Interval-Valued Complex Fuzzy Logic. In Proceedings of the IEEE International Conference on Fuzzy Systems (FUZZ), Vancouver, BC, Canada, 24–29 July 2016; pp. 1–6. [CrossRef]

49. Yazdanbakhsh, O.; Dick, S. A systematic review of complex fuzzy sets and logic. *Fuzzy Sets Syst.* **2018**, *338*, 1–22. [CrossRef]

50. Alkouri, A.; Salleh, A. Complex Intuitionistic Fuzzy Sets. In Proceedings of the 2nd International Conference on Fundamental and Applied Sciences, Kuala Lumpur, Malaysia, 12–14 June 2012; Volume 1482, pp. 464–470.

51. Alkouri, A.U.M.; Salleh, A.R. Complex Atanassov's intuitionistic fuzzy relation. *Abstr. Appl. Anal.* **2013**, *2013*, 287382. [CrossRef]

52. Rani, D.; Garg, H. Distance measures between the complex intuitionistic fuzzy sets and its applications to the decision-making process. *Int. J. Uncertain. Quantif.* **2017**, *7*, 423–439. [CrossRef]

53. Kumar, T.; Bajaj, R.K. On complex intuitionistic fuzzy soft sets with distance measures and entropies. *J. Math.* **2014**, *2014*, 972198. [CrossRef]

54. Selvachandran, G.; Majib, P.; Abed, I.E.; Salleh, A.R. Complex vague soft sets and its distance measures. *J. Intell. Fuzzy Syst.* **2016**, *31*, 55–68. [CrossRef]

55. Selvachandran, G.; Maji, P.K.; Abed, I.E.; Salleh, A.R. Relations between complex vague soft sets. *Appl. Soft Comput.* **2016**, *47*, 438–448. [CrossRef]

56.   Selvachandran, G.; Garg, H.; Alaroud, M.H.S.; Salleh, A.R. Similarity measure of complex vague soft sets and its application to pattern recognition. *Int. J. Fuzzy Syst.* **2018**, 1–14. [CrossRef]
57.   Singh, P.K. Complex vague set based concept lattice. *Chaos Solitons Fractals* **2017**, *96*, 145–153. [CrossRef]
58.   Hu, D.; Hong, Z.; Wang, Y. A new approach to entropy and similarity measure of vague soft sets. *Sci. World J.* **2014**, *2014*, 610125. [CrossRef] [PubMed]
59.   Arora, R.; Garg, H. A robust correlation coefficient measure of dual hesitant fuzzy soft sets and their application in decision making. *Eng. Appl. Artif. Intell.* **2018**, *72*, 80–92. [CrossRef]
60.   Qiu, D.; Lu, C.; Zhang, W.; Lan, Y. Algebraic properties and topological properties of the quotient space of fuzzy numbers based on mares equivalence relation. *Fuzzy Sets Syst.* **2014**, *245*, 63–82. [CrossRef]
61.   Garg, H. Generalised Pythagorean fuzzy geometric interactive aggregation operators using Einstein operations and their application to decision making. *J. Exp. Theor. Artif. Intell.* **2018**. [CrossRef]
62.   Garg, H.; Arora, R. Novel scaled prioritized intuitionistic fuzzy soft interaction averaging aggregation operators and their application to multi criteria decision making. *Eng. Appl. Artif. Intell.* **2018**, *71*, 100–112. [CrossRef]
63.   Qiu, D.; Zhang, W. Symmetric fuzzy numbers and additive equivalence of fuzzy numbers. *Soft Comput.* **2013**, *17*, 1471–1477. [CrossRef]
64.   Garg, H.; Kumar, K. Distance measures for connection number sets based on set pair analysis and its applications to decision making process. *Appl. Intell.* **2018**, 1–14. [CrossRef]
65.   Qiu, D.; Zhang, W.; Lu, C. On fuzzy differential equations in the quotient space of fuzzy numbers. *Fuzzy Sets Syst.* **2016**, *295*, 72–98. [CrossRef]
66.   Garg, H.; Kumar, K. An advanced study on the similarity measures of intuitionistic fuzzy sets based on the set pair analysis theory and their application in decision making. *Soft Comput.* **2018**. [CrossRef]
67.   Garg, H.; Nancy. Linguistic single-valued neutrosophic prioritized aggregation operators and their applications to multiple-attribute group decision-making. *J. Ambient Intell. Hum. Comput.* **2018**, 1–23. [CrossRef]
68.   Garg, H.; Kumar, K. Some aggregation operators for Linguistic intuitionistic fuzzy set and its application to group decision-making process using the set pair analysis. *Arab. J. Sci. Eng.* **2018**, *43*, 3213–3227. [CrossRef]

entropy

MDPI

Article

A Novel Belief Entropy for Measuring Uncertainty in Dempster-Shafer Evidence Theory Framework Based on Plausibility Transformation and Weighted Hartley Entropy

Qian Pan [1,*], Deyun Zhou [1], Yongchuan Tang [1], Xiaoyang Li [1] and Jichuan Huang [2]

[1] School of Electronics and Information, Northwestern Polytechnical University, Xi'an 710072, China; dyzhounpu@nwpu.edu.cn (D.Z.); tangyongchuan@mail.nwpu.edu.cn (Y.T.); lixiaoyang@mail.nwpu.edu.cn (X.L.)
[2] First Military Representative Office of Air Force Equipment Department, People's Liberation Army Air Force, Chengdu 610013, China; jichuan1980@163.com
[*] Correspondence: panq@mail.nwpu.edu.cn; Tel.: +86-29-8843-1267

Received: 22 January 2019; Accepted: 7 February 2019; Published: 10 February 2019

Abstract: Dempster-Shafer evidence theory (DST) has shown its great advantages to tackle uncertainty in a wide variety of applications. However, how to quantify the information-based uncertainty of basic probability assignment (BPA) with belief entropy in DST framework is still an open issue. The main work of this study is to define a new belief entropy for measuring uncertainty of BPA. The proposed belief entropy has two components. The first component is based on the summation of the probability mass function (PMF) of single events contained in each BPA, which are obtained using plausibility transformation. The second component is the same as the weighted Hartley entropy. The two components could effectively measure the discord uncertainty and non-specificity uncertainty found in DST framework, respectively. The proposed belief entropy is proved to satisfy the majority of the desired properties for an uncertainty measure in DST framework. In addition, when BPA is probability distribution, the proposed method could degrade to Shannon entropy. The feasibility and superiority of the new belief entropy is verified according to the results of numerical experiments.

Keywords: Dempster-Shafer evidence theory; uncertainty of basic probability assignment; belief entropy; plausibility transformation; weighted Hartley entropy; Shannon entropy

---

1. Introduction

Dempster-Shafer evidence theory (DST) [1,2], which was initially introduced by Dempster in the context of statistical inference and then extended by Shafer into a general framework, has drawn great and continued attention in recent years [3–6]. The DST could be regarded as an extension of probability theory (PT). In DST, the probabilities are assigned to basic probability assignments (BPAs), which is presented to generalize the BPA in probability distribution in PT. The DST has shown its effectiveness and advantages in wide applications with uncertainty in terms of decision making, such as knowledge reasoning [7–9], sensor fusion [10–13], reliability analysis [14,15], fault diagnosis [16–18], assessment and evaluation [19–21], image recognition [22,23], and others [24–26].

Decision making in the framework of DST is based on the combination results of BPAs. Nonetheless, how to measure the uncertainty of BPA is still an open issue, which has not been completely solved [27]. The uncertainty of BPA mainly contains discord uncertainty and non-specificity uncertainty. Working out the uncertainty of BPA is the groundwork and precondition of applying DST to applications [28]. Entropy was initially proposed to measure the uncertainty in statistical thermodynamics [29]. Then Claude Shannon extended this concept to solve the problem of information theory, namely Shannon entropy [30]. Although the Shannon entropy is admitted as an efficient way for measuring uncertainty in PT framework, it is unavailable to be used directly in the DST as the BPA described by sets of probabilities rather than single events [31]. For the sake of better standardizing the uncertainty measure in the framework of DST, Klir and Wierman defined a list of five basic required properties that an uncertainty measure should verify in DST [32]. Many attempts have been made to extend the Shannon entropy for measuring the uncertainty of BPA in the framework of DST, including Dubois and Prade's weighted Hartley entropy [33], Höhle's confusion uncertainty measure [34], Yager's dissonance uncertainty measure [35], Klir and Ramer's discord uncertainty measure [36], Klir and Parviz's strife uncertainty measure [37], Jousselme's ambiguity uncertainty measure [38], and Deng entropy [39]. Generally speaking, these approaches could degenerate to Shannon entropy if the probability values are assigned to single events. A belief entropy following Deng entropy is proposed by Pan and Deng to measure uncertainty in DST [40]. The method borrows from the idea of Deng entropy and is based on the probability interval, which is composed of the belief function and plausibility function. Although this Deng entropy-based method contains more information and could effectively measure the uncertainty in numerical cases, it does not satisfy most of the desired properties. Moreover, the expression of discord uncertainty measure in the method just considers the central values of the lower and upper bounds of the interval, which lacks explicit practical significance. Recently, Jiroušek and Shenoy added four new properties to the set of basic requirements. Thereafter, they define a belief entropy, which could verify six desired properties [41]. Their approach uses the probability mass function (PMF) transformed by plausibility transformation and weighted Hartley entropy to measure the discord and non-specificity uncertainty, respectively. However, the PMF used in the discord uncertainty measure may cause information loss when it is converted from BPA [42]. Hence, the discord uncertainty measure used in Jiroušek and Shenoy's belief entropy needs to be improved.

In this study, inspired by Pan and Deng's uncertainty measure [32] and Jiroušek and Shenoy's uncertainty measure [41], a novel belief entropy is proposed to measure the uncertainty in DST framework. The novel belief entropy has two components, the discord uncertainty measure and non-specificity uncertainty measure. The non-specificity uncertainty measure is the same as Dubois and Prade's weighted Hartley entropy, which could efficiently reflect the scale of each BPA. The discord uncertainty measure is based on the sum of PMFs transformed by plausibility transformation of single events, which are contained in each BPA. The sum of PMFs could be seen as the representative of probability interval with a practical significance. The discord uncertainty measure in the proposed method could capture sufficient information. In addition, the proposed method could satisfy six basic required properties.

The rest of this study is organized as follows. In Section 2, the preliminaries of DST, probability transformation of BPA, and Shannon entropy are briefly introduced. In Section 3, we discuss the desired properties of uncertainty measure in DST framework. Section 4 presents the exiting belief entropies and the proposed belief entropy. The property analysis of the proposed belief entropy is also conducted in this section. In Section 5, some significant numerical experiments are carried out to illustrate the feasibility and effectiveness of the proposed belief entropy. Finally, in Section 6, the conclusion and future work are summarized.

2. Preliminaries

Some basic concepts are briefly introduced in this section, including Dempster-Shafer evidence theory [1,2], probability transformation of transforming a BPA to a PMF [43,44], and Shannon entropy [30].

2.1. Dempster-Shafer Evidence Theory

Let $\Theta = \{x_1, x_2, \ldots, x_n\}$ be a nonempty finite set of mutually exclusive and collectively exhaustive alternatives. The Θ is called the frame of discernment (FOD). The power set of Θ is denoted by 2^Θ, namely

$$2^\Theta = \{\varnothing, \{x_1\}, \{x_2\}, \ldots, \{x_n\}, \{x_1, x_2\}, \ldots, \{x_1, x_2, \ldots, x_i\}, \ldots, \Theta\}, \tag{1}$$

A BPA is a mapping m from power set 2^Θ to $[0, 1]$, which satisfies the condition:

$$m(\varnothing) = 0 \quad and \quad \sum_{A \in 2^\Theta} m(A) = 1. \tag{2}$$

A is called a focal element such that $m(A) > 0$. The BPA is also known as mass function.

There are two functions associated with each BPA called belief function $Bel(A)$ and plausibility function $Pl(A)$, respectively. The two functions are defined as follows:

$$Bel(A) = \sum_{B \subseteq A} m(A),$$
$$Pl(A) = \sum_{A \cap B \neq \varnothing} m(A). \tag{3}$$

The plausibility function $Pl(A)$ denotes the degree of BPA that potentially supports A, while the belief function $Bel(A)$ denotes the degree of BPA that definitely supports A. Thus, $Bel(A)$ and $Pl(A)$ could be seen as the lower and upper probability of A.

Suppose m_1 and m_2 are two independent BPAs in the same FOD Θ, and they can be combined by using the Dempster-Shafer combination rule as follows:

$$m(A) = m_1 \oplus m_2 = \begin{cases} \frac{\sum_{B \cap C = A} m_1(B) m_2(C)}{1-k}, A \neq \varnothing \\ 0, A = \varnothing \end{cases}, \tag{4}$$

with

$$k = \sum_{B \cap C = \varnothing} m_1(B) m_2(C), \tag{5}$$

where the k is the conflict coefficient to measure the degree of conflict among BPAs. The operator $\oplus$ denotes the Dempster-Shafer combination rule. Please note that the Dempster-Shafer combination rule is unavailable for combining BPAs such that $k > 0$.

2.2. Probability Transformation

There are many ways to transform a BPA m to a PMF. Here, the pignistic transformation and the plausibility transformation are introduced.

Let m be a BPA on FOD Θ. Its associated probabilistic expression of PMF on Θ is defined as follows:

$$BetP(x) = \sum_{A \in 2^\Theta, x \in A} \frac{m(A)}{|A|}, \tag{6}$$

where the $|A|$ is the cardinality of A. The transformation between m and $BetP(x)$ is called the pignistic transformation.

$Pt(x)$ is a probabilistic expression of PMF that is obtained from m by using plausibility transformation as follows:

$$Pt(x) = \frac{Pl(x)}{\sum_{x \in \Theta} Pl(x)}, \tag{7}$$

where the $Pl(x)$ is the plausibility function of specific element x in Θ. The transformation between m and $Pt(x)$ is called the plausibility transformation.

2.3. Shannon Entropy

Let Ω be a FOD with possible values $\{w_1, w_2, \ldots, w_n\}$. The Shannon entropy is explicitly defined as:

$$H_s = \sum_{w_i \in \Omega} p(w_i) \log_2 \left[\frac{1}{p(w_i)} \right], \tag{8}$$

where the $p(w_i)$ is the probability of alternative w_i, which satisfies $\sum_{i=1}^{n} p(w_i) = 1$. If some $p(w_i) = 0$, we follow the convention that $p(w_i) \log_2 \left[\frac{1}{p(w_i)} \right] = 0$ as $\lim_{x \to 0^+} x \log_2(x) = 0$. Please note that we will simply use log for $\log_2$ in the rest of this paper.

3. Desired Properties of Uncertainty Measures in The DS Theory

In the research of Klir and Wierman [32], Klir and Lewis [45], and Klir [46], five basic required properties are defined for uncertainty measure in DST framework, namely probabilistic consistency, set consistency, range, sub-additivity, and additivity. These requirements are detailed as follows.

- *Probability consistency.* Let m be a BPA on FOD X. If m is a Bayesian BPA, then $H(m) = \sum_{x \in X} m(x) \log \left[\frac{1}{m(x)} \right]$.

- *Additivity.* Let m_X and m_Y be distinct BPAs for FOD X and FOD Y, respectively. The combined BPA $m_X \oplus m_Y$ using Dempster-Shafer combination rules must satisfy the following equality:

$$H(m_X \oplus m_Y) = H(m_X) + H(m_Y), \tag{9}$$

where the $m_X \oplus m_Y$ is a BPA for $\{X, Y\}$. For all $A \times B \in 2^{\{X,Y\}}$, where $A \in 2^X$ and $B \in 2^Y$, we have:

$$(m_X \oplus m_Y)(A \times B) = m_X(A)m_Y(B) \tag{10}$$

- *Sub-additivity.* Let m be a BPA on the space $X \times Y$, with marginal BPAs $m^{\downarrow X}$ and $m^{\downarrow Y}$ on FOD X and FOD Y, respectively. The uncertainty measure must satisfy the following inequality:

$$H(m) \le H\left(m^{\downarrow X}\right) + H\left(m^{\downarrow Y}\right) \tag{11}$$

- *Set consistency.* Let m be a BPA on FOD X. If there exists a focal element $A \in X$ and $m(A) = 1$, then an uncertainty measure must degrade to Hartley measure:

$$H(m) = \log |A|. \tag{12}$$

- *Range.* Let m be a BPA on FOD X. The range of an uncertainty measure $H(m)$ must be $[0, \log |X|]$.

These properties illuminated in DST framework start from the verification by Shannon entropy in PT. In DST, there exist more situations of uncertainty than in PT framework [47]. Therewith, by analyzing shortcomings of these properties, Jiroušek and Shenoy add four other desired properties for measuring uncertainty in DST framework, including consistency with DST semantics, non-negativity, maximum entropy, monotonicity [41].

The uncertainty measure for BPA in DST must agree on the DST semantics [48]. Many uncertainty measures are based on the PMFs which are transformed from BPA [49–51]. However, only the plausibility transformation is compatible with the Dempster-Shafer combination rule [41,44]. Therefore, the property of consistency with DST semantics is presented to require the uncertainty measure to satisfy the tenets in DST framework.

- *Consistency with DST semantics.* Let m_1 and m_2 be two BPAs in the same FOD. If an uncertainty measure is based on a probability transformation of BPA, which transforms a BPA m to a PMF P_m, then the PMFs of m_1 and m_2 must satisfy the following condition:

$$P_{m_1 \oplus m_2} = P_{m_1} \otimes P_{m_2}, \tag{13}$$

where $\otimes$ denotes the Bayesian combination rule [41], i.e., pointwise multiplication followed by normalization. Notice that this property is not presupposing the use of probability transformation in the uncertainty measure.

The property of additivity is easy to satisfy by most definitions of uncertainty measure [41]. The property of consistency with DST semantics is regarded as reinforcement of the additivity property, which makes sure that any uncertainty measure in DST framework follows the Dempster-Shafer combination rule.

Since the number of uncertainty type in DST framework is larger than that in PT framework. One can find that uncertainty measures in DST framework prefer a wider range than that in PT framework, namely $[0, \log |X|]$. Thus, in Jiroušek and Shenoy's opinion, the properties of non-negativity, maximum entropy, and monotonicity are pivotal to uncertainty measure in DST framework.

- *Non-negativity.* Let m be a BPA on FOD X. The uncertainty measure $H(m)$ must satisfy the following inequality:

$$H(m) \geq 0, \tag{14}$$

where the equality holds up if and only if m is Bayesian and $m\{(x)\} = 1$ with $x \in X$.

- *Maximum entropy.* Let m be a BPA on FOD X. The vacuous BPA m_v should have the most uncertainty, then the uncertainty measure must satisfy the following inequality:

$$H(m_v) \geq H(m), \tag{15}$$

where the equality holds up if and only if $m = m_v$.

- *Monotonicity.* Let v_X and v_Y be the vacuous BPAs of FOD X and FOD Y, respectively. If $|X| < |Y|$, then $H(v_X) < H(v_Y)$.

The property of set consistency entails that the uncertainty of a vacuous BPA m_v for FOD X is $\log |X|$. The probability consistency entails that the uncertainty of a Bayesian BPA m_e, which has the equally likely probabilities for X, is $\log |X|$ too. However, these two requirements are contradictory as the property of maximum entropy consider $H(m_v) > H(m_e)$. About this contradiction, there is a debatable open issue. Some researchers suggest the uncertainty of these two kinds of BPA should be equal and be the maximum possible uncertainty as we cannot get information to help us make a determinate decision [52,53]. Some other researchers deem the uncertainty of a vacuous BPA to be greater than a Bayesian uniform BPA, which is demonstrated by Ellsberg paradox phenomenon [54–56]. To provide a comprehensive understanding for our definition of uncertainty measure, all the above-mentioned properties are taken into account.

4. The Belief Entropy for Uncertainty Measure in DST Framework

4.1. The Existing Definitions of Belief Entropy of BPAs

The majority of the uncertainty measures have the Shannon entropy as the start point, which plays an important role to address the uncertainty in PT framework. Nevertheless, the Shannon entropy has inherent limitations to handle the uncertainty in DST as there are more types of uncertainty [27,57]. This is reasonable because the BPA includes more information than probabilistic distribution [4]. In the earlier literatures, the definitions of belief entropy only focus on one aspect of discord uncertainty or non-specificity uncertainty in the BPAs. Then, Yager makes a contribution to distinction between the discord uncertainty and non-specificity uncertainty [35]. Thereafter, the discord and non-specificity are taken into consideration in most of the definitions of belief entropy. Some representative belief entropies and their definitions are listed as follows:

Höhel. One of the earliest uncertainty measures in DST is presented by Höhel as shown [34]:

$$H_o(m) = \sum_{A \in 2^X} m(A) \log \left[\frac{1}{Bel(A)} \right], \tag{16}$$

where the $Bel(A)$ is the belief function of proposition A. $H_o(m)$ only considers the discord uncertainty measure.

Nguyen defines the belief entropy of BPA m using the original BPAs [58]:

$$H_n(m) = \sum_{A \in 2^X} m(A) \log \left[\frac{1}{m(A)} \right]. \tag{17}$$

As the definition of $H_o(m)$, $H_n(m)$ only captures the discord part of uncertainty.

Dubois and Prade define the belief entropy using the cardinality of BPAs [33]:

$$H_d(m) = \sum_{A \in 2^X} m(A) \log |A|. \tag{18}$$

$H_d(m)$ considers only the non-specificity portion of the uncertainty. Dubois and Prade's definition could be regarded as the weighted Hartley entropy $H_h(m)$, where $H_h(m) = \log |A|$.

Pal et al. define a belief entropy as [59]:

$$H_p(m) = \sum_{A \in 2^X} m(A) \log \left[\frac{1}{m(A)} \right] + \sum_{A \in 2^X} m(A) \log (|A|). \tag{19}$$

In $H_p(m)$, the first component is the measure of discord uncertainty, and the second component is the measure of non-specificity uncertainty.

Jousselme et al. define a belief entropy based on the pignistic transformation [38]:

$$H_j(m) = \sum_{x \in X} BetP(x) \log \left[\frac{1}{BetP(x)} \right], \tag{20}$$

where the $BetP(x)$ is the PMF of pignistic transformation. The $H_j(m)$ using the Shannon entropy of $BetP(x)$

Deng defines a belief entropy, namely Deng entropy, as follows [39]:

$$H_{deng}(m) = \sum_{A \in 2^X} m(A) \log \left[\frac{1}{m(A)} \right] + \sum_{A \in 2^X} m(A) \log \left[2^{|A|} - 1 \right]. \tag{21}$$

The $H_{deng}(m)$ is very similar to the definition of $H_p(m)$, while $H_{deng}(m)$ employs the $2^{|A|} - 1$ instead of $|A|$ to measure the non-specificity uncertainty of the BPA.

Pan and Deng develop Deng entropy $H_{deng}(m)$ with the definition [40]:

$$H_{pd}(m) = \sum_{A \in 2^X} \frac{1}{2} [Bel(A) + Pl(A)] \log \left\{ \frac{1}{\frac{1}{2} [Bel(A) + Pl(A)]} \right\} + \sum_{A \in 2^X} m(A) \log \left[2^{|A|} - 1 \right], \tag{22}$$

where the $Bel(A)$ and $Pl(A)$ are the belief function and plausibility function, respectively. $H_{pd}(m)$ uses the central value of the probability interval $[Bel(A), Pl(A)]$ to measure the discord uncertainty of BPA.

It is obvious that all these uncertainty measures are the extension of the Shannon entropy in DST. Apart from the aforementioned methods of belief entropy, there are, of course, some other entropy-based uncertainty measures for BPAs in DST framework. One can find an expatiatory and detailed introduction to these methods in the literature [41,47].

Jiroušek and Shenoy define a concept for measuring uncertainty, as follows [41]:

$$H_{JS}(m) = \sum_{x \in X} Pt(x) \log \left[\frac{1}{Pt(x)} \right] + \sum_{A \in 2^X} m(A) \log (|A|). \tag{23}$$

The $H_{JS}(m)$ consists of two components. The first part is Shannon entropy of a PMF based on the plausibility transformation, which is associated with discord uncertainty. The second part is the entropy of Dubois and Prade for measuring non-specificity in BPAs. The $H_{JS}(m)$ satisfies the six desired properties, including consistency with DST semantics, non-negativity, maximum entropy, monotonicity, probability consistency, and additivity. Moreover, the properties of range and set consistency are expanded.

4.2. The Proposed Belief Entropy

Although the $H_{JS}(m)$ can better meet the requirement of the basic properties for uncertainty measure, it has an intrinsic defect. The first part in $H_{JS}(m)$ using Shannon entropy captures only the probability of plausibility transformation, which may lead to information loss. As argued in $H_{pd}(m)$, the probability interval $[Bel(A), Pl(A)]$ can provide more information according to the BPAs in each proposition. However, the $H_{pd}(m)$ considers only the numerical average of the probability interval, which lacks the piratical physical significance. In this study, by combining the merit of $H_{JS}(m)$ and $H_{pd}(m)$, a new definition of belief entropy-based uncertainty measure in DST framework is proposed as follows:

$$H_{PQ}(m) = \sum_{A \in 2^X} m(A) \log \left[\frac{1}{Pm(A)} \right] + \sum_{A \in 2^X} m(A) \log (|A|), \tag{24}$$

where the $Pm(A) = \sum_{x \in A} Pt(x)$ is the summation of plausibility transformation-based PMFs of x contained in A.

Similar to most of the belief entropies, the first component $\sum_{A \in 2^X} m(A) \log \left[Pm^{-1}(A) \right]$ in $H_{PQ}(m)$ is designed to measure the discord uncertainty of BPA. The information contained in not only BPAs but also the plausibility function based on $Pt(x)$ is taken into consideration. Since the $Pt(x)$ reflects the support degree of different propositions to element x, it could provide more information than $m(A)$. Furthermore, the $Pm(A) = \sum_{x \in A} Pt(x)$ satisfies the $Bel(A) \leq Pm(A) \leq Pl(A)$, which could be seen as a representative of the probability interval. At length, the second component $\sum_{A \in 2^X} m(A) \log (|m(A)|)$ in H_{PQ} is the same as the $H_d(m)$ to measure the non-specificity uncertainty of BPA. Therefore, we believe that the new proposed belief entropy can be more effective to measure the uncertainty of BPAs in DST framework. The property analysis of $H_{PQ}(m)$ is explored as follows.

(1) Consistency with DST semantics. The first part in $H_{PQ}(m)$ uses $Pt(x)$ based on the plausibility transformation, which is compatible with the definition of the property. The second part is not a Shannon entropy based on probability transformation. Thus, $H_{PQ}(m)$ satisfies the consistency with DST semantics property.

(2) Non-negativity. As $Pm(A) \in [0,1]$, $m(A) \in [0,1]$ and $0 < |m(A)|$, thus, $H_{PQ}(m) \geq 0$. If and only if the m is a Bayesian BPA and $m(x) = 1$, $H_{PQ}(m) = 0$. Thus, $H_{PQ}(m)$ satisfies the non-negativity property.

(3) Maximum entropy. Let m_e and m_v be a uniform Bayesian BPA and a vacuous BPA in the same FOD X, respectively. We could obtain $H_{PQ}(m_e) = H_{PQ}(m_v) = \log (|X|)$, therefore $H_{PQ}(m)$ dissatisfies the maximum entropy property.

(4) Monotonicity. Since $H_{PQ}(m_v) = \log (|X|)$, $H_{PQ}(m_v)$ is monotonic in $|X|$. Therefore $H_{PQ}(m)$ satisfies the monotonicity property.

(5) Probability consistency. If m is a Bayesian BPA, then $Pm(x) = Pt(x) = m(x)$ and $H_d(m) = 0$. Hence, $H_{PQ}(m) = \sum_{x \in X} m(x) \log \left[\frac{1}{m(x)}\right]$. Therewith, we know that the $H_{PQ}(m)$ satisfies the probability consistency property.

(6) Set consistency. If a focal element has the whole support degree such that $m(A) = 1$, $H_{PQ}(m) = \log(|X|)$. Hence, the $H_{PQ}(m)$ satisfies the set consistency property.

(7) Range. As $Pm(A)$ includes the support from the other propositions, thus, $m(A) \leq Pm(A)$. Therefore, $\sum_{A \in 2^X} m(A)\log\left[Pm^{-1}(A)\right] \leq \sum_{A \in 2^X} m(A)\log\left[m^{-1}(A)\right] = H_n(m)$. The range of $H_n(m)$ and $H_d(m)$ both are $[0, \log(|X|)]$. Thus, the range of $H_{PQ}(m)$ is $[0, 2\log(|X|)]$, which means the $H_{PQ}(m)$ dissatisfies the range property.

(8) Additivity. Let m_X and m_Y be two BPAs of FOD X and FOD Y, respectively, $A \subseteq 2^X$, and $B \subseteq 2^Y$. Let $C = A \times B$ be the corresponding joint focal element on $X \times Y$, $x \in X$, and $y \in Y$. Let m be a joint BPA defined on $X \times Y$ which is obtained by using Equation (10). Thus, $m(C) = (m_X \oplus m_Y)(A \times B) = m_X(A)m_Y(B)$. Then the new belief entropy for m is:

$$H_{PQ}(m) = H_{PQ}(m_X \oplus m_Y) = \sum_{C \in 2^{\{X \times Y\}}} m(C) \log \left[\frac{|C|}{Pm(C)}\right],$$

where

$$m(C) = m_X(A)m_Y(B),$$

$$Pm(C) = Pm(A \times B) = \sum_{(x,y) \in \{A \times B\}} Pt(x,y) = \sum_{(x,y) \in \{A \times B\}} \frac{Pl(x,y)}{\sum_{(x,y) \in \{X \times Y\}} Pl(x,y)}.$$

As proved in [33], we have $Pl(A \times B) = Pl(A)Pl(B)$. Thus, we know

$$\sum_{(x,y) \in \{A \times B\}} \frac{Pl(x,y)}{\sum_{(x,y) \in \{X \times Y\}} Pl(x,y)} = \sum_{x \in A} \sum_{y \in B} \frac{Pl(x)Pl(y)}{\sum_{x \in X} \sum_{y \in Y} Pl(x)Pl(y)} = \sum_{x \in A} \frac{Pl(x)}{\sum_{x \in X} Pl(x)} \sum_{y \in B} \frac{Pl(y)}{\sum_{y \in Y} Pl(y)}$$

and

$$Pm(C) = \sum_{x \in A} \frac{Pl(x)}{\sum_{x \in X} Pl(x)} \sum_{y \in B} \frac{Pl(y)}{\sum_{y \in Y} Pl(y)} = Pm(A)Pm(B).$$

Consequently,

$$H_{PQ}(m_X \oplus m_Y) = \sum_{C \in 2^{\{X \times Y\}}} m(C) \log \left[\frac{|A||B|}{Pm(A)Pm(B)}\right]$$

$$= \sum_{A \in 2^{\{X\}}} \sum_{B \in 2^{\{Y\}}} m_X(A)m_Y(B) \log \left[\frac{|A|}{Pm(A)}\right] + \sum_{A \in 2^{\{X\}}} \sum_{B \in 2^{\{Y\}}} m_X(A)m_Y(B) \log \left[\frac{|B|}{Pm(B)}\right]$$

$$= \sum_{A \in 2^{\{X\}}} m_X(A) \sum_{B \in 2^{\{Y\}}} m_Y(B) \log \left[\frac{|A|}{Pm(A)}\right] + \sum_{A \in 2^{\{X\}}} m_X(A) \sum_{B \in 2^{\{Y\}}} m_Y(B) \log \left[\frac{|B|}{Pm(B)}\right]$$

$$= \sum_{A \in 2^{\{X\}}} m_X(A) \log \left[\frac{|A|}{Pm(A)}\right] + \sum_{B \in 2^{\{Y\}}} m_Y(B) \log \left[\frac{|B|}{Pm(B)}\right]$$

$$= H_{PQ}(m_X) + H_{PQ}(m_Y).$$

Hence, the $H_{PQ}(m)$ satisfies the additivity property.

(9) Sub-additivity. An example of binary-valued variables is given to check whether the $H_{PQ}(m)$ satisfies the sub-additivity as follows with masses

$$m(z_{11}) = m(z_{12}) = 0.1, m(z_{21}) = m(z_{22}) = 0.3, m(X \times Y) = 0.2,$$

where $z_{ij} = (x_i, y_j)$. The marginal BPAs of for X and Y are $m^{\downarrow X}$ and $m^{\downarrow Y}$, respectively, shown as following ones.

$$m^{\downarrow X}(x_1) = 0.2, m^{\downarrow X}(x_2) = 0.6, m^{\downarrow X}(X) = 0.2$$
$$m^{\downarrow Y}(y_1) = 0.4, m^{\downarrow Y}(y_2) = 0.4, m^{\downarrow Y}(Y) = 0.2$$

Thus,

$$Pl(x_1) = 0.4, Pl(x_2) = 0.8, Pl(y_1) = 0.5, Pl(y_2) = 0.5,$$
$$Pl(Z_{11}) = 0.3, Pl(Z_{12}) = 0.3, Pl(Z_{21}) = 0.5, Pl(Z_{22}) = 0.5$$
$$Pt(x_1) = 0.333, Pt(x_2) = 0.667, Pt(y_1) = 0.5, Pt(y_2) = 0.5,$$
$$Pt(Z_{11}) = 0.188, Pt(Z_{12}) = 0.188, Pt(Z_{21}) = 0.312, Pt(Z_{22}) = 0.312$$
$$H_{PQ}(m) = 1.8899, H_{PQ}(m^{\downarrow X}) + H_{PQ}(m^{\downarrow Y}) = 0.8678 + 1 = 1.8687.$$

Obviously, $H_{PQ}(m) > H_{PQ}(m^{\downarrow X}) + H_{PQ}(m^{\downarrow Y})$, thus the $H_{PQ}(m)$ dissatisfies the sub-additivity property.

In summary, the new belief entropy $H_{PQ}(m)$ for uncertainty measure in DST framework satisfies the properties of consistency with DST semantics, non-negativity, set consistency, probability consistency, additivity, monotonicity, and does not satisfy the properties of sub-additivity, maximum entropy, range. An overview of the properties of existing belief entropies for uncertainty measure are listed in Table 1.

Table 1. An overview of the properties of existing belief entropies and the proposed method.

Definition	Cons.w DST	Non-neg	Max. ent	Monoton	Prob. cons	Add	Subadd	Range	Set. cons
Höhle	yes	no	no	no	yes	yes	no	yes	no
Smets	yes	no	no	no	no	yes	no	yes	no
Yager	yes	no	no	no	yes	yes	no	yes	no
Nguyen	yes	no	no	no	yes	yes	no	yes	no
Dubois-Prade	yes	no	yes	yes	no	yes	yes	yes	yes
Klir-Ramer	yes	yes	no	yes	yes	yes	no	no	yes
Klir-Parviz	yes	yes	no	yes	yes	yes	no	no	yes
Pal et al.	yes	yes	no	yes	yes	yes	no	no	yes
George-Pal	yes	no	no	no	no	no	no	no	yes
Maeda-Ichihashi	no	yes	yes	yes	yes	yes	yes	no	yes
Harmanec-Klir	no	yes	no	yes	yes	yes	yes	no	no
Abellán-Moral	no	yes	yes	yes	yes	yes	yes	no	yes
Jousselme et al.	no	yes	no	yes	yes	yes	no	yes	yes
Pouly et al.	no	yes	no	yes	yes	yes	no	no	yes
Jiroušek-Shenoy	yes	yes	yes	yes	yes	yes	no	no	no
Deng	yes	yes	no	yes	yes	no	no	no	no
Pan-Deng	yes	yes	no	yes	yes	no	no	no	no
Proposed method	yes	yes	no	yes	yes	yes	no	no	yes

Additionally, based on combining the advantages of the definition of Jiroušek-Shenoy and Pan-Deng, the new belief entropy involves more information, which can better meet the requirements. The properties of maximum entropy and range that the new belief entropy dissatisfies need further discussion. For maximum entropy properties, we think that the uncertainty of a vacuous BPA and an equally likely Bayesian BPA should be equivalent. There is a classical example.

Assume a bet on a race conducted by four cars, A, B, C, and D. Two experts give their opinion. Expert-1 suggests that the ability of the four drivers and the performance of the four cars are almost the same. Expert-2 has no idea about the traits of each car and driver. The opinion of the Expert-1 could be regarded as a uniform probability distribution with $m(A) = m(B) = m(C) = m(D) = \frac{1}{4}$. while the Expert-2 produces a vacuous BPA with $m(A, B, C, D) = 1$. Based on only one piece of these two pieces of evidence, we have no information to support us to make a certain bet. Besides, it is very convincing that the range property is not suitable for uncertain measure. The range $[0, \log(|X|)]$

can only reflect one aspect of uncertainty, which lacks consideration for multiple uncertainties of a BPA in DST framework. As a consequence, the properties of maximum entropy and range should be extended.

5. Numerical Experiment

In this section, several numerical experiments are verified to demonstrate the reasonability and effectiveness of our proposed new belief entropy.

5.1. Example 1

Let $\Theta = \{x\}$ be the FOD. Given a BPA with $m(x) = 1$, we can obtain the $Pt(x)$ and $Pm(x)$ with:

$$Pt(x) = 1, Pm(x) = 1.$$

Then, the associated Shannon entropy $H_s(m)$ and the proposed belief entropy $H_{PQ}(m)$ are calculated as follows:

$$H_s(m) = 1 \times \log(1) = 0, H_{PQ}(m) = 1 \times \log\left(\frac{1}{1}\right) = 0.$$

Obviously, the above example shows that the Shannon entropy and the proposed belief entropy are equal when the FOD has only one single element, where exits no uncertainty.

5.2. Example 2

Let $\Theta = \{x_1, x_2, x_3, x_4, x_5\}$ be the FOD. A uniform BPA of FOD is given as $m(x_1) = m(x_2) = m(x_3) = m(x_4) = m(x_5) = \frac{1}{5}$. Then,

$$Pt(x_1) = Pt(x_2) = Pt(x_3) = Pt(x_4) = Pt(x_5) = \frac{1}{5},$$

$$Pm(x_1) = Pm(x_2) = Pm(x_3) = Pm(x_4) = Pm(x_5) = \frac{1}{5},$$

$$H_s(m) = \frac{1}{5} \times \log 5 + \frac{1}{5} \times \log 5 + \frac{1}{5} \times \log 5 + \frac{1}{5} \times \log 5 + \frac{1}{5} \times \log 5 = 2.3219,$$

$$H_{PQ}(m) = \frac{1}{5} \times \log 5 + \frac{1}{5} \times \log 5 + \frac{1}{5} \times \log 5 + \frac{1}{5} \times \log 5 + \frac{1}{5} \times \log 5 + \dots$$

$$\dots + \frac{1}{5} \times \log 1 + \frac{1}{5} \times \log 1 + \frac{1}{5} \times \log 1 + \frac{1}{5} \times \log 1 + \frac{1}{5} \times \log 1 = 2.3219.$$

As shown above, the proposed belief entropy is the same as the Shannon entropy when the BPA is the probability distribution. Sections 5.1 and 5.2 verify that the proposed belief entropy will degenerate into the Shannon entropy when the belief is assigned to singleton elements.

5.3. Example 3

Let $\Theta = \{x_1, x_2, x_3, x_4, x_5\}$ be the FOD. A vacuous BPA of FOD is given as $m(x_1, x_2, x_3, x_4, x_5) = 1$. Then,

$$Pt(x_1) = Pt(x_2) = Pt(x_3) = Pt(x_4) = Pt(x_5) = \frac{1}{5},$$

$$Pm(x_1) = Pm(x_2) = Pm(x_3) = Pm(x_4) = Pm(x_5) = \frac{1}{5},$$

$$H_{PQ}(m) = 1 \times \log 1 + 1 \times \log 5 = 2.3219.$$

Compared to Section 5.2, we know that the uncertainty of this example is the same as the Section 5.2. This is reasonable. As discussed in Section 4.2, neither the uniform BPA nor the vacuous

BPA in the same FOD could provide more information for a determinate single element. Thus, their uncertainty should be equal.

5.4. Example 4

Two experiments in [40] are recalled in this example. Let $\Theta = \{x_1, x_2, x_3, x_4\}$ be the FOD. Two BPAs are given as m_1 and m_2. The detailed BPAs are:

$$m_1(x_1) = \frac{1}{4}, m_1(x_2) = \frac{1}{3}, m_1(x_3) = \frac{1}{6}, m_1(x_1, x_2, x_3) = \frac{1}{6}, m_1(x_4) = \frac{1}{12},$$

$$m_2(x_1) = \frac{1}{4}, m_2(x_2) = \frac{1}{3}, m_2(x_3) = \frac{1}{6}, m_2(x_1, x_2) = \frac{1}{6}, m_2(x_4) = \frac{1}{12}.$$

The corresponding $H_{PQ}(m_1)$ and $H_{PQ}(m_2)$ are calculated as follows:

$$Pl_{m_1}(x_1) = \frac{5}{12}, Pl_{m_1}(x_2) = \frac{1}{2}, Pl_{m_1}(x_3) = \frac{1}{3}, Pl_{m_1}(x_4) = \frac{1}{12},$$

$$Pl_{m_2}(x_1) = \frac{5}{12}, Pl_{m_2}(x_2) = \frac{1}{2}, Pl_{m_2}(x_3) = \frac{1}{6}, Pl_{m_2}(x_4) = \frac{1}{12},$$

$$Pt_{m_1}(x_1) = \frac{5}{16}, Pt_{m_1}(x_2) = \frac{6}{16}, Pt_{m_1}(x_3) = \frac{4}{16}, Pt_{m_1}(x_4) = \frac{1}{16},$$

$$Pt_{m_2}(x_1) = \frac{5}{14}, Pt_{m_2}(x_2) = \frac{6}{14}, Pt_{m_2}(x_3) = \frac{2}{14}, Pt_{m_2}(x_4) = \frac{1}{14},$$

$$H_{PQ}(m_1) = \frac{1}{4} \times \log(\frac{1}{5/16}) + \frac{1}{3} \times \log(\frac{1}{6/16}) + \frac{1}{6} \times \log(\frac{1}{4/16}) + \ldots$$

$$\ldots + \frac{1}{6} \times \log(\frac{3}{15/16}) + \frac{1}{12} \times \log(\frac{1}{1/16}) = 1.8375,$$

$$H_{PQ}(m_2) = \frac{1}{4} \times \log(\frac{1}{5/14}) + \frac{1}{3} \times \log(\frac{1}{6/14}) + \frac{1}{6} \times \log(\frac{1}{2/14}) + \ldots$$

$$\ldots + \frac{1}{6} \times \log(\frac{2}{13/14}) + \frac{1}{12} \times \log(\frac{1}{1/14}) = 1.7485.$$

It can be seen from the results of $H_{PQ}(m_1)$ and $H_{PQ}(m_2)$, the belief entropy of m_1 is larger than the m_2. This is logical because the $m_1(x_1, x_2, x_3) = \frac{1}{6}$ has one more single element than $m_2(x_1, x_2) = \frac{1}{6}$, which implies that the $m_1(x_1, x_2, x_3)$ contains more information. Thus, the m_1 should be more uncertain.

5.5. Example 5

Consider a target recognition problem in [60]. Target detection results provided by two independent sensors. Let A, B, C, and D be the potential target types. The results are represented by BPAs shown as follows.

$$m_1(A, B) = 0.4, m_1(C, D) = 0.6,$$

$$m_2(A, C) = 0.4, m_2(B, C) = 0.6.$$

Then the corresponding uncertainty measure with $H_{deng}(m)$, $H_{pd}(m)$ and $H_{PQ}(m)$ are calculated as:

$$Bel_{m_1}(A,B) = 0.4, Pl_{m_1}(A,B) = 0.4, Bel_{m_1}(C,D) = 0.6, Pl_{m_1}(C,D) = 0.6,$$

$$Bel_{m_2}(A,C) = 0.4, Pl_{m_2}(A,C) = 1.0, Bel_{m_2}(B,C) = 0.6, Pl_{m_2}(B,C) = 1.0,$$

$$H_d(m_1) = 0.4\log\frac{2^{|2|}-1}{0.4} + 0.6\log\frac{2^{|2|}-1}{0.6} = 2.5559,$$

$$H_d(m_2) = 0.4\log\frac{2^{|2|}-1}{0.4} + 0.6\log\frac{2^{|2|}-1}{0.6} = 2.5559,$$

$$H_{pd}(m_1) = \frac{0.4+0.4}{2}\log\frac{2^{|2|}-1}{(0.4+0.4)/2} + \frac{0.6+0.6}{2}\log\frac{2^{|2|}-1}{(0.6+0.6)/2} = 2.5559,$$

$$H_{pd}(m_2) = \frac{0.4+1.0}{2}\log\frac{2^{|2|}-1}{(0.4+1.0)/2} + \frac{0.6+1.0}{2}\log\frac{2^{|2|}-1}{(0.6+1.0)/2} = 2.9952,$$

$$Pl_{m_1}(A) = 0.4, Pl_{m_1}(B) = 0.4, Pl_{m_1}(C) = 0.6, Pl_{m_1}(D) = 0.6,$$

$$Pl_{m_2}(A) = 0.4, Pl_{m_2}(B) = 0.6, Pl_{m_2}(C) = 1.0,$$

$$Pt_{m_1}(A) = 0.2, Pt_{m_1}(B) = 0.2, Pt_{m_1}(C) = 0.3, Pt_{m_1}(D) = 0.3,$$

$$Pt_{m_2}(A) = 0.2, Pt_{m_2}(B) = 0.3, Pt_{m_2}(C) = 0.5,$$

$$H_{PQ}(m_1) = 0.4 \times \log\frac{2}{0.4} + 0.6 \times \log\frac{2}{0.6} = 1.9710,$$

$$H_{PQ}(m_2) = 0.4 \times \log\frac{2}{0.7} + 0.6 \times \log\frac{2}{0.8} = 1.3390.$$

Though the two BPAs have the same value, the BPA m_1 has four potential targets, namely A, B, C, D, while the BPA m_2 has just three potential targets, namely A, B, C. As verified in [60], it is intuitively expected that m_1 has a larger uncertainty than m_2. According to the above calculation results, the $H_{deng}(m)$ illustrates that the two BPAs have the same uncertainty, and the $H_{pd}(m)$ suggests that the m_2 has a larger uncertainty. Therefore, both $H_{deng}(m)$ and $H_{pd}(m)$ are unable to reflect the prospective difference. The proposed belief entropy can effectively quantify this divergence by considering not only the information contained in each focal element but also the mutual support degree among different focal elements. Therefore, it is safe to say that the capability of the proposed belief entropy $H_{PQ}(m)$ is unavailable in the $H_{deng}(m)$ and $H_{pd}(m)$.

5.6. Example 6

Let $\Theta = \{x_1, x_2, x_3, x_4, x_5, x_6\}$ be the FOD. Two BPAs are given as follows.

$$m_1(x_1,x_2) = \frac{1}{3}, m_1(x_3,x_4) = \frac{1}{3}, m_1(x_5,x_6) = \frac{1}{3},$$

$$m_2(x_1,x_2,x_3) = \frac{1}{2}, m_2(x_4,x_5,x_6) = \frac{1}{2}.$$

According to the Jiroušek-Shenoy entropy $H_{JS}(m)$ in Equation (23) and the proposed belief entropy $H_{PQ}(m)$ in Equation (24), both kinds of entropy consists of the discord uncertainty measure and the non-specificity uncertainty measure. Then the $H_{JS}(m)$ and $H_{PQ}(m)$ are calculated as follows.

$$Pl_{m_1}(x_1) = \frac{1}{3}, Pl_{m_1}(x_2) = \frac{1}{3}, Pl_{m_1}(x_3) = \frac{1}{3}, Pl_{m_1}(x_4) = \frac{1}{3}, Pl_{m_1}(x_5) = \frac{1}{3}, Pl_{m_1}(x_6) = \frac{1}{3},$$

$$Pl_{m_2}(x_1) = \frac{1}{2}, Pl_{m_2}(x_2) = \frac{1}{2}, Pl_{m_2}(x_3) = \frac{1}{2}, Pl_{m_2}(x_4) = \frac{1}{2}, Pl_{m_2}(x_5) = \frac{1}{2}, Pl_{m_2}(x_6) = \frac{1}{2},$$

$$Pt_{m_1}(x_1) = \frac{1}{6}, Pt_{m_1}(x_2) = \frac{1}{6}, Pt_{m_1}(x_3) = \frac{1}{6}, Pt_{m_1}(x_4) = \frac{1}{6}, Pt_{m_1}(x_5) = \frac{1}{6}, Pt_{m_1}(x_6) = \frac{1}{6},$$

$$Pt_{m_2}(x_1) = \frac{1}{6}, Pl_{m_2}(x_2) = \frac{1}{6}, Pl_{m_2}(x_3) = \frac{1}{6}, Pl_{m_2}(x_4) = \frac{1}{6}, Pl_{m_2}(x_5) = \frac{1}{6}, Pl_{m_2}(x_6) = \frac{1}{6},$$

$$H_{JS}(m_1) = H_{JS}^{dis}(m_1) + H_{JS}^{nos-spe}(m_1) = 6 \times \frac{1}{6} \times \log(\frac{1}{1/6}) + 3 \times \frac{1}{3} \times \log(2) = 3.5850,$$

$$H_{JS}(m_2) = H_{JS}^{dis}(m_2) + H_{JS}^{nos-spe}(m_2) = 6 \times \frac{1}{6} \times \log(\frac{1}{1/6}) + 2 \times \frac{1}{2} \times \log(3) = 4.1699,$$

$$H_{PQ}(m_1) = H_{PQ}^{dis}(m_1) + H_{PQ}^{nos-spe}(m_1) = 3 \times \frac{1}{3} \times \log(\frac{1}{1/3}) + 3 \times \frac{1}{3} \times \log(2) = 2.5850,$$

$$H_{PQ}(m_2) = H_{PQ}^{dis}(m_2) + H_{PQ}^{nos-spe}(m_2) = 2 \times \frac{1}{2} \times \log(\frac{1}{1/2}) + 2 \times \frac{1}{2} \times \log(3) = 2.5850.$$

The results calculated by the $H_{JS}^{dis}(m_1)$ and $H_{JS}^{dis}(m_2)$ are the same, which are equal to $\log(6)$. This outcome is counterintuitive. The BPAs in m_1 are completely different from that in m_2, thus the $H_{JS}^{dis}(m)$ and $H_{JS}^{non-spe}(m)$ of m_1 are expected to be distinguished from those ones of m_2. However, only the $H_{JS}^{non-spe}(m_1)$ and $H_{JS}^{non-spe}(m_2)$ are different. The reason for this situation is that the discord uncertainty measure $H_{JS}^{dis}(m)$ in $H_{JS}(m)$ overly concerns the conflict involved in single elements and ignores the information contained in the original BPAs. The discord uncertainty measure $H_{PQ}^{dis}(m)$ in $H_{PQ}(m)$ combines the original BPAs with the probability distribution of single elements included in the BPA can better resolve the limitations. In short, this example indicates the effectiveness for measuring the discord uncertainty of the proposed belief entropy.

5.7. Example 7

Let $\Theta = (1, 2, \ldots, 14, 15)$ be a FOD with 15 elements. The mass functions of Θ is denoted as:

$$m(3,4,5) = 0.05, \quad m(7) = 0.05, \quad m(A) = 0.8, \quad m(\Theta) = 0.1$$

The proposition A is a variable subset of 2^Θ with the number of single elements changing from 1 to 14. To verify the merit and effectiveness of the proposed belief entropy, eight uncertainty measures listed in Table 1 are selected for comparison, including Dubois and Prade's weighted Hartley entropy [33], Höhle's confusion uncertainty measure [34], Yager's dissonance uncertainty measure [35], Klir and Ramer's discord uncertainty measure [36], Klir and Parviz's strife uncertainty measure [37], George and Pal's conflict uncertainty measure [61], Pan and Deng's uncertainty measure [40], Jiroušek and Shenoy's uncertainty measure [41]. The experimental results are shown in Table 2. The Höhle's confusion uncertainty measure (M_2), Yager's dissonance uncertainty measure (M_3), Klir and Ramer's discord uncertainty measure (M_4), Klir and Parviz's strife uncertainty measure (M_5), and George and Pal's conflict uncertainty measure (M_6) are plotted in Figure 1. The Dubois and Prade's weighted Hartley entropy (M_1), Pan and Deng's uncertainty measure (M_7), Jiroušek and Shenoy's uncertainty measure (M_8), and proposed belief entropy (M_9) are plotted in Figure 2.

Table 2. The value of different uncertainty measures.

Cases	M_1	M_2	M_3	M_4	M_5	M_6	M_7	M_8	M_9
A={1}	0.4699	0.6897	0.3953	6.4419	3.3804	0.3317	16.1443	3.8322	1.9757
A={1,2}	1.2699	0.6897	0.3953	5.6419	3.2956	0.3210	17.4916	4.4789	2.3362
A={1,2,3}	1.7379	0.6897	0.1997	4.2823	2.9709	0.2943	19.8608	4.8870	2.5232
A={1,2,3,4}	2.0699	0.6897	0.1997	3.6863	2.8132	0.2677	20.8229	5.2250	2.7085
A={1,2,3,4,5}	2.3275	0.6198	0.1997	3.2946	2.7121	0.2410	21.8314	5.5200	2.8749
A={1,2,...,6}	2.5379	0.6198	0.1997	3.2184	2.7322	0.2383	22.7521	5.8059	3.0516
A={1,2,...,7}	2.7158	0.5538	0.0074	2.4562	2.5198	0.2220	24.1131	6.0425	3.0647
A={1,2,...,8}	2.8699	0.5538	0.0074	2.4230	2.5336	0.2170	25.0685	6.2772	3.2042
A={1,2,...,9}	3.0059	0.5538	0.0074	2.3898	2.5431	0.2108	26.0212	6.4921	3.3300
A={1,2,...,10}	3.1275	0.5538	0.0074	2.3568	2.5494	0.2037	27.1947	6.6903	3.4445
A={1,2,...,11}	3.2375	0.5538	0.0074	2.3241	2.5536	0.1959	27.9232	6.8743	3.5497
A={1,2,...,12}	3.3379	0.5538	0.0074	2.2920	2.5562	0.1877	29.1370	7.0461	3.6469
A={1,2,...,13}	3.4303	0.5538	0.0074	2.2605	2.5577	0.1791	30.1231	7.2071	3.7374
A={1,2,...,14}	3.5158	0.5538	0.0074	2.2296	2.5582	0.1701	31.0732	7.3587	3.8219

M_1 is the Dubois and Prade's weighted Hartley entropy; M_2 is the Höhle's confusion uncertainty measure; M_3 is the Yager's dissonance uncertainty measure; M_4 is the Klir and Ramer's discord uncertainty measure; M_5 is the Klir and Parviz's strife uncertainty measure; M_6 is the George and Pal's conflict uncertainty measure; M_7 is the Pan and Deng's uncertainty measure; M_8 is the Jiroušek and Shenoy's uncertainty measure; M_9 is the proposed belief entropy.

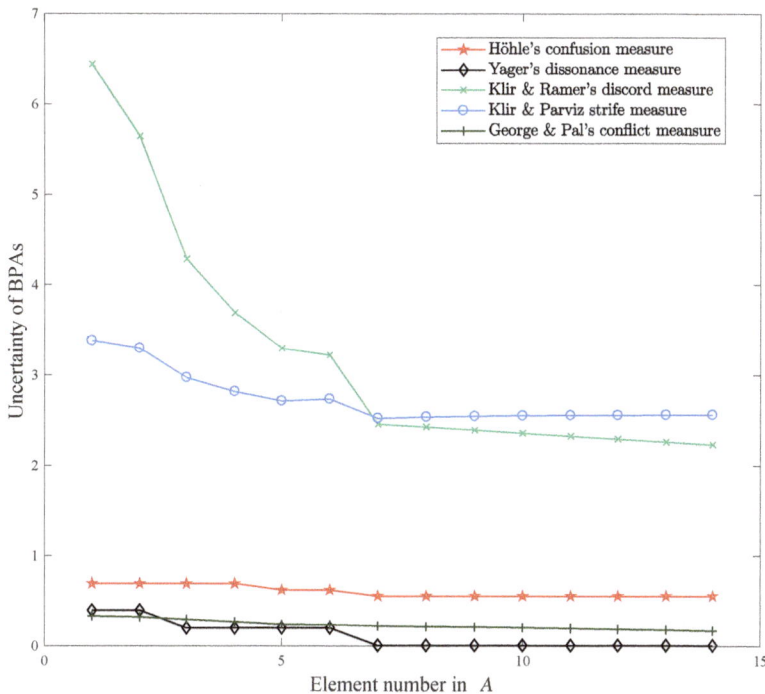

Figure 1. Results comparison of M_2, M_3, M_4, M_5, and M_6 in DST (Dempster-Shafer evidence theory).

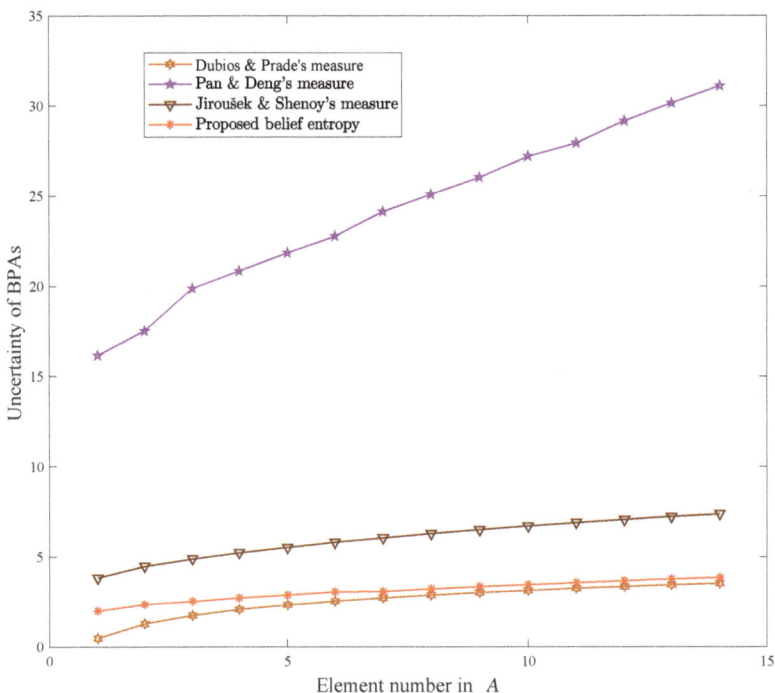

Figure 2. Results comparison of M_1, M_7, M_8, and M_9 in DST.

As shown in Figure 1, it is obvious that the uncertain degree measured by the George and Pal's conflict measure is almost unchanged when the element number increase in proposition A. Similarly, the Höhle's confusion uncertainty measure and Yager's dissonance uncertainty measure have the same situation to reflect the variation on uncertain degree in this case. Thus, these three uncertainty measures cannot detect the change in proposition A. Although the uncertainty degrees obtained by the Klir and Ramer's discord uncertainty measure and Klir and Parviz's strife uncertainty measure change with the growth of element number in A, the variation trends of both methods are contrary to expectation that the uncertainty degree increases with the augment of the element number in A. These methods only measure the discord uncertainty of the BPAs, but ignore the non-specificity uncertainty of the BPAs. Besides, from the Table 2, we can find that the Yager's dissonance uncertainty measure has the minimum uncertainty degree. This is because this method uses the plausibility function to measure the discord uncertainty. The plausibility function contains all the support degree to the single events from other propositions, which could lead to information redundant and uncertainty reducing incorrectly. To sum up, the uncertainty degree obtained by George and Pal's method, Höhle's method, Yager's method, Klir and Ramer's method, and Klir and Parviz's method are unreasonable and counterintuitive, which means these methods cannot measure the uncertainty in this case aright.

From Figure 2, it can be seen that the uncertainty degrees measured by Dubois and Prade's weighted Hartley entropy, Pan and Deng's uncertainty measure, Jiroušek and Shenoy's uncertainty measure, and the proposed belief entropy are increasing visibly with the rising of the element number in A. These methods consider not only the discord uncertainty but also the non-specificity uncertainty. Furthermore, Pan and Deng's uncertainty measure is the largest among all the methods in Table 2. This is understandable. The non-specificity uncertainty measure in Pan and Deng's method is exponential, while the others are linear. As the number of elements in A increases, the uncertainty degree of Pan and Deng's method increases faster than the other methods. Non-specificity uncertainty measure using exponential form may cause the possible uncertainty degree from the discord part to be significantly smaller than the ones from the non-specificity part. Additionally, Jiroušek and Shenoy's uncertainty measure is larger than the proposed belief entropy. Compared to the Jiroušek and Shenoy's uncertainty measure, which uses the probability distribution of single element obtained by plausibility transformation to measure the discord uncertainty, the proposed belief entropy measure that one by using the information of each mass function and the single element each BPA contains. The redundant information is removed, and the possible values of discord uncertainty is decreased notably in the proposed method. More importantly, except for the proposed method, the other three uncertainty measures have shortcomings. The Dubois and Prade's weighted Hartley entropy does not consider the discord uncertainty of BPAs. The Pan and Deng's uncertainty measure cannot measure accurately two similar BPAs in Section 5.5. The discord uncertainty measure of Jiroušek and Shenoy's uncertainty measure is irrational in Section 5.6. Thus, the proposed belief entropy is the only effective approach for uncertainty measure among these given methods in this case. Therefore, the proposed belief entropy, which considers the information contained in BPAs and single elements, is reasonable and effective for uncertainty measure in Dempster-Shafer framework.

6. Conclusions

How to measure the uncertainty of BPA in the framework of DST is an open issue. In this study, the main contribution is that a new belief entropy is proposed to quantify the uncertainty of BPA. The proposed belief entropy is comprised of the discord uncertainty measurement and the non-specificity uncertainty measurement. In particular, in the discord uncertainty measure component, the idea of probability interval and conversion BPA to probability using the plausibility transformation are combined. The new method takes advantage of the information of not only the BPAs, but also the total support degree of the single events contained in the BPAs. By addressing appropriate information in a BPA, which means less information loss and less information redundancy, the proposed belief entropy could measure the uncertainty of BPA efficiently. In addition, the proposed belief entropy could satisfy six desired properties of consistency with DST semantics, non-negativity, set consistency, probability consistency, additivity, and monotonicity. The results of numerical experiments demonstrate that the proposed belief entropy can be more effective and accurate when compared to the existing uncertainty measures in the framework of DST. Future work of this study will be focused on extending the proposed method to open-world assumptions and applying it to solve problems in real applications.

Author Contributions: Conceptualization, Q.P. and D.Z.; Data curation, Q.P. and J.H.; Formal analysis, Q.P.; Methodology, Q.P.; Software, Q.P. and Y.T.; Validation, X.L.; Writing–original draft, Q.P.; Writing–review & editing, Q.P.

Funding: This work was supported in part by the National Natural Science Foundation of China (Grant Nos. 61603299 and 61602385), Innovation Foundation for Doctor Dissertation of Northwestern Polytechnical University (Grant CX201705).

Acknowledgments: The authors greatly appreciate the reviews' valuable suggestions and the editor's encouragement. The authors greatly appreciate Boya Wei, the research assistant in Shanxi Normal Unicersity, for her advice on manuscript writing.

Conflicts of Interest: The authors declare no conflicts of interest.

References

1. Dempster, A.P. Upper and lower probabilities induced by a multivalued mapping. In *Classic Works of the Dempster-Shafer Theory of Belief Functions*; Springer: Berlin, Germany, 2008; pp. 57–72.
2. Shafer, G. *A Mathematical Theory of Evidence*; Princeton University Press: Princeton, NJ, USA, 1976; Volume 42.
3. Wang, J.; Qiao, K.; Zhang, Z. An improvement for combination rule in evidence theory. *Futur. Gener. Comput. Syst.* **2019**, *91*, 1–9. [CrossRef]

4.   Jiao, Z.; Gong, H.; Wang, Y. A DS evidence theory-based relay protection system hidden failures detection method in smart grid. *IEEE Trans. Smart Grid* **2018**, *9*, 2118–2126. [CrossRef]

5.   Jiang, W.; Zhan, J. A modified combination rule in generalized evidence theory. *Appl. Intell.* **2017**, *46*, 630–640. [CrossRef]

6.   de Oliveira Silva, L.G.; de Almeida-Filho, A.T. A multicriteria approach for analysis of conflicts in evidence theory. *Inf. Sci.* **2016**, *346*, 275–285. [CrossRef]

7.   Liu, Z.G.; Pan, Q.; Dezert, J.; Martin, A. Combination of classifiers with optimal weight based on evidential reasoning. *IEEE Trans. Fuzzy Syst.* **2018**, *26*, 1217–1230. [CrossRef]

8.   Mi, J.; Li, Y.F.; Peng, W.; Huang, H.Z. Reliability analysis of complex multi-state system with common cause failure based on evidential networks. *Reliab. Eng. Syst. Saf.* **2018**, *174*, 71–81. [CrossRef]

9.   Zhao, F.J.; Zhou, Z.J.; Hu, C.H.; Chang, L.L.; Zhou, Z.G.; Li, G.L. A new evidential reasoning-based method for online safety assessment of complex systems. *IEEE Trans. Syst. Man Cybern. Syst.* **2018**, *48*, 954–966. [CrossRef]

10.   Xiao, F. Multi-sensor data fusion based on the belief divergence measure of evidences and the belief entropy. *Inf. Fusion* **2019**, *46*, 23–32. [CrossRef]

11.   Tang, Y.; Zhou, D.; Chan, F.T.S. An Extension to Deng's Entropy in the Open World Assumption with an Application in Sensor Data Fusion. *Sensors* **2018**, *18*, 1902. [CrossRef]

12.   Zhu, J.; Wang, X.; Song, Y. Evaluating the Reliability Coefficient of a Sensor Based on the Training Data within the Framework of Evidence Theory. *IEEE Access* **2018**, *6*, 30592–30601. [CrossRef]

13.   Xiao, F.; Qin, B. A Weighted Combination Method for Conflicting Evidence in Multi-Sensor Data Fusion. *Sensors* **2018**, *18*, 1487. [CrossRef] [PubMed]

14.   Kang, B.; Chhipi-Shrestha, G.; Deng, Y.; Hewage, K.; Sadiq, R. Stable strategies analysis based on the utility of Z-number in the evolutionary games. *Appl. Math. Comput.* **2018**, *324*, 202–217. [CrossRef]

15.   Zheng, X.; Deng, Y. Dependence assessment in human reliability analysis based on evidence credibility decay model and IOWA operator. *Ann. Nuclear Energy* **2018**, *112*, 673–684. [CrossRef]

16.   Lin, Y.; Li, Y.; Yin, X.; Dou, Z. Multisensor Fault Diagnosis Modeling Based on the Evidence Theory. *IEEE Trans. Reliab.* **2018**, *67*, 513–521. [CrossRef]

17.   Song, L.; Wang, H.; Chen, P. Step-by-step Fuzzy Diagnosis Method for Equipment Based on Symptom Extraction and Trivalent Logic Fuzzy Diagnosis Theory. *IEEE Trans. Fuzzy Syst.* **2018**, *26*, 3467–3478. [CrossRef]

18.   Gong, Y.; Su, X.; Qian, H.; Yang, N. Research on fault diagnosis methods for the reactor coolant system of nuclear power plant based on DS evidence theory. *Ann. Nucl. Energy* **2018**, *112*, 395–399. [CrossRef]

19.   Zheng, H.; Deng, Y. Evaluation method based on fuzzy relations between Dempster–Shafer belief structure. *Int. J. Intell. Syst.* **2018**, *33*, 1343–1363.

20.   Song, Y.; Wang, X.; Zhu, J.; Lei, L. Sensor dynamic reliability evaluation based on evidence theory and intuitionistic fuzzy sets. *Appl. Intell.* **2018**, *48*, 3950–3962. [CrossRef]

21.   Ruan, Z.; Li, C.; Wu, A.; Wang, Y. A New Risk Assessment Model for Underground Mine Water Inrush Based on AHP and D–S Evidence Theory. *Mine Water Environ.* **2019**, 1–9, doi:10.1007/s10230-018-00575-0. [CrossRef]

22.   Ma, X.; Liu, S.; Hu, S.; Geng, P.; Liu, M.; Zhao, J. SAR image edge detection via sparse representation. *Soft Comput.* **2018**, *22*, 2507–2515. [CrossRef]

23.   Moghaddam, H.A.; Ghodratnama, S. Toward semantic content-based image retrieval using Dempster–Shafer theory in multi-label classification framework. *Int. J. Multimed. Inf. Retr.* **2017**, *6*, 317–326. [CrossRef]

24.   Torous, J.; Nicholas, J.; Larsen, M.E.; Firth, J.; Christensen, H. Clinical review of user engagement with mental health smartphone apps: evidence, theory and improvements. *Evid. Ment. Health* **2018**, *21*, 116–119. [CrossRef] [PubMed]

25.   Liu, T.; Deng, Y.; Chan, F. Evidential supplier selection based on DEMATEL and game theory. *Int. J. Fuzzy Syst.* **2018**, *20*, 1321–1333. [CrossRef]

26.   Orient, G.; Babuska, V.; Lo, D.; Mersch, J.; Wapman, W. A Case Study for Integrating Comp/Sim Credibility and Convolved UQ and Evidence Theory Results to Support Risk Informed Decision Making. In *Model Validation and Uncertainty Quantification*; Springer: Berlin, Germany, 2019; Volume 3, pp. 203–208.

27.   Li, Y.; Xiao, F. Bayesian Update with Information Quality under the Framework of Evidence Theory. *Entropy* **2019**, *21*, 5. [CrossRef]

28. Dietrich, C.F. *Uncertainty, Calibration and Probability: the Statistics of Scientific and Industrial Measurement*; Routledge: London, NY, USA, 2017.

29. Rényi, A. *On Measures of Entropy and Information*; Technical Report; Hungarian Academy of Sciences: Budapest, Hungary, 1961.

30. Shannon, C. A mathematical theory of communication. *ACM SIGMOBILE Mob. Comput. Commun. Rev.* **2001**, *5*, 3–55. [CrossRef]

31. Zhou, M.; Liu, X.B.; Yang, J.B.; Chen, Y.W.; Wu, J. Evidential reasoning approach with multiple kinds of attributes and entropy-based weight assignment. *Knowl. Syst.* **2019**, *163*, 358–375. [CrossRef]

32. Klir, G.J.; Wierman, M.J. *Uncertainty-Based Information: Elements of Generalized Information Theory*; Springer: Berlin, Germany, 2013; Volume 15.

33. Dubois, D.; Prade, H. Properties of measures of information in evidence and possibility theories. *Fuzzy Sets Syst.* **1987**, *24*, 161–182. [CrossRef]

34. Hohle, U. Entropy with respect to plausibility measures. In Proceedings of the 12th International Symposium on Multiple-Valued Logic, Paris, France, 25–27 May 1982.

35. Yager, R.R. Entropy and specificity in a mathematical theory of evidence. *Int. J. Gen. Syst.* **1983**, *9*, 249–260. [CrossRef]

36. Klir, G.J.; Ramer, A. Uncertainty in the Dempster-Shafer theory: a critical re-examination. *Int. J. Gen. Syst.* **1990**, *18*, 155–166. [CrossRef]

37. Klir, G.J.; Parviz, B. A note on the measure of discord. In *Uncertainty in Artificial Intelligence*; Elsevier: Amsterdam, The Netherlands, 1992; pp. 138–141.

38. Jousselme, A.L.; Liu, C.; Grenier, D.; Bossé, É. Measuring ambiguity in the evidence theory. *IEEE Trans. Syst. Man Cybern.-Part A Syst. Hum.* **2006**, *36*, 890–903. [CrossRef]

39. Deng, Y. Deng entropy. *Chaos, Solitons & Fractals* **2016**, *91*, 549–553.

40. Pan, L.; Deng, Y. A New Belief Entropy to Measure Uncertainty of Basic Probability Assignments Based on Belief Function and Plausibility Function. *Entropy* **2018**, *20*, 842. [CrossRef]

41. Jiroušek, R.; Shenoy, P.P. A new definition of entropy of belief functions in the Dempster–Shafer theory. *Int. J. Approx. Reason.* **2018**, *92*, 49–65. [CrossRef]

42. Yang, Y.; Han, D. A new distance-based total uncertainty measure in the theory of belief functions. *Knowl. Syst.* **2016**, *94*, 114–123. [CrossRef]

43. Smets, P. Decision making in the TBM: the necessity of the pignistic transformation. *Int. J. Approx. Reason.* **2005**, *38*, 133–148. [CrossRef]

44. Cobb, B.R.; Shenoy, P.P. On the plausibility transformation method for translating belief function models to probability models. *Int. J. Approx. Reason.* **2006**, *41*, 314–330. [CrossRef]

45. Klir, G.J.; Lewis, H.W. Remarks on "Measuring ambiguity in the evidence theory". *IEEE Trans. Syst. Man Cybern.-Part A Syst. Hum.* **2008**, *38*, 995–999. [CrossRef]

46. Klir, G.J. *Uncertainty and Information: Foundations of Generalized Information Theory*; John Wiley & Sons: Hoboken, NJ, USA, 2005.

47. Abellán, J. Analyzing properties of Deng entropy in the theory of evidence. *Chaos Solitons Fractals* **2017**, *95*, 195–199. [CrossRef]

48. Smets, P. Decision making in a context where uncertainty is represented by belief functions. In *Belief Functions in Business Decisions*; Springer: Berlin, Germany, 2002; pp. 17–61.

49. Daniel, M. On transformations of belief functions to probabilities. *Int. J. Intell. Syst.* **2006**, *21*, 261–282. [CrossRef]

50. Cuzzolin, F. On the relative belief transform. *Int. J. Approx. Reason.* **2012**, *53*, 786–804. [CrossRef]

51. Shahpari, A.; Seyedin, S. A study on properties of Dempster-Shafer theory to probability theory transformations. *Iran. J. Electr. Electron. Eng.* **2015**, *11*, 87.

52. Jaynes, E.T. Where do we stand on maximum entropy? *Maximum Entropy Formalism* **1979**, *15*, 15–118.

53. Klir, G.J. Principles of uncertainty: What are they? Why do we need them? *Fuzzy Sets Syst.* **1995**, *74*, 15–31.

54. Ellsberg, D. Risk, ambiguity, and the Savage axioms. *Q. J. Econ.* **1961**, *75*, 643–669. [CrossRef]

55. Dubois, D.; Prade, H. Properties of measures of information in evidence and possibility theories. *Fuzzy Sets Syst.* **1999**, *100*, 35–49.

56. Abellan, J.; Moral, S. Completing a total uncertainty measure in the Dempster-Shafer theory. *Int. J. Gen. Syst.* **1999**, *28*, 299–314.

57. Li, Y.; Deng, Y. Generalized Ordered Propositions Fusion Based on Belief Entropy. *Int. J. Comput. Commun. Control* **2018**, *13*. [CrossRef]
58. Nguyen, H.T. On entropy of random sets and possibility distributions. *Anal. Fuzzy Inf.* **1987**, *1*, 145–156.
59. Pal, N.R.; Bezdek, J.C.; Hemasinha, R. Uncertainty measures for evidential reasoning II: A new measure of total uncertainty. *Int. J. Approx. Reason.* **1993**, *8*, 1–16. [CrossRef]
60. Zhou, D.; Tang, Y.; Jiang, W. An improved belief entropy and its application in decision-making. *Complexity* **2017**, *2017*. [CrossRef]
61. George, T.; Pal, N.R. Quantification of conflict in Dempster-Shafer framework: A new approach. *Int. J. Gen. Syst.* **1996**, *24*, 407–423. [CrossRef]

entropy

MDPI

Article

On the Information Content of Coarse Data with Respect to the Particle Size Distribution of Complex Granular Media: Rationale Approach and Testing

Carlos García-Gutiérrez [1,*], Miguel Ángel Martín [1] and Yakov Pachepsky [2]

[1] Department of Applied Mathematics, Universidad Politécnica de Madrid, 28040 Madrid , Spain;
    miguelangel.martin@upm.es
[2] USDA-ARS Environmental Microbial and Food Safety Laboratory, Beltsville, MD 20705, USA;
    Yakov.Pachepsky@ARS.USDA.GOV
*   Correspondence: carlos.garciagutierrez@upm.es

Received: 3 June 2019; Accepted: 17 June 2019; Published: 17 June 2019

Abstract: The particle size distribution (PSD) of complex granular media is seen as a mathematical measure supported in the interval of grain sizes. A physical property characterizing granular products used in the Andreasen and Andersen model of 1930 is re-interpreted in Information Entropy terms leading to a differential information equation as a conceptual approach for the PSD. Under this approach, measured data which give a coarse description of the distribution may be seen as initial conditions for the proposed equation. A solution of the equation agrees with a selfsimilar measure directly postulated as a PSD model by Martín and Taguas almost 80 years later, thus both models appear to be linked. A variant of this last model, together with detailed soil PSD data of 70 soils are used to study the information content of limited experimental data formed by triplets and its ability in the PSD reconstruction. Results indicate that the information contained in certain soil triplets is sufficient to rebuild the whole PSD: for each soil sample tested there is always at least a triplet that contains enough information to simulate the whole distribution.

Keywords: information entropy; particle size distribution; selfsimilar measure; simulation

1. Introduction

Granular media resulting from sedimentation and/or fragmentation processes are of great interest in different fields of science, technology and industry. The particle size distribution (PSD) is a main characteristic of these granular media since it has a crucial influence on their physical properties. These media are formed by an enormous amount of particles and, in fact, any particle size within the size interval might potentially be represented in a sample, so that, the PSD may be considered as a continuous distribution. In spite of this, experimental data on this distribution is usually very limited. In the case of soil, a paradigmatic natural granular media, the distribution information is commonly reduced to three classical size fractions, clay, silt and sand [1]. A first natural question arising at this point is if there is a theoretical framework supporting the supposed information value of so limited experimental data. While several mathematical distributions have been used as PSD models [2], the PSD reconstruction from this extremely poor information needs a rationale based in some driving idea different from empirical fitting procedures.

To address this challenge, this work is focused on characterizing the PSD by a specific property which could satisfy a simple equation (differential, difference or dynamical). An outstanding example of this

Entropy **2019**, *21*, 601; doi:10.3390/e21060601   www.mdpi.com/journal/entropy

kind of approach is the pioneering work [3] in which a differential equation is proposed as a semiempirical model for the cumulative mass-size distribution $Q(x)$ of certain granular media with grain size below a given limit x.

$$\frac{dQ}{d\left(\log x\right)} = \alpha Q$$

The differential equation is formulated for granular materials whose grain distribution is arranged in the same statistical manner for both the smaller and for the greater sizes and conformed in such a way that adding a portion of greater grains, the resulting distribution is geometrically similar to the previous one; using the terminology of the authors, they have the same *granulography*.

Interestingly, behind this *old* model swarm features nowadays recognized in many complex dissipative systems. Indeed the formation processes of some granular media have certain aspects that are present in the dynamics of dissipative systems. Fragmentation of particles together with other coupled processes, suggest that the use of energy and its storage takes place in the form of "information" or disorder in the particle sizes. There are two constrains: the available energy for fragmentation is limited and also the energy needed to fragment a particle has a power law dependence of the size of the particle [4]. The maximum entropy principle [5] states that, under certain rules of optimality and randomness, the system thus would reach the maximum level of disorder conditional to the constrains imposed on the process. Entropy maximization methods have already been used to explain the power-scaling nature of size distributions caused by sudden breakage [6]. According to Prigogine [7], the balance of entropy production in dissipative systems should produce a characteristic organization level in a stationary state. In the context of this paper, this corresponds with a characteristic PSD heterogeneity. Notably the term *granulography* used in [3] may be interpreted in a very similar way. These features suggest the use certain elements of Information and Complex Systems Theories in the study of complex granular media, with the goal of establishing a rational basis under which one can evaluate and test the information content of a small number of wide ranges from the distribution.

The paper is organized as follows: in Section 2 a differential information equation is proposed as a conceptual approach for the PSD of complex granular media. Under this approach, experimental data may be seen as initial conditions of the above differential information equation. In Section 3 the use of detailed soil PSD data, together with methods based in the above mathematical approach are used to test the ability of limited experimental data to generate a full reconstruction of the PSD.

2. The Differential Information Equation for the PSD

Instead of the differential equation for the cumulative distribution proposed in [3], we present a rather different type of differential equation framed in a typical quantity used in the description of complex systems: the information entropy (IE).

In mathematical terms, the PSD of granular media may be seen as a mass particle-size distribution μ supported in the interval I of grain sizes.

Limited information on PSD is usually provided as a list of size ranges that cover I. Grains sorted according to their size thus appear distributed in a partition of size classes $P = \{I_1, I_2, \ldots, I_k\}$ defined by those ranges on the list. If the corresponding mass fractions are $p_1 = \mu(I_1), p_2 = \mu(I_2), \ldots, p_k = \mu(I_k)$, respectively, the IE of the partition P is defined by [8]

$$H_\mu(P) = -\sum_{i=1}^{k} p_i \log p_i, \tag{1}$$

provided $p_i \log p_i = 0$ if $p_i = 0$.

The number $H_\mu(P)$ is expressed in information units (bits) and its extreme values are $\log k$, which corresponds to the most even case, when all the intervals have the same cumulative mass; and 0, which corresponds to the most uneven case, when the whole mass is concentrated in a single interval.

The number $H_\mu(P)$ can be interpreted as a measure of heterogeneity. In fact, in [9] it is shown that any measure of heterogeneity having the natural properties for this purpose, must be a multiple of $H_\mu(P)$. Both, the physical hypothesis and the differential equation in [3] have an implicit recognition of scale invariant features. Also the term *granulography* used there agrees with the concept of heterogeneity, which has a precise formulation in mathematical terms, as it was said above. Thus, instead of the differential equation proposed in [3] for the cumulative distribution, we propose a different type of differential equation involving the IE.

If we consider all the partitions $P = \{I_1, I_2, \ldots, I_k\}$ of the size interval I that support the mass particle-size distribution μ, we define

$$H_\mu(r) = \inf\{H_\mu(P) : diam\ P \le r\} \tag{2}$$

where $r > 0$ and *diam P* is the diameter of P, this is, the length of the greater subinterval of P.

The use of IE allows to formulate a natural property of many multiparticular granular media similar to the master property proposed in [3]: after an arbitrary sieving at a characteristic size scale r, the amount of information received is related to that received at an "inmediately previous" sieving. This relation can be encapsulated in the following initial value problem

$$\begin{cases} \frac{d\,H_{\mu(r)}}{d\,(\log r)} = D, \\ H_\mu(r_0) = H_0 \end{cases} \tag{3}$$

where D is constant and $H_\mu(r_0)$ is the information received at an initial sieving of characteristic size r_0. This is the model we propose for the quantitative description of the PSD.

Although formally this differential equation resembles the proposed in [3], it involves different variables and has a complete different meaning. The physical hypothesis stated by [3], in these new terms, signify that when one travels through the scales using the logarithmic transformation, i.e., changes the size scale, the information content increases on the multiplicative scale. In particular, the equation implies that the information is conserved through the scales.

Under the theoretical point of view, for any partition $P = \{I_1, I_2, \ldots, I_k\}$, the coarse information content of the corresponding empiric data

$$p_1 = \mu(I_1), p_2 = \mu(I_2), \ldots, p_k = \mu(I_k),$$

may be used to provide the initial condition

$$H_0 = -\sum_{i=1}^{k} p_i \log p_i.$$

A first issue is to find out if there is a solution of the Equation (3) corresponding to these initial conditions. Theoretical results from Fractal Geometry [10,11] assure that for each set of empiric data there exists a unique selfsimilar measure, which is a particular solution of (3) using those measured data as initial condition.

Initial data has an static information content that can be calculated with Shannon's entropy. But the information potential of initial data is to suppose that this static information content is mantained, at least statistically, across the scales, which is exactly what the model (3) implies. Also, there is a scaling behaviour in every natural granular media.

Moreover, it turns out, that in the general case ($p_1 = \mu(I_1)$, $p_2 = \mu(I_2)$, ..., $p_k = \mu(I_k)$) this measure agrees with the proposed as a model for soil mass size particle-size soil distribution in [12]. The latter now appears founded under this approach, and also gives the possibility of simulation. Furthermore, this result links the fractal PSD model proposed in [12] with the model given more than seventy five years earlier in [3].

Nevertheless, the multiplicative cascade associated to each set of experimental data is unique, so a second issue is to study for which initial conditions the corresponding solution μ better reconstructs the real PSD of a certain granular media. This study will take account of which experimental initial conditions (fraction contents) storage greater information content and are most useful to retrieve the actual PSD. There rest of this work is devoted to the study of this problem in the particular case of soil.

3. Materials and Methods

3.1. Data

A total of 70 soil samples from the provinces of Jaen and Segovia in Spain were used. Samples were selected so that they covered the biggest possible part of the USDA textural triangle (Figure 1).

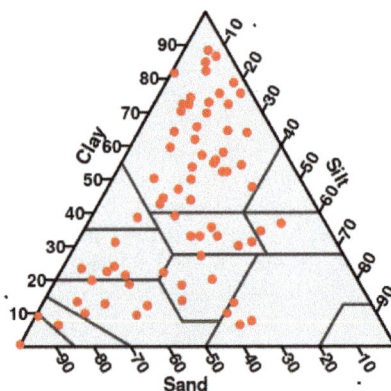

Figure 1. Representation of the 70 soil samples in the USDA textural triangle.

They belong to 10 different soil textural classes from the USDA textural classification [1], being the clay class the most represented, with 38 of the samples belonging to it. From a soil classification point of view, these soils belong to to 10 different soil classes, being Calcic Cambisol the most frequent. A complete description of these soils can be found in [13] and references within.

The particle size distribution of these samples was measured using the laser diffraction method [13] with the Longbench Mastersizer S (Malvern Instruments) with a He-Ne laser of 5 mW and a wavelength of 632.8 nm. This apparatus yields a set of data of the form

$$\{I_i = [\varphi_i, \varphi_{i+1}], v_i\}_{i=1}^N,$$

where N is the total number of size intervals I_i, and v_i is the percentage of total volume of particles whose sizes belong to the size interval I_i. Sizes are given in μm. Let I be the total size interval, i.e., $I = \cup_{i=1}^N I_i$.

In this case $I = [0.59 - 3473.45]$ Assuming a constant particle density the probability associated to each interval I_i can be calculated as

$$p_i = \frac{v_i}{\sum_{i=1}^{N} v_i}.$$

The length of the size intervals, I_i, was not constant. The first interval is 0.12 μm and the last one is 574.36 μm. Nevertheless, when using a logarithmic scale, the interval sizes become even and the endpoints of the intervals verify that the quotient $\log \frac{\varphi_{i+1}}{\varphi_i}$ remains equal. Thus, we considered the following size intervals instead:

$$\{J_i = [\phi_i, \phi_{i+1}]\}_{i=1}^{N}, \quad \phi_j = \log_{10} \varphi_j, \quad j = 1, 2, \ldots, N+1,$$

and J is the new total size interval in the logarithmic scale.

3.2. Simulation and Testing

Soil is a paradigmatic essentially complex granular system whose PSD is usually given in terms of the mass of only three size fractions, clay, silt and sand. These few data are are used as proxy for deriving many soil properties. Therefore it seemed natural to test sets of triplets of intervals along with their masses as initial conditions for the equation, simulate the whole distribution and compare it to a detailed description of the PSD. Indeed, different initial conditions lead to different PSD reconstructions. Testing differrent triplets would allow to find out if any of the them contains enough information to recover the whole PSD.

3.2.1. Triplet Description

The hypothesis was tested using only three mass size intervals in the input partition.

By collapsing the detailed PSD description obtained through the experimental analysis different triplets of mass-size intervals were obtained to use as input data. With 48 available data intervals $\{J_1, \ldots, J_{48}\}$ along their respective masses $\{q_1, \ldots, q_{48}\}$ ($\sum q_i = 1$), the number of possible combinations of those intervals into a valid triplet $\{T_1, T_2, T_3\}$ was 1081. The mass of each input interval is the sum of all the masses from the data intervals that it comprises.

As an example, let $T_1 = J_1 \cup J_2$, $T_2 = J_3$ and finally $T_3 = \cup_{i=4}^{64} J_i$. Thus, the corresponding masses are $p_1 = q_1 + q_2$, $p_2 = q_3$ and $p_3 = \sum_{i=4}^{64} q_i$.

The geometric description of the triplets is the three intervals it comprises, i.e., $T_1 = [a, b]$, $T_2 = [b, c]$ and $T_3 = [c, d]$. As a and d are the same for all triplets, the transformation

$$\alpha_1 = \frac{b-a}{d-a}, \quad \alpha_2 = \frac{c-a}{d-a},$$

allows for a succint representation of any triplet in the plane. The values of α_1 ranged from 3.504×10^{-5} to 0.697. The values from α_2 ranged from 7.375×10^{-5} to 0.835.

3.2.2. Simulation Algorithm

To simulate the distribution using limited inputs, an iterated function system (IFS) was used [12]. This algorithm can simulate the mass within any interval $A \subset J$, being I the size interval provided by the laser diffraction analysis.

Once the initial interval is divided into the three size fractions that are going to be used as inputs: T_i, where $J = \cup_{i=1}^{3} T_i$, with their respective masses p_i, we shall calculate the linear transformations, ξ_i, $i = 1, 2, 3$, that map J into T_i. Then the simulation algorithm is as follows:

1. take any $x_0 \in I$ as a starting point,

2. choose randomly, with probability p_i, one of the three linear transformations ξ_i, $i = 1, 2, 3$ and calculate the next point in the simulation $\xi_i(x_0) = x_1$,
3. repeat step (2), obtaining a sequence $\{x_k\}$ with $x_k = \xi_i(x_{k-1})$ with probability p_i, chosen randomly, $i = 1, 2, 3$.

This process defines a limit measure. The measure of any interval $A \subset J$, $\mu(A)$ can be calculated as

$$\mu(A) = \lim_{n \to \infty} \frac{m(n)}{n+1},$$

$m(n)$ being the number of points of the orbit $\{x_i\}$ of n points that fall within the interval A.

The simulated distribution was statistically compared to the experimental one using a Kolmogorov-Smirnov (KS) test [14], and thus the distributions are statistically similar.

The convergence of the algorithm is fast. We performed simulations using increasing powers of ten points in the simulation. Only 0.34% of the triplets had a different KS test result when changing from 10^5 to 10^6 points in the simulation.

4. Results and Discussion

There was total of 1081 possible input triplets to simulate the whole PSD. At least 28 triplets (2.6%) passed the KS test for all soils in the database. On average 16.9% (~182) of the available triplets passed the test for each soil. The number of triplets that passed the test was above 400 (>37%) for soils labeled 6 and 44. Example of simulation results is shown in Figure 2. Simulation with two different triplets is shown. The first one with input intervals $I_1 = [0.59 - 1.46]$, $I_2 = [1.46 - 1406.77]$ and $I_3 = [1406.77 - 3473.45]$ in µm. The values of α_1 and α_2 for this triplet were 0.0003 and 0.4049.

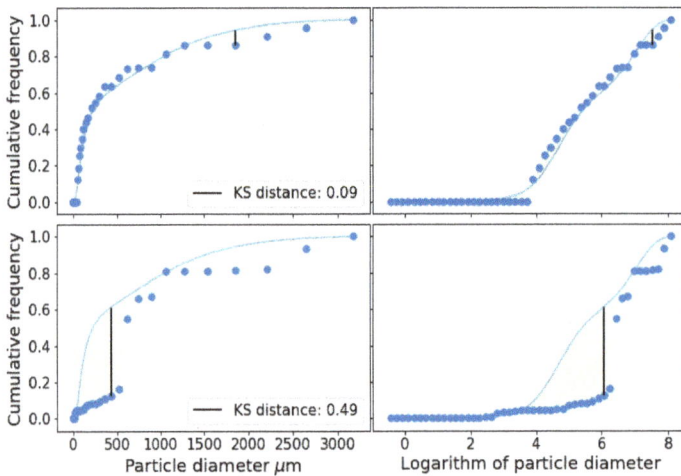

Figure 2. Actual (continuous line) and simulated (dots) PSD for soil 44. Top row shows the simulation using triplet $I_1 = [0.59 - 1.46]$, $I_2 = [1.46 - 1406.77]$ and $I_3 = [1406.77 - 3473.45]$ µm. On the left the x scale is the particle diameter in µm, while on the right, for visualization purposes, it is on the logarithmic scale. For the bottom row, the input triplet used was $I_1 = [0.59 - 37.84]$, $I_2 = [37.84 - 1174.13]$ and $I_3 = [1174.13 - 3473.45]$ µm. In both cases, the maximum allowed distance for the acceptance of the KS test at a 0.05 level was 0.28.

The other triplet was $I_1 = [0.59 - 37.84]$, $I_2 = [37.84 - 1174.13]$ and $I_3 = [1174.13 - 3473.45]$ µm, with $\alpha_1 = 0.0107$ and $\alpha_2 = 0.3379$.

Figure 3 shows the Kolmogorov-Smirnov statistics for all triplets represented in the (α_1, α_2) plane. The horizontal plane at the height of acceptance of the null hypothesis is shown.

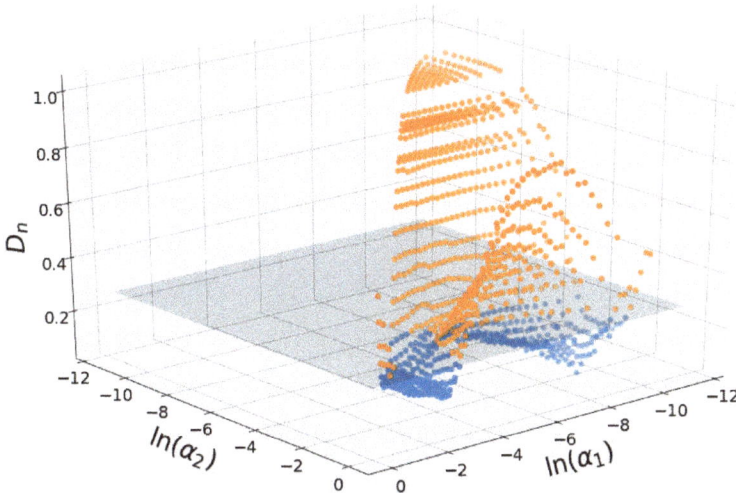

Figure 3. Representation of the KS distance, D_n, for all possible triplets, in the (α_1, α_2) plane, for soil 44. The values of α_i are in the log scale. The horizontal plane, at height, 0.28, is the limit value for D_n for the acceptance region at the 0.05 level. Blue points, below the plane, are the ones that passed the test, while orange ones do not pass it. For this soil, 403 triplets (37.28%) pass the test.

The surface has butterfly-like shape with maximum values found at the extremes of the α_2 values and intermediate α_1 values. Either fine fraction (small α_2) or coarse fraction (large α_2) have to dominate to provide better modeling results.

Total of 536 (49.6%) of all the 1,081 available input triplets passed the KS test for at least one of the soils. The triplet that passed the KS test for most soils was $I_1 = [0.59 - 192.61]$, $I_2 = [192.61 - 979.85]$, $I_3 = [979.85 - 3473.45]$ µm, with $\alpha_1 = 0.055$ and $\alpha_2 = 0.282$. The number of soils for which this triplet passed the KS test was 65 (93%). In terms of the USDA textural classification, this triplet consists of "clay + silt + fine sand", "medium sand + coarse sand", and "very coarse sand + gravel". Total of 22 triplets passed the KS test for 55 soils (79%) or more. The percentage of soils for which each triplet (α_1, α_2) passes the KS test is shown in Figure 4.

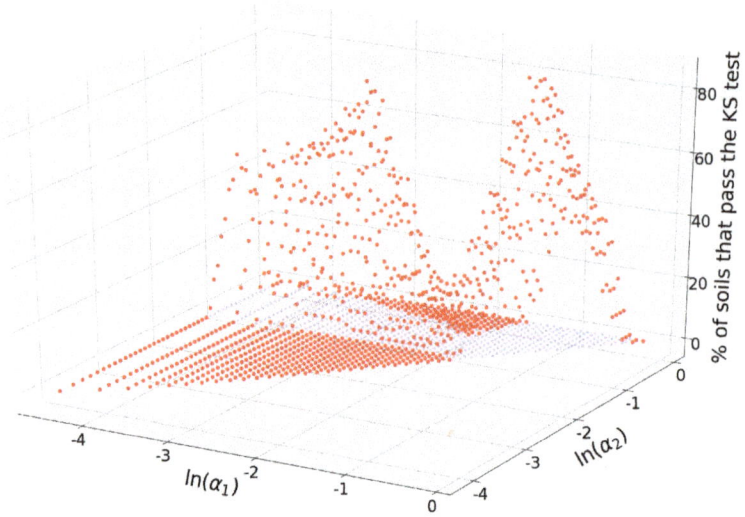

Figure 4. Red dots represent the percentage of samples that pass the KS test for a given triplet (α_1, α_2). For visualization purposes, the projection of the percentages on the plane have been added (blue dots).

Triplets with maximum acceptance rates are concentrated in the neighborhood of the point $(\alpha_1, \alpha_2) = (0.055, 0.282)$, but there are other combinations of alpha values with relatively large acceptances. Figure 5 shows a heatmap of the acceptance percentage for the (α_1, α_2) plane.

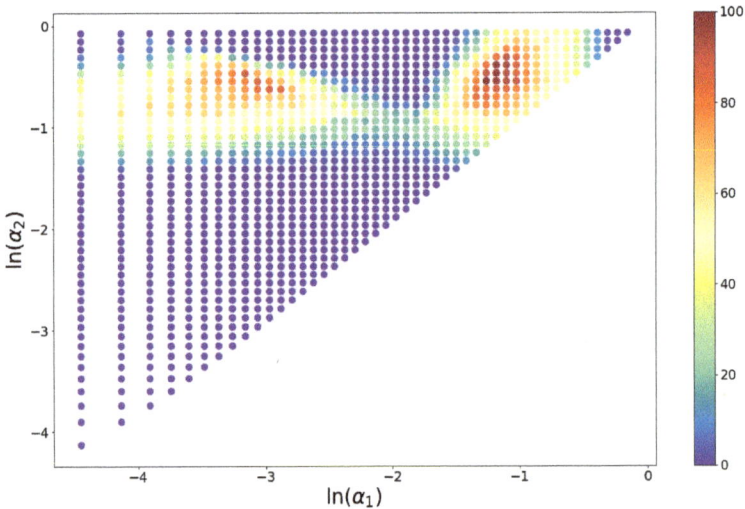

Figure 5. Heatmap for the percentage of samples that pass the KS test for a given triplet (α_1, α_2).

Entropy **2019**, *21*, 601

Whether these points attractors have a general significance, are soil database specific or have relations with conditions of soil formation presents an interesting avenue for further research.

5. Conclusions

A new model is proposed for the PSD of granular systems. This model links two works that are separated almost 70 years in time. The link between both models is made by interpreting the driving idea of the first one by means of a differential information equation which leads to the second one. This model provides a theoretical based way to simulate the entire PSD using coarse textural data as initial conditions. This advantage allows to investigate which triplets (as coarse date) stores more information content, this measured by evaluating its potential in reconstructing the real PSD.

Any coarse description of the PSD can be used as an initial condition for the model. In particular, results indicate that the information contained in certain soil triplets is sufficient to reconstruct the whole PSD. For each soil tested there is always at least one triplet that contains enough information to simulate the whole distribution.

Author Contributions: Conceptualization C.G.-G., M.Á.M. and Y.P.; methodology, C.G.-G., M.Á.M. and Y.P.; software, C.G.-G.; formal analysis, C.G.-G.; investigation, C.G.-G., M.Á.M., and Y.P.; validation, C.G.-G.; resources, M.Á.M.; data curation, C.G.-G.; writing—original draft preparation, C.G.-G., M.Á.M.; writing—review and editing, C.G.-G., M.Á.M. and Y.P.; visualization, C.G.-G.; funding acquisition, M.Á.M.

Funding: This research work was funded by Spain's Plan Nacional de Investigación Científica, Desarrollo e Innovación Tecnológica (I+D+I), under ref. AGL2015-69697-P.

Conflicts of Interest: The authors declare no conflict of interest.

Abbreviations

The following abbreviations are used in this manuscript:

PSD	Particle Size Distribution
IE	Information Entropy
KS	Kolmogorov-Smirnov
USDA	United States Department of Agriculture

References

1. Ditzler, C.; Scheffe, K.; Monger, H.C. (Eds.) *Soil Science Division Staff. Soil Survey Manual. Agriculture Handbook 18*; Government Printing Office: Washington, DC, USA, 2017.
2. Bayat, H.; Rastgou, M.; Nemes, A.; Mansourizadeh, M.; Zamani, P. Mathematical models for soil particle-size distribution and their overall and fraction-wise fitting to measurements. *Eur. J. Soil Sci.* **2017**, *68*, 345–364. [CrossRef]
3. Andreasen, A.H.M.; Andersen, J. Über die Beziehung zwischen Kornabstufung und Zwischenraum in Produkten aus losen Körnern. *Kolloid-Zeitschrift* **1930**, *50*, 217–228. [CrossRef]
4. Bertoin, J.; Martínez, S. Fragmentation Energy. *Adv. Appl. Probab.* **2005**, *37*, 553–570. [CrossRef]
5. Jaynes, E.T. Information theory and statistical mechanics. *Phys. Rev.* **1957**, *106*, 620–630. [CrossRef]
6. Englman, R.; Rivier, N.; Jaeger, Z. Size-distribution in sudden breakage by the use of entropy maximization. *J. Appl. Phys.* **1988**, *63*, 4766. [CrossRef]
7. Prigogine, I. Modération et transformations irreversibles des systèmes ouverts. *Bull. Classe Sci. Acad. R. Belg.* **1945**, *31*, 600–606.
8. Shannon, C.E. A mathematical theory of communication. *Bell Syst. Tech. J.* **1948**, *27*, 379–423. [CrossRef]
9. Khinchin, A.Y. *Mathematical Foundation of Information Theory*; Dover Publications: Mineola, NY, USA, 1957; ISBN 978-048-660-434-3.

10. Deliu, A.; Geronimo, J.S.; Shonkwiler, R.; Hardin, D. Dimensions associated with recurrent self-similar sets. *Math. Proc. Camb. Philos. Soc.* **1991**, *110*, 327–336. [CrossRef]

11. Morán, M.; Rey, J.-M. Singularity of self-similar measures with respect to Hausdorff measures. *Trans. Am. Math. Soc.* **1998**, *350*, 2297–2310. [CrossRef]

12. Martín, M.A.; Taguas, F.J. Fractal modelling, characterization and simulation of particle-size distributions in soil. *Proc. R. Soc. Lond. A* **1998**, *454*, 1457–1468. [CrossRef]

13. Montero, E. Aplicacion de Técnicas de Análisis Multifractal a Distribuciones de Tamaño-Volumen de Partículas de Suelo Obenidas Mediante análisis por Difracción de Láser. Ph.D. Thesis, Universidad Politécnica de Madrid, Madrid, Spain, 2003.

14. DeGroot, M.H. *Probability and Statistics*, 2nd ed.; Addison-Wesley Publ. Co.: Reading, MA, USA, 1986; ISBN 978-013-468-700-1.

entropy

MDPI

Article

A Novel Hybrid Meta-Heuristic Algorithm Based on the Cross-Entropy Method and Firefly Algorithm for Global Optimization

Guocheng Li [1,2,*], Pei Liu [3], Chengyi Le [4] and Benda Zhou [1,2]

1   School of Finance and Mathematics, West Anhui University, Lu'an 237012, China; bendazhou@wxc.edu.cn
2   Institute of Financial Risk Intelligent Control and Prevention, West Anhui University, Lu'an 237012, China
3   College of Computer Science, Sichuan University, Chengdu 610065, China; peiliu0408@outlook.com
4   School of Economic & Management, East China Jiaotong University, Nanchang 330013, China;
     ncycy@126.com
*   Correspondence: liguocheng@wxc.edu.cn; Tel.: +86-564-3305031

Received: 1 April 2019; Accepted: 5 May 2019; Published: 14 May 2019

Abstract: Global optimization, especially on a large scale, is challenging to solve due to its nonlinearity and multimodality. In this paper, in order to enhance the global searching ability of the firefly algorithm (FA) inspired by bionics, a novel hybrid meta-heuristic algorithm is proposed by embedding the cross-entropy (CE) method into the firefly algorithm. With adaptive smoothing and co-evolution, the proposed method fully absorbs the ergodicity, adaptability and robustness of the cross-entropy method. The new hybrid algorithm achieves an effective balance between exploration and exploitation to avoid falling into a local optimum, enhance its global searching ability, and improve its convergence rate. The results of numeral experiments show that the new hybrid algorithm possesses more powerful global search capacity, higher optimization precision, and stronger robustness.

Keywords: global optimization; meta-heuristic; firefly algorithm; cross-entropy method; co-evolution

1. Introduction

In many tasks or applications, global optimization plays a vital role, such as in power systems, industrial design, image processing, biological engineering, job-shop scheduling, economic dispatch and financial markets. In this paper, we focus our attention on unconstrained optimization problems which can be formulated as $\min f(x) : x \in \mathbb{R}^n$, where $f : \mathbb{R}^n \mapsto \mathbb{R}$ and n refers to the problems' dimension [1]. Traditional optimization methods such as the gradient-based methods usually struggle to deal with these challenging problems due to the objective function $f(x)$ can be nonlinearity, multimodality and non-convexity [2,3]. Thus, for decades, researchers have explored many derivative-free optimization methods to solve them. Generally, these optimization methods can be divided into two main classes: deterministic algorithms and stochastic algorithms [3,4]. The former, such as the Hill-Climbing [5], Newton–Raphson [6], DIRECT Algorithm [7], and Geometric and Information Global Optimization Methods with local tuning or local improvement [8,9], can get the same final results if the same set of initial values are used at the beginning [10]. However, the latter such as two well-known algorithms—Genetic Algorithm (GA) [11] and Particle Swarm Optimization (PSO) [12]—often use some randomness in their strategies which can enable the algorithm to escape from the local optima to search more regions on a global scale [10], and which have become very popular for solving real-life problems [3].

In the past two decades, meta-heuristics based on evolutionary computation and swarm intelligence have emerged and become prevalent, such as Ant Colony Optimization (ACO) [13],

Differential Evolution (DE) [14], Harmony Search (HS) [15], Bacterial Foraging Optimization Algorithm (BFOA) [16], Honey Bees Mating Optimization (HBMO) [17], Artificial Bee Colony (ABC) [18], Biogeography-Based Optimization (BBO) [19], Gravitational Search Algorithm (GSA) [20], Firefly Algorithm (FA) [21], Cuckoo Search (CS) [22], Bat Algorithm (BA) [23], Grey Wolf Optimizer (GWO) [24], Ant Lion Optimizer (ALO) [25], Moth Flame Optimizer (MFO) [26], Dragonfly Algorithm (DA) [27], Whale Optimization Algorithm (WOA) [28], Salp Swarm Algorithm (SSA) [29], Crow Search Algorithm (CSA) [30], Polar Bear Optimization (PBO) [31], Tree Growth Algorithm (TGA) [32], and Butterfly Optimization Algorithm (BOA) [33]. Meta-heuristic algorithms have been widely adopted to deal with global optimization and engineering optimization problems, and have attracted much attention as effective tools for optimization.

However, superior performance for any meta-heuristic algorithm is a target. They perform well when dealing with certain optimization problems but are not ideal in most cases [34]. In order to overcome this shortcoming, many hybrid meta-heuristic algorithms trying to combine meta-heuristics and exact algorithms or other meta-heuristics have been proposed to solve more complicated optimization problems, such as Hybrid Genetic Algorithm with Particle Swarm Optimization [35], Hybrid Particle Swarm and Ant Colony Optimization [36], Hybrid Particle Swarm Optimization with Gravitational Search Algorithm [37], Hybrid Evolutionary Firefly Algorithm [38], Hybrid Artificial Bee Colony with Firefly Algorithm [39], Hybrid Firefly-Genetic Algorithm [40], Hybrid Firefly Algorithm with Differential Evolution [10], Simulated Annealing Gaussian Bat Algorithm [41], Hybrid Harmony Search with Cuckoo Search [42], Hybrid Harmony Search with Artificial Bee Colony Algorithm [43], and Hybrid Whale Optimization Algorithm with Simulated Annealing [44]. These hybrid meta-heuristic algorithms have been successfully applied in function optimization, engineering optimization, portfolio selection, shop scheduling optimization, and feature selection.

Based on co-evolution, this paper explores a new hybrid meta-heuristic algorithm combining the cross-entropy (CE) method and the firefly algorithm (FA). The cross-entropy method was proposed by Rubinstein [45] in 1997 to solve rare event probability estimation in complex random networks, while the firefly algorithm (FA) was developed by Yang [21] and inspired by the flashing pattern of tropical fireflies in nature for multimodal optimization. The motivation of our new proposed hybrid algorithm is to improve the global search ability by embedding the cross-entropy method into the firefly algorithm to obtain an effective balance between exploration and exploitation.

The rest of the paper is organized as follows. In Section 2, CE and FA are briefly introduced, and their hybridization study is presented in Section 3. Numeral experiments and results are given in Section 4. Further analysis and a discussion of the performance of the new method are conducted in Section 5. In Section 6, the conclusions of the paper are presented.

2. Preliminaries

2.1. The Cross-Entropy Method

The cross-entropy (CE) method was proposed by Rubinstein [45] in 1997 based on Monte Carlo technology and uses Kullback–Leibler divergence to measure the cross-entropy between two sampling distributions, solve an optimization problem by minimizing them, and obtain the optimal probability distribution parameters. CE has excellent global optimization capability, good adaptability, and strong robustness. Thus, Yang regards it as a meta-heuristic algorithm [4]. However, due to the large sample size, it has the disadvantages of large computational cost and slow convergence rate. CE not only solves rare event probability estimation problems. It can also be used to solve complex optimization problems such as combination optimization [46–48], function optimization [46,48,49], engineering design [50], vehicle routing problems [51], and problems from other fields [52–54].

Let us consider the optimization problem as follows:

$$\min S(x) : X \in R^n \to R, \tag{1}$$

where S is a real-valued performance function on X.

Now, we associate the above problem with a probability distribution estimation problem, and the auxiliary problem is obtained:

$$l(\gamma) = P_\mu(S(X) \leq \gamma) = E_\mu[I_{S(X)\leq\gamma}], \tag{2}$$

where E_μ is the expectation operator, γ is a threshold or level parameter, and I is the indicator function, whose value is 1 if $S(X) \leq \gamma$ and 0, otherwise. In order to reduce the number of samples, the importance sampling method is introduced in CE. Consequently, we can rewrite Equation (2) as

$$l(\gamma) = \frac{1}{N} \sum_{i=1}^{N} I_{S(X)\leq\gamma} \frac{f(x^i; v)}{g(x^i)}, \tag{3}$$

where x^i is a random sample from $f(x; v)$ with importance sampling density $g(x)$. In order to obtain the optimal importance sampling density, the Kullback–Leibler divergence is employed to measure the distance between two densities, i.e., the cross-entropy, and the Kullback–Leibler divergence is minimized to obtain the optimal density $g^*(x)$, which is equivalent to solving the minimization problem [45]

$$\min_v \frac{1}{N} \sum_{i=1}^{N} I_{S(X)\leq\gamma} \ln f(x^i; v). \tag{4}$$

The main CE algorithm for optimization problems is summarized in Algorithm 1.

Algorithm 1: CE for Optimization Problems

Begin

Set $t = 0$. Initialize the value of the probability distribution parameter $\hat{v}_k$.

while ($t < MaxGeneration$)

Generate $Y_1, Y_2, ..., Y_L \sim_{iid} f(x; \hat{v}_k)$. Evaluate and rank the sample.

Use the sample $Y_1, Y_2, ..., Y_L$ to solve the problem given in Equation (4). Denote the solution by $\tilde{v}$.

Adaptive smoothing $\hat{v}_k$ is demoted by $\tilde{v}$.

$$\hat{v}_{k+1} = \alpha\tilde{v} + (1 - \alpha)\hat{v}_k, \tag{5}$$

where $0 \leq \alpha \leq 1$ is a smoothing parameter.

Set $t = t + 1$.

end while

Output the best solution.

End

2.2. Firefly Algorithm

The firefly algorithm (FA) was proposed by Yang [21] and inspired by the unique light signal system of fireflies in nature. Fireflies use radiance as a signal to locate and attract the opposite sex, even to forage. Based on idealizing the flashing characteristics of fireflies, the firefly algorithm was formulated for solving optimization problems. Using this algorithm, random search and optimization can be performed within a certain range, such as the solution space. Through the movement of fireflies and the constant renewal of brightness and attraction, they are constantly approaching the best position and ultimately get the best solution to the problem. FA has attracted much attention and has been applied to many applications such as global optimization [55], multimodal optimization [21], multi-objective optimization [56], engineering design problems [57], scheduling problems [58], and other fields [59–62].

In order to design FA properly, two important issues need to be defined: the variation of light intensity and formulation of the attractiveness [21]. The light intensity of a firefly can be approximated as follows:

$$I = I_0 \times e^{-\gamma r_{ij}^2},$$ (6)

where I_0 represents the original light intensity and γ is a fixed light absorption coefficient. r_{ij} indicates the distance between firefly i and firefly j and is defined as follows:

$$r_{ij} = \|x_i - x_j\| = \sqrt{\sum_{k=1}^{d} (x_{ik} - x_{jk})^2}.$$ (7)

The attractiveness of a firefly can be formulated as follows:

$$\beta = \beta_0 \times e^{-\gamma r_{ij}^2},$$ (8)

where β_0 represents the attractiveness at $r = 0$, which is the maximum attractiveness. Due to the attractiveness from firefly j, the position of firefly i is updated as follows:

$$s_i = s_i + \beta \times (s_j - s_i) + \lambda \times (rand - 0.5),$$ (9)

where s_i and s_j are the positions of fireflies i and j, respectively. The step factor λ is a constant and satisfies $0 < \lambda < 1$, and $rand$ is a random number generator uniformly distributed in $[0, 1]$, which was later replaced by Lévy flight [55].

Based on the above, the main FA can be summarized in pseudo-code as Algorithm 2.

Algorithm 2: Firefly Algorithm

Begin
    Objective function $f(x), x = (x_1, x_2, ..., x_d)^T$.
    Initialize a population of fireflies $pop_i(i = 1, 2, ..., n)$.
    Calculate the fitness value $f(pop_i)$ to determine the light intensity I_i at pop_i.
    Define light absorption coefficient γ.
    while ($t < MaxGeneration$)
        for $i = 1 : n$ all n fireflies
            for $j = 1 : n$ all n fireflies
                if ($I_j > I_i$)
                    Move firefly i towards j in all d-dimensions via Lévy flight.
                end if
                Attractiveness varies with distance r via $-e^{-\gamma r^2}$.
                Evaluate new solutions and update light intensity.
            end for j
        end for i
        Rank the fireflies and find the current best.
    end while
    Output the best solution.
End

---

3. Novel Hybrid Cross-Entropy Method and Firefly Algorithm

In this section, the details of the new hybrid algorithm are presented. A meta-heuristic algorithm should have two main exploration and exploitation functions, and an excellent meta-heuristic algorithm should try to effectively balance them and achieve better performance [63]. The cross-entropy method based on the Monte Carlo technique has the advantages of strong global optimization ability, good adaptability, and robustness [46]. It also has obvious disadvantages of large sample size, high computational cost, and slow convergence. At the same time, the firefly algorithm based on bionics has the advantages of strong local search ability and fast convergence, but it tends to fall into a local optimum rather than obtaining a global optimal solution [21]. Based on a co-evolutionary technique, this paper explores constructing a new hybrid meta-heuristic algorithm, named the Cross-Entropy Firefly Algorithm (CEFA), by embedding the cross-entropy method into the firefly algorithm. The new method contains two optimization operators—the CE operator and FA operator—which implement information sharing between the CE sample and the FA population through co-evolution in each iteration. While the FA operator updates its population using the elite sample from CE to improve the population diversity, the CE operator uses the FA population to calculate the initial probability distribution parameters in order to speed up its convergence.

The new hybrid meta-heuristic algorithm based on a co-evolutionary technique preserves the advantage of fast convergence of the swarm intelligent bionic algorithm in local search. At the same time, it also makes full use of the global optimization ability of the cross-entropy stochastic optimization method. The introduction of a co-evolutionary technique not only makes the meta-heuristic algorithms from different backgrounds complement each other but also enhances their respective advantages. Therefore, it has strong global exploration capability and local exploitation capability, and can quickly converge to global optimal solution, which provides powerful algorithm support for complex function optimization or engineering optimization problems.

The pseudo-code of CEFA is described in Algorithm 3.

In order to more clearly show the co-evolutionary process between the FA operator and the CE operator, the flow chart of CEFA is presented in Figure 1.

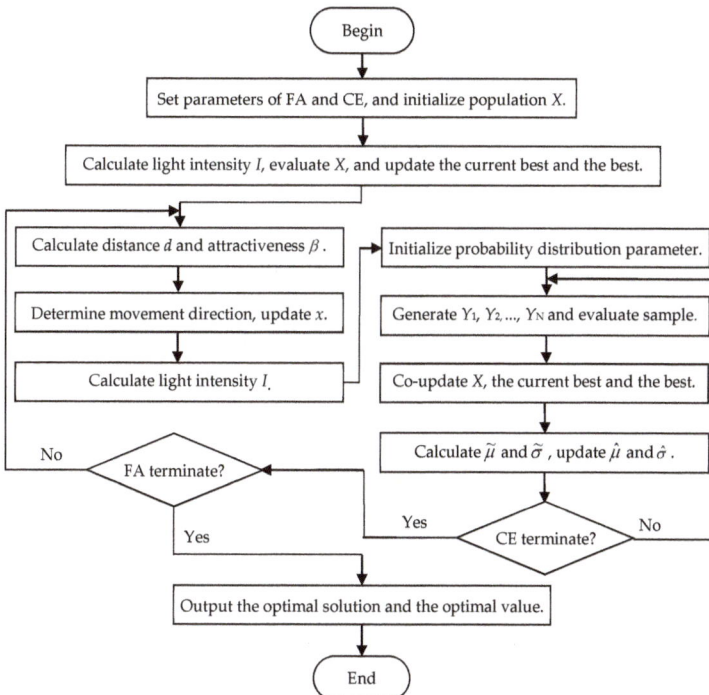

Figure 1. The flow chart of the Cross-Entropy Firefly Algorithm (CEFA).

Algorithm 3: Cross-Entropy Firefly Algorithm

Begin

  Objective function $f(x), x = (x_1, x_2, ..., x_d)^T$.

  Initialize a population of fireflies $X_i (i = 1, 2, ..., n)$.

  Calculate the fitness value $f(X_i)$ to determine the light intensity I_i at X_i.

  Define light absorption coefficient γ.

  while ($t < MaxGeneration\_FA$)

    for $i = 1 : n$ all n fireflies

      for $j = 1 : n$ all n fireflies

        if ($I_j > I_i$)

          Move firefly i towards j in all d-dimensions via Lévy flight.

        end if

        Attractiveness varies with distance r via $-e^{-\gamma r^2}$.

        Evaluate new solutions and update light intensity.

      end for j

    end for i

    Rank the fireflies and find the current best.

    Initialize the probability distribution parameter $\hat{v}_k$ by the population **X**.

    for $k = 1 : MaxGeneration\_CE$

      Generate $Y_1, Y_2, ..., Y_N \sim_{iid} f(x; \hat{v}_k)$. Evaluate the sample **Y**.

      Rank the population **X** and the sample **Y** together, update the current best.

      Update the population **X** of FA and the elite sample $\mathbf{Y}_e$ of CE.

      Calculate the probability distribution parameter $\tilde{v}$ by the elite sample $\mathbf{Y}_e$.

      Update the probability distribution parameter via Equation (5).

    end for k

  end while

  Output the best solution.

End

---

4. Experiment and Results

4.1. Benchmark Functions

In this section, 23 standard testing functions utilized by many researchers [20,24,25,27–29] were employed to evaluate the performance of the proposed hybrid algorithm CEFA for numerical optimization problems. The benchmark functions including seven unimodal functions, six multimodal functions and ten fixed-dimension multimodal functions are described in Appendix A (Table A1). The unimodal functions were used to evaluate the exploitation and convergence of an algorithm, while the multimodal functions were used to benchmark the performance of exploration and local optima avoidance [25,27]. Further information on all the benchmark functions can be found in Yao et al. (1999) [64].

4.2. Experiment Setting

Three test experiments were performed using the proposed CEFA method, and the obtained numerical solutions were compared with those from FA [21], CE [45], GA [11], PSO [12], SSA [29], BOA [31], and Hybrid Firefly Algorithm (HFA) [10] on the benchmark functions. Further information on the experiments is shown in Table 1. For these experiments, the variants were coded in MATLAB R2018b, running on a PC with an Intel Core i7-8700 machine (Gainesville, FL, USA), 3.19 GHz CPU, and 16 GB of RAM.

Table 1. Information about the three test experiments.

Name	Functions	Dimension	Comparisons
Test 1	F1–F23	2–30	FA, CE, GA, PSO, SSA, BOA, HFA, CEFA
Test 2	F1–F13	50	GA, PSO, SSA, BOA, HFA, CEFA
Test 3	F1–F13	100	GA, PSO, SSA, BOA, HFA, CEFA

Test experimental conditions and settings: (1) The population size of the FA operator in CEFA was set to 60 for Test 1 and 100 for Tests 2 and 3, while the sample size of the CE operator was 98. The maximum number of iterations of the FA operator in CEFA was 50, while the CE operator's was 30 for Test 1 and 50 for Tests 2 and 3. (2) The population sizes of other algorithms for comparison were 100, and the maximum number of iterations were 1500 for Test 1 and 2500 for Tests 2 and 3. (3) All the other parameters of each algorithm were set to be as the same as the original reference. This experimental setup ensures fairness in comparison because the numbers of functional evaluations (NFEs) for all algorithms were the same in the same test.

It is well known that all the intelligent methods are based on a certain stochastic distribution, so 30 independent runs were carried out for each method on each test function in order to statistically evaluate the proposed hybrid algorithm. The average value and standard deviation of the best approximated solution in the last iteration are introduced to compare the overall performance of the algorithms.

4.3. Results and Comparisons

The results of Test 1 are shown in Table 2. The winner (best value) is identified in bold. Among the results, the average value was used to evaluate the overall quality of the solution, reflecting the average solution accuracy of the algorithm, and the standard deviation was used to evaluate the stability of the algorithm. From Table 2, we can see the following: (1) The proposed algorithm outperforms FA, CE, GA, PSO, and SSA on almost all seven unimodal functions and six multimodal functions, while it is superior to BOA and HFA for the majority of them. This indicates that CEFA has good performance in terms of exploitation, exploration and local optima avoidance. (2) CEFA provides very competitive results in most of the ten fixed-dimension multimodal functions and tends to outperform other algorithms. The advantages of CEFA have not been fully demonstrated when solving low-dimensional function optimization problems.

The progress of the average best value over 30 runs for the benchmark functions F1, F2, F6, F10, F12, and F13 is shown in Figure 2; it shows that the proposed CEFA tends to find the global optimum significantly faster than other algorithms and has a higher convergence rate. This is due to the employed co-evolutionary mechanisms adopted between CE and FA to place emphasis on the local search and exploitation as the iteration number increases, which highly accelerate the convergence towards the optimum in the final steps of the iterations.

Tests 2 and 3 were intended to further explore the advantages of the CEFA algorithm in solving large-scale optimization problems. The test results are shown in Tables 3 and 4. Both of them show that the proposed algorithm outperforms GA, PSO, and SSA on all test problems, except for one problem with a slight difference from GA or PSO and provides very competitive results compared to BOA and HFA on the majority of multimodal functions. The superior performance of the proposed method in solving large-scale optimization problems is attributed to a good balance between exploration and exploitation, which also enhances CEFA's exploration and exploitation capabilities to focus on the high-performance areas of the search space.

Table 2. Comparison of the optimization results obtained in Test 1 ($d = 2$–30).

Fun.	Meas.	FA	CE	GA	PSO	SSA	BOA	HFA	CEFA
F1	Aver.	1.23×10^{-03}	5.45×10^{-01}	1.10×10^{-09}	3.18×10^{-23}	5.92×10^{-09}	3.09×10^{-16}	1.64×10^{-63}	$\mathbf{3.04 \times 10^{-68}}$
	Stdev.	4.35×10^{-03}	6.72×10^{-02}	3.48×10^{-09}	8.40×10^{-23}	8.80×10^{-10}	1.40×10^{-17}	1.91×10^{-64}	$\mathbf{1.58 \times 10^{-68}}$
F2	Aver.	4.36×10^{-02}	6.00×10^{-01}	3.84×10^{-05}	1.76×10^{-15}	5.24×10^{-06}	2.26×10^{-13}	1.57×10^{-32}	$\mathbf{4.18 \times 10^{-33}}$
	Stdev.	4.80×10^{-02}	6.86×10^{-02}	1.22×10^{-04}	2.33×10^{-15}	6.71×10^{-07}	9.69×10^{-15}	1.55×10^{-33}	$\mathbf{1.14 \times 10^{-33}}$
F3	Aver.	$8.59 \times 10^{+01}$	$5.19 \times 10^{+02}$	2.39×10^{-02}	$3.48 \times 10^{+00}$	7.03×10^{-10}	3.27×10^{-16}	$\mathbf{5.02 \times 10^{-18}}$	$1.08 \times 10^{+02}$
	Stdev.	4.44×10^{-01}	$1.24 \times 10^{+02}$	7.74×10^{-02}	$2.56 \times 10^{+00}$	2.17×10^{-10}	1.19×10^{-17}	$\mathbf{6.98 \times 10^{-18}}$	$9.61 \times 10^{+01}$
F4	Aver.	8.45×10^{-01}	$1.07 \times 10^{+00}$	1.96×10^{-01}	2.63×10^{-01}	1.07×10^{-05}	2.51×10^{-13}	$\mathbf{3.51 \times 10^{-14}}$	4.53×10^{-02}
	Stdev.	$1.01 \times 10^{+00}$	8.09×10^{-02}	6.15×10^{-01}	1.12×10^{-01}	1.86×10^{-06}	$\mathbf{1.34 \times 10^{-14}}$	1.31×10^{-13}	1.77×10^{-01}
F5	Aver.	$3.85 \times 10^{+01}$	$3.89 \times 10^{+01}$	$\mathbf{7.45 \times 10^{-01}}$	$3.81 \times 10^{+01}$	$3.37 \times 10^{+01}$	$2.89 \times 10^{+01}$	$6.02 \times 10^{+00}$	$2.73 \times 10^{+01}$
	Stdev.	$1.27 \times 10^{+01}$	$1.25 \times 10^{+01}$	$6.90 \times 10^{+00}$	$2.69 \times 10^{+01}$	$6.96 \times 10^{+01}$	$\mathbf{3.01 \times 10^{-02}}$	$2.40 \times 10^{+00}$	2.22×10^{-01}
F6	Aver.	7.08×10^{-04}	5.69×10^{-01}	1.14×10^{-09}	2.45×10^{-23}	4.48×10^{-10}	$4.93 \times 10^{+00}$	0	0
	Stdev.	3.05×10^{-03}	9.65×10^{-02}	3.65×10^{-09}	6.62×10^{-23}	1.60×10^{-10}	6.58×10^{-01}	0	0
F7	Aver.	4.61×10^{-02}	1.20×10^{-03}	4.26×10^{-02}	3.20×10^{-03}	1.32×10^{-03}	$\mathbf{2.74 \times 10^{-04}}$	5.55×10^{-04}	3.09×10^{-03}
	Stdev.	1.88×10^{-02}	2.94×10^{-04}	1.32×10^{-01}	1.19×10^{-03}	9.73×10^{-04}	$\mathbf{9.29 \times 10^{-05}}$	1.65×10^{-04}	7.28×10^{-04}
F8	Aver.	$-4.08 \times 10^{+03}$	$-4.39 \times 10^{+03}$	$-1.07 \times 10^{+03}$	$-6.76 \times 10^{+03}$	$-2.96 \times 10^{+03}$	$-4.39 \times 10^{+03}$	$\mathbf{-1.04 \times 10^{+04}}$	$-5.21 \times 10^{+03}$
	Stdev.	$2.53 \times 10^{+02}$	$3.47 \times 10^{+02}$	$3.25 \times 10^{+03}$	$7.70 \times 10^{+02}$	$\mathbf{2.25 \times 10^{+02}}$	$3.04 \times 10^{+02}$	$5.77 \times 10^{+02}$	$2.06 \times 10^{+03}$
F9	Aver.	$1.49 \times 10^{+02}$	$1.57 \times 10^{+02}$	1.99×10^{-01}	$3.28 \times 10^{+01}$	$1.30 \times 10^{+01}$	$\mathbf{5.69 \times 10^{-15}}$	$2.47 \times 10^{+01}$	$5.44 \times 10^{+00}$
	Stdev.	$1.17 \times 10^{+01}$	$8.45 \times 10^{+00}$	9.38×10^{-01}	1.09×10^{-01}	$5.82 \times 10^{+00}$	$\mathbf{1.80 \times 10^{-14}}$	$5.94 \times 10^{+00}$	$2.27 \times 10^{+00}$
F10	Aver.	$\mathbf{4.44 \times 10^{-15}}$	3.64×10^{-01}	7.97×10^{-06}	6.98×10^{-13}	$2.20 \times 10^{+00}$	1.92×10^{-13}	6.93×10^{-15}	$\mathbf{4.44 \times 10^{-15}}$
	Stdev.	0	2.80×10^{-02}	2.43×10^{-05}	1.14×10^{-12}	7.19×10^{-01}	4.17×10^{-14}	1.66×10^{-15}	0
F11	Aver.	2.84×10^{-03}	7.13×10^{-01}	9.86×10^{-05}	1.12×10^{-02}	3.04×10^{-01}	0	0	0
	Stdev.	1.32×10^{-03}	4.16×10^{-02}	9.86×10^{-04}	1.25×10^{-02}	1.58×10^{-01}	0	0	0
F12	Aver.	5.50×10^{-05}	5.51×10^{-03}	4.15×10^{-03}	2.03×10^{-26}	1.04×10^{-01}	3.51×10^{-01}	1.57×10^{-32}	$\mathbf{1.57 \times 10^{-32}}$
	Stdev.	7.78×10^{-05}	8.54×10^{-04}	2.52×10^{-02}	5.27×10^{-26}	3.20×10^{-01}	9.59×10^{-02}	3.16×10^{-02}	$\mathbf{5.57 \times 10^{-48}}$
F13	Aver.	5.46×10^{-03}	6.17×10^{-02}	4.39×10^{-04}	1.10×10^{-03}	7.32×10^{-04}	$1.98 \times 10^{+00}$	$\mathbf{1.35 \times 10^{-32}}$	$\mathbf{1.35 \times 10^{-32}}$
	Stdev.	7.45×10^{-03}	9.98×10^{-03}	2.16×10^{-03}	3.35×10^{-03}	2.79×10^{-03}	3.36×10^{-01}	$\mathbf{5.57 \times 10^{-48}}$	$\mathbf{5.57 \times 10^{-48}}$
F14	Aver.	1.0037	1.0970	0.3948	1.3280	$\mathbf{0.9980}$	0.9983	$\mathbf{0.9980}$	1.2470
	Stdev.	3.12×10^{-02}	5.42×10^{-01}	$1.56 \times 10^{+00}$	9.47×10^{-01}	1.51×10^{-16}	1.34×10^{-03}	0	9.44×10^{-01}
F15	Aver.	6.89×10^{-04}	$\mathbf{3.07 \times 10^{-04}}$	3.83×10^{-04}	3.69×10^{-04}	6.94×10^{-04}	3.18×10^{-04}	3.07×10^{-04}	7.31×10^{-04}
	Stdev.	1.73×10^{-04}	3.30×10^{-10}	2.05×10^{-03}	2.32×10^{-04}	3.64×10^{-04}	8.14×10^{-06}	7.67×10^{-20}	2.37×10^{-05}
F16	Aver.	-1.0316	-1.0316	-0.1032	-1.0316	-1.0316	-1.0316	-1.0316	-1.0316
	Stdev.	$\mathbf{6.78 \times 10^{-16}}$	6.20×10^{-07}	3.11×10^{-01}	$\mathbf{6.78 \times 10^{-16}}$	1.04×10^{-15}	5.73×10^{-06}	$\mathbf{6.78 \times 10^{-16}}$	$\mathbf{6.78 \times 10^{-16}}$
F17	Aver.	0.3979	0.3979	0.3979	0.4665	0.3979	0.3979	0.3979	0.3979
	Stdev.	0	9.80×10^{-06}	1.20×10^{-01}	1.27×10^{-01}	2.63×10^{-15}	9.97×10^{-05}	0	0

Table 2. *Cont.*

Fun.	Meas.	FA	CE	GA	PSO	SSA	BOA	HFA	CEFA
F18	Aver.	3.9000	6.4068	**3.0000**	**3.0000**	**3.0000**	3.0020	**3.0000**	3.9000
	Stdev.	$4.93 \times 10^{+00}$	$1.09 \times 10^{+01}$	1.245×10^{-10}	1.31×10^{-15}	3.80×10^{-14}	1.41×10^{-03}	$\mathbf{1.76 \times 10^{-15}}$	$4.93 \times 10^{+00}$
F19	Aver.	**−3.8628**	−3.8593	−0.3863	−3.7727	**−3.8628**	−3.8619	**−3.8628**	−3.8064
	Stdev.	$\mathbf{2.71 \times 10^{-15}}$	1.20×10^{-02}	$1.16 \times 10^{+00}$	6.63×10^{-02}	2.84×10^{-15}	1.17×10^{-03}	$\mathbf{2.71 \times 10^{-15}}$	1.96×10^{-01}
F20	Aver.	−3.2784	−3.2863	−0.3251	−2.3324	−3.2190	−3.1088	−3.27	**−3.2900**
	Stdev.	5.83×10^{-02}	5.54×10^{-02}	9.80×10^{-01}	3.16×10^{-01}	$\mathbf{4.11 \times 10^{-02}}$	7.21×10^{-02}	5.92×10^{-02}	5.33×10^{-02}
F21	Aver.	**−10.1532**	−6.1882	−0.638	−2.3449	−9.0573	−9.1254	**−10.1532**	−6.7096
	Stdev.	$\mathbf{6.63 \times 10^{-15}}$	$3.77 \times 10^{+00}$	$2.18 \times 10^{+00}$	9.81×10^{-01}	$2.27 \times 10^{+00}$	9.23×10^{-01}	$1.90 \times 10^{+00}$	$3.75 \times 10^{+00}$
F22	Aver.	−9.5164	−10.1479	−0.7815	−2.2815	−9.8742	−9.7991	**−10.4029**	**−10.4029**
	Stdev.	$2.58 \times 10^{+00}$	$1.40 \times 10^{+00}$	$2.57 \times 10^{+00}$	9.73×10^{-01}	$1.61 \times 10^{+00}$	5.03×10^{-01}	1.75×10^{-15}	$\mathbf{1.65 \times 10^{-15}}$
F23	Aver.	−10.3130	**−10.5364**	−0.8559	−2.3258	−9.919	−10.0764	**−10.5364**	**−10.5364**
	Stdev.	$2.88 \times 10^{+00}$	2.22×10^{-09}	$2.76 \times 10^{+00}$	9.13×10^{-01}	$1.91 \times 10^{+00}$	2.96×10^{-01}	$\mathbf{1.62 \times 10^{-15}}$	1.81×10^{-15}

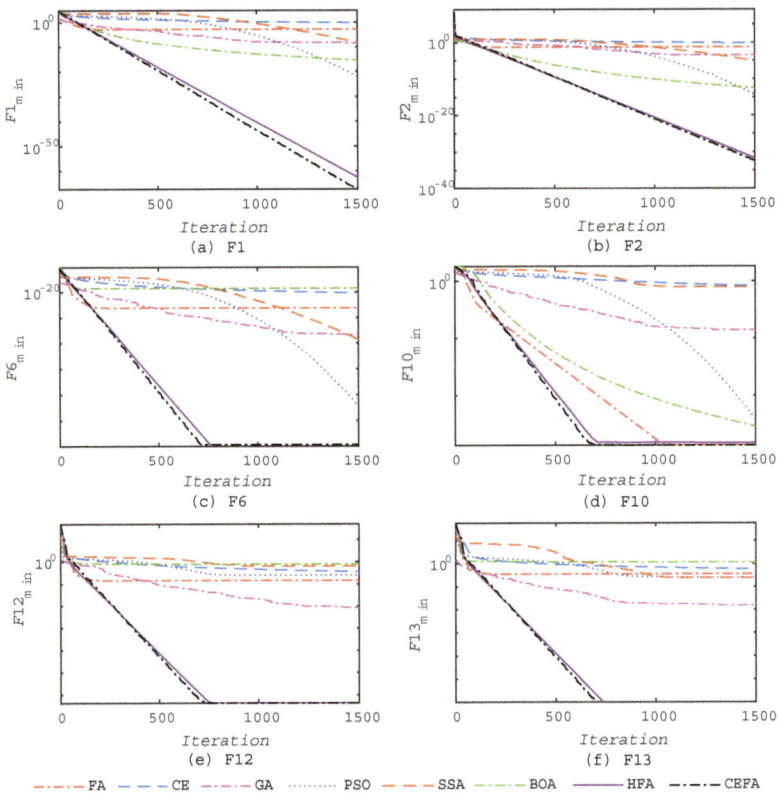

Figure 2. Convergence of algorithms on some of the benchmark functions in Test 1.

In addition, the good convergence speed of the proposed CEFA algorithm could be concluded from Figures 3 and 4 when solving large-scale optimization problems, in which the same six functions, F1, F2, F6, F10, F12, and F13, were selected from the benchmark functions for comparison. From these, we can see that the local optima avoidance of this algorithm is satisfactory since it is able to avoid all of the local optima and approximate the global optima on the majority of the multimodal test functions. These results reaffirm that the operators of CEFA appropriately balance exploration and exploitation to handle difficulty in a challenging and high-dimensional search space.

Table 3. Comparison of the optimization results obtained in Test 2 ($d = 50$).

F	Meas.	GA	PSO	SSA	BOA	HFA	CEFA
F1	Aver.	4.92×10^{-09}	5.32×10^{-19}	4.68×10^{-09}	2.28×10^{-18}	$\mathbf{3.24 \times 10^{-106}}$	2.05×10^{-65}
	Stdev.	1.91×10^{-08}	8.24×10^{-19}	7.90×10^{-10}	8.51×10^{-20}	$\mathbf{1.64 \times 10^{-59}}$	7.97×10^{-66}
F2	Aver.	1.16×10^{-02}	1.96×10^{-12}	4.58×10^{-06}	$3.47 \times 10^{+20}$	$\mathbf{4.08 \times 10^{-54}}$	1.25×10^{-31}
	Stdev.	5.01×10^{-02}	4.69×10^{-12}	9.23×10^{-07}	$1.90 \times 10^{+21}$	$\mathbf{3.36 \times 10^{-55}}$	2.66×10^{-32}
F3	Aver.	2.28×10^{-01}	$1.58 \times 10^{+02}$	5.06×10^{-10}	$\mathbf{2.33 \times 10^{-18}}$	3.38×10^{-09}	$5.61 \times 10^{+02}$
	Stdev.	7.19×10^{-01}	$5.26 \times 10^{+01}$	1.55×10^{-10}	$\mathbf{7.58 \times 10^{-20}}$	2.92×10^{-09}	$2.98 \times 10^{+02}$
F4	Aver.	2.34×10^{-01}	$2.48 \times 10^{+00}$	1.03×10^{-05}	$\mathbf{1.98 \times 10^{-15}}$	1.43×10^{-02}	$1.91 \times 10^{+00}$
	Stdev.	7.34×10^{-01}	4.73×10^{-01}	1.58×10^{-06}	$\mathbf{5.10 \times 10^{-17}}$	1.86×10^{-02}	$1.89 \times 10^{+00}$
F5	Aver.	$\mathbf{2.07 \times 10^{+00}}$	$7.89 \times 10^{+01}$	$6.51 \times 10^{+01}$	$4.89 \times 10^{+01}$	$2.55 \times 10^{+01}$	$3.92 \times 10^{+01}$
	Stdev.	$1.14 \times 10^{+01}$	$3.40 \times 10^{+01}$	$6.03 \times 10^{+01}$	$\mathbf{3.00 \times 10^{-02}}$	$2.27 \times 10^{+01}$	$5.17 \times 10^{+00}$
F6	Aver.	1.28×10^{-08}	5.97×10^{-19}	3.36×10^{-10}	$9.52 \times 10^{+00}$	2.47×10^{-33}	$\mathbf{0}$
	Stdev.	5.90×10^{-08}	1.20×10^{-18}	1.01×10^{-10}	7.24×10^{-01}	5.63×10^{-33}	$\mathbf{0}$
F7	Aver.	1.39×10^{-01}	8.66×10^{-03}	9.23×10^{-04}	$\mathbf{1.79 \times 10^{-04}}$	1.59×10^{-03}	3.76×10^{-03}
	Stdev.	4.26×10^{-01}	2.33×10^{-03}	8.29×10^{-04}	$\mathbf{6.52 \times 10^{-05}}$	4.09×10^{-04}	9.56×10^{-04}
F8	Aver.	$-1.67 \times 10^{+03}$	$-1.13 \times 10^{+04}$	$-3.01 \times 10^{+03}$	$-5.98 \times 10^{+03}$	$\mathbf{-1.61 \times 10^{+04}}$	$-7.42 \times 10^{+03}$
	Stdev.	$5.05 \times 10^{+03}$	$1.22 \times 10^{+03}$	$\mathbf{2.30 \times 10^{+02}}$	$4.52 \times 10^{+02}$	$7.73 \times 10^{+02}$	$4.08 \times 10^{+03}$
F9	Aver.	1.99×10^{-01}	$5.91 \times 10^{+01}$	$1.23 \times 10^{+01}$	$\mathbf{0}$	$6.83 \times 10^{+01}$	$1.34 \times 10^{+01}$
	Stdev.	7.75×10^{-01}	$1.34 \times 10^{+01}$	$4.20 \times 10^{+00}$	$\mathbf{0}$	$1.55 \times 10^{+01}$	$3.20 \times 10^{+00}$
F10	Aver.	1.76×10^{-02}	1.20×10^{-10}	2.26×10^{-01}	4.20×10^{-15}	8.70×10^{-15}	$\mathbf{1.98 \times 10^{-15}}$
	Stdev.	1.24×10^{-01}	1.54×10^{-10}	6.37×10^{-01}	9.01×10^{-16}	2.17×10^{-15}	$\mathbf{1.79 \times 10^{-16}}$
F11	Aver.	1.48×10^{-04}	7.55×10^{-03}	2.80×10^{-01}	$\mathbf{0}$	1.15×10^{-03}	$\mathbf{0}$
	Stdev.	1.48×10^{-03}	8.76×10^{-03}	1.20×10^{-01}	$\mathbf{0}$	3.09×10^{-02}	$\mathbf{0}$
F12	Aver.	3.73×10^{-03}	8.29×10^{-03}	2.07×10^{-02}	6.73×10^{-01}	2.08×10^{-02}	$\mathbf{9.42 \times 10^{-33}}$
	Stdev.	1.48×10^{-02}	2.70×10^{-02}	7.89×10^{-02}	1.04×10^{-01}	1.03×10^{-01}	$\mathbf{2.78 \times 10^{-48}}$
F13	Aver.	7.69×10^{-04}	3.30×10^{-03}	1.65×10^{-11}	$4.06 \times 10^{+00}$	2.56×10^{-03}	$\mathbf{1.35 \times 10^{-32}}$
	Stdev.	4.75×10^{-03}	5.12×10^{-03}	5.72×10^{-12}	7.17×10^{-01}	4.73×10^{-03}	$\mathbf{5.57 \times 10^{-48}}$

Table 4. Comparison of the optimization results obtained in Test 3 ($d = 100$).

F	Meas.	GA	PSO	SSA	BOA	HFA	CEFA
F1	Aver.	1.02×10^{-02}	1.40×10^{-05}	4.65×10^{-09}	2.34×10^{-18}	6.00×10^{-44}	$\mathbf{1.93 \times 10^{-44}}$
	Stdev.	3.45×10^{-02}	1.18×10^{-05}	8.94×10^{-10}	6.49×10^{-20}	9.67×10^{-44}	$\mathbf{6.54 \times 10^{-45}}$
F2	Aver.	6.88×10^{-01}	7.58×10^{-04}	4.69×10^{-06}	$3.76 \times 10^{+46}$	$\mathbf{1.81 \times 10^{-29}}$	1.70×10^{-21}
	Stdev.	$2.22 \times 10^{+00}$	2.11×10^{-03}	1.02×10^{-06}	$8.32 \times 10^{+46}$	$\mathbf{1.52 \times 10^{-29}}$	2.88×10^{-22}
F3	Aver.	$2.95 \times 10^{+00}$	$7.67 \times 10^{+03}$	4.73×10^{-10}	$\mathbf{2.39 \times 10^{-18}}$	$3.03 \times 10^{+03}$	$7.50 \times 10^{+03}$
	Stdrv.	$9.16 \times 10^{+00}$	$1.70 \times 10^{+03}$	1.95×10^{-10}	$\mathbf{6.96 \times 10^{-20}}$	$4.07 \times 10^{+03}$	$1.84 \times 10^{+03}$
F4	Aver.	2.53×10^{-01}	$8.39 \times 10^{+00}$	1.01×10^{-05}	$\mathbf{2.00 \times 10^{-15}}$	$5.89 \times 10^{+01}$	$1.51 \times 10^{+01}$
	Stdev.	7.75×10^{-01}	7.99×10^{-01}	1.51×10^{-06}	$\mathbf{6.00 \times 10^{-17}}$	$5.13 \times 10^{+00}$	$4.30 \times 10^{+00}$
F5	Aver.	$\mathbf{1.65 \times 10^{+01}}$	$2.38 \times 10^{+02}$	$1.70 \times 10^{+02}$	$9.89 \times 10^{+01}$	$1.34 \times 10^{+02}$	$1.04 \times 10^{+02}$
	Stdev.	$5.37 \times 10^{+01}$	$9.49 \times 10^{+01}$	$7.30 \times 10^{+01}$	$\mathbf{2.74 \times 10^{-02}}$	$5.26 \times 10^{+01}$	$2.47 \times 10^{+01}$
F6	Aver.	$3.08 \times 10^{+00}$	8.74×10^{-06}	3.57×10^{-10}	$2.23 \times 10^{+01}$	2.17×10^{-31}	$\mathbf{0}$
	Stdev.	1.06×10^{-01}	8.32×10^{-06}	1.39×10^{-10}	9.59×10^{-01}	2.40×10^{-31}	$\mathbf{0}$
F7	Aver.	3.57×10^{-01}	6.37×10^{-02}	8.10×10^{-04}	$\mathbf{1.79 \times 10^{-04}}$	1.26×10^{-02}	9.36×10^{-03}
	Stdev.	$1.11 \times 10^{+00}$	1.09×10^{-02}	5.87×10^{-04}	$\mathbf{6.49 \times 10^{-05}}$	2.91×10^{-03}	1.46×10^{-03}
F8	Aver.	$-2.81 \times 10^{+03}$	$-2.10 \times 10^{+04}$	$-3.06 \times 10^{+03}$	$-8.52 \times 10^{+03}$	$\mathbf{-3.00 \times 10^{+04}}$	$-9.24 \times 10^{+03}$
	Stdev.	$8.47 \times 10^{+03}$	$2.19 \times 10^{+04}$	$\mathbf{3.55 \times 10^{+02}}$	$7.06 \times 10^{+02}$	$1.32 \times 10^{+03}$	$3.90 \times 10^{+02}$
F9	Aver.	$2.59 \times 10^{+00}$	$1.29 \times 10^{+02}$	$4.71 \times 10^{+01}$	$\mathbf{0}$	$2.28 \times 10^{+02}$	$3.91 \times 10^{+01}$
	Stdev.	$8.08 \times 10^{+00}$	$2.03 \times 10^{+01}$	$1.43 \times 10^{+01}$	$\mathbf{0}$	$4.64 \times 10^{+01}$	$5.09 \times 10^{+00}$
F10	Aver.	1.12×10^{-01}	1.54×10^{-02}	2.66×10^{-01}	4.44×10^{-15}	2.43×10^{-01}	$\mathbf{4.44 \times 10^{-15}}$
	Stdev.	3.44×10^{-01}	6.60×10^{-02}	6.26×10^{-01}	5.32×10^{-16}	5.12×10^{-01}	$\mathbf{4.01 \times 10^{-16}}$
F11	Aver.	2.30×10^{-04}	7.21×10^{-03}	3.02×10^{-01}	$\mathbf{0}$	3.37×10^{-03}	$\mathbf{0}$
	Stdev.	7.63×10^{-04}	1.39×10^{-02}	1.03×10^{-01}	$\mathbf{0}$	5.62×10^{-03}	$\mathbf{0}$
F12	Aver.	2.18×10^{-03}	1.66×10^{-02}	6.22×10^{-02}	9.51×10^{-01}	7.61×10^{-02}	$\mathbf{2.92 \times 10^{-04}}$
	Stdev.	8.86×10^{-03}	3.03×10^{-02}	1.51×10^{-01}	7.96×10^{-02}	1.16×10^{-01}	$\mathbf{1.60 \times 10^{-03}}$
F13	Aver.	2.71×10^{-03}	7.72×10^{-03}	7.32×10^{-04}	$9.98 \times 10^{+00}$	9.08×10^{-02}	$\mathbf{1.35 \times 10^{-32}}$
	Stdev.	9.66×10^{-03}	1.04×10^{-02}	2.79×10^{-03}	6.30×10^{-03}	3.28×10^{-01}	$\mathbf{5.57 \times 10^{-48}}$

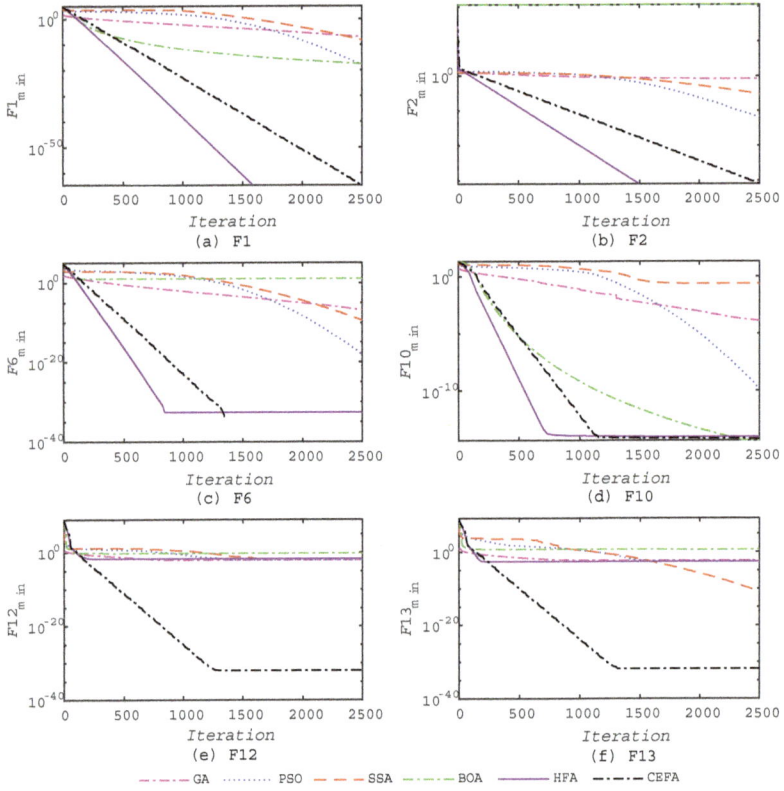

Figure 3. Convergence of algorithms on some of the benchmark functions in Test 2.

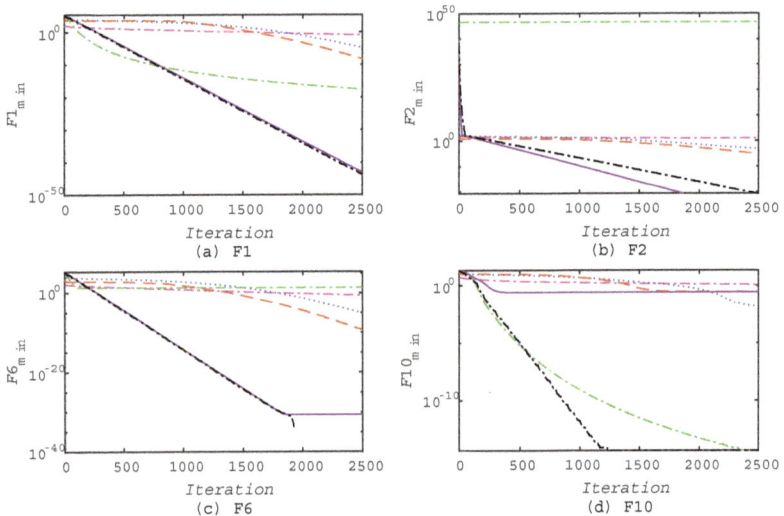

Figure 4. *Cont.*

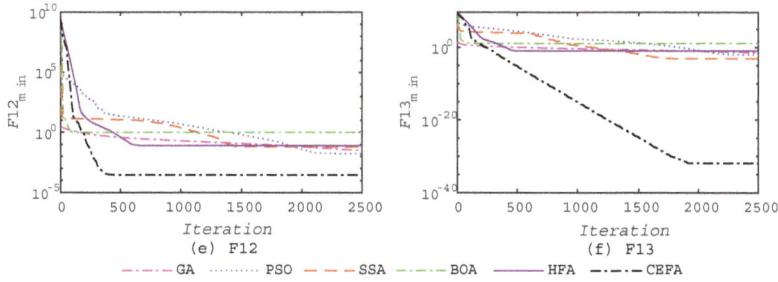

Figure 4. Convergence of algorithms on some of the benchmark functions in Test 3.

5. Discussion

5.1. Advantage Analysis of CEFA

The main reasons for the superior performance of the proposed hybrid meta-heuristic algorithm based on CE and FA in solving complex numerical optimization problems may be summarized as follows:

- CE is a global stochastic optimization method based on Monte Carlo technology, and has the advantages of randomness, adaptability, and robustness; this makes the FA population in the hybrid algorithm have good diversity so that it can effectively overcome its tendency to fall into a local optimum and improve its global optimization ability.
- FA mimicking the flashing mechanism of fireflies in nature has the advantage of fast convergence. With co-evolution, CEFA uses the superior individuals obtained by the FA operator to update the probability distribution parameters in the CE operator during the iterative process, which improves the convergence rate of the CE operator.
- The hybrid meta-heuristic algorithm CEFA introduces the co-evolutionary technique to collaboratively update the FA population and the probability distribution parameters in CE, which obtains a good balance between exploration and exploitation, and has excellent performance in terms of exploitation, exploration, and local optima avoidance in solving complex numerical optimization problems. In addition, the proposed CEFA can effectively solve complex high-dimensional optimization problems due to the superior performance of CE in solving them.

5.2. Efficiency Analysis of Co-Evolution

The proposed hybrid meta-heuristic algorithm CEFA employs co-evolutionary technology to achieve a good balance between exploration and exploitation. The application of this co-evolutionary technology can be summarized by three aspects: (1) The CE operator and the FA operator collaboratively update the optimal solution and optimal value. (2) The initial probability distribution parameters of the CE operator during the iterative process are updated with the population of the FA operator. (3) The result of each iteration of the CE operator updates the current population of the FA operator to obtain the best population.

Figure 5 shows the specific process of co-evolution when the hybrid algorithm is used to solve F1 and F9 selected from the benchmark functions, where "o" is the optimal function value updated by the FA operator and "." is updated by the CE operator. This fully demonstrates that the co-evolutionary technology can be well implemented in the proposed method and the optimal function value is collaboratively updated by the two operators FA and CE during the iterative process.

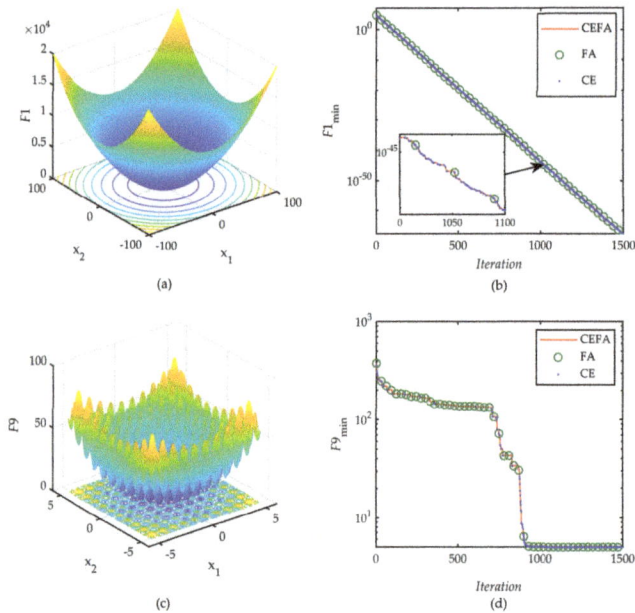

Figure 5. Efficiency analysis of co-evolution: (**a,c**) two-dimensional versions of F1 and F9; (**b,d**) FA and CE co-update the current best in CEFA's iterative process.

5.3. Parameter Analysis of CEFA

In the proposed hybrid meta-heuristic algorithm, the numbers of iterations of the operators CE and FA are two key parameters that affect its performance in solving numerical optimization problems. To this end, this paper took F1 (dimension $d = 30$) as an example, and used the experimental method to explore the influence of their different combinations on the optimization results. The specific experiment was set as follows: the number of iterations N_1 of the CE operator was set to 1, 5, 10, 30, 50, 100, 200, or 300, while the number of iterations of the FA operator N_2 took values of 30, 50, 100, 200, 500, or 1000, and all the other parameters were the same as before. The results were averaged over 30 runs and the average optimal function value and time consumption are reported in Table 5.

Table 5. Experimental results of different numbers of iterations for FA and CE operators in CEFA.

N_1	N_2	30	50	100	200	500	1000
1	$F1_{min}$	$6.59 \times 10^{+00}$	8.94×10^{-02}	3.32×10^{-04}	6.03×10^{-07}	4.95×10^{-15}	3.62×10^{-28}
	T	0.01	0.02	0.05	0.10	0.24	0.48
5	$F1_{min}$	8.49×10^{-11}	6.98×10^{-21}	3.08×10^{-45}	4.43×10^{-95}	0	0
	T	0.03	0.04	0.08	0.15	0.34	0.76
10	$F1_{min}$	7.16×10^{-20}	2.80×10^{-36}	4.30×10^{-76}	0	0	0
	T	0.04	0.05	0.11	0.31	0.76	2.10
30	$F1_{min}$	8.83×10^{-40}	6.80×10^{-69}	0	0	0	0
	T	0.14	0.21	0.42	0.79	2.16	4.70
50	$F1_{min}$	6.84×10^{-51}	1.32×10^{-87}	0	0	0	0
	T	0.24	0.31	0.57	1.15	3.43	6.35
100	$F1_{min}$	8.11×10^{-68}	0	0	0	0	0
	T	0.35	0.64	1.19	2.19	6.57	12.03
200	$F1_{min}$	3.10×10^{-86}	0	0	0	0	0
	T	0.54	1.07	2.10	4.61	12.75	24.10
300	$F1_{min}$	8.97×10^{-98}	0	0	0	0	0
	T	1.13	1.48	3.01	6.80	18.56	38.75

Table 5 shows that the hybrid algorithm can adjust the number of iterations N_1 and N_2 of the two operators in solving the specific optimization problem to achieve higher accuracy. The values of N_1 and N_2 are determined by the characteristics and complexity of the given optimization problem, and they are generally between 30 and 100.

5.4. Performance of CEFA for High-Dimensional Function Optimization Problems

In order to further explore the influence of search space dimension on the optimization performance and convergence rate of CEFA when solving high-dimensional function optimization problems, this paper selected the standard GA, PSO, SSA, BOA, and HFA as comparison objects to test F1 from the benchmark functions. The dimension of the search space was increased from 10 to 200 in steps of 10.

It can be seen from Figure 6 that the accuracy of the proposed CEFA is not greatly affected by the increase of the dimension of the search space, which is obviously different from GA, PSO, and SSA. It can be also seen that BOA has the same advantage, but its solution accuracy is not as high as that of CEFA. As the dimensions of the search space increase, for example, it is greater than 70 for F1, CEFA obtains more accurate results than HFA. This may provide a new and effective way for solving high-dimensional function optimization problems.

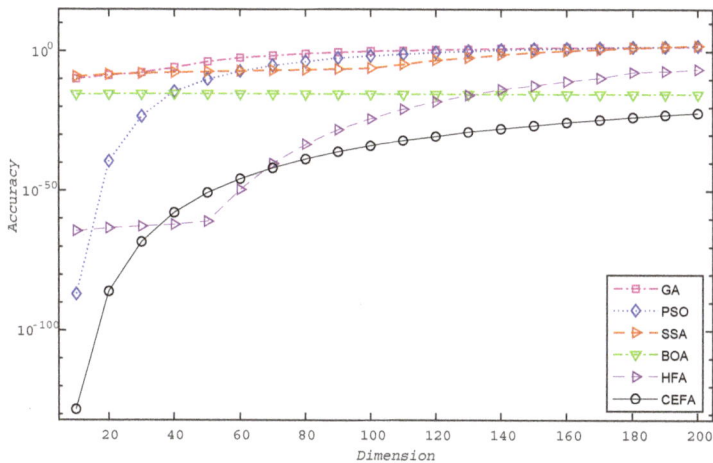

Figure 6. Comparison of optimization accuracy of different search space dimensions.

6. Conclusions

Global optimization problems are challenging to solve due to their nonlinearity and multimodality. In this paper, based on the firefly algorithm and the cross-entropy method, a novel hybrid meta-heuristic algorithm was constructed. In order to enhance the global search ability of the proposed method, the co-evolutionary technique was introduced to obtain an efficient balance between exploration and exploitation. The benchmark functions are employed to evaluate the performance of the proposed hybrid algorithm CEFA for numerical optimization problems. The results of the numeral experiments show that the new method provides very competitive results and possesses more powerful global search capacity, higher optimization precision, and stronger robustness. Furthermore, the new method exhibits excellent performance in solving high-dimensional function optimization problems. In addition, for future research, a discrete version of CEFA will be developed to solve combinatorial optimization problems.

Author Contributions: G.L. designed the algorithm, conducted all experiments, analyzed the results, and wrote the manuscript. P.L. analyzed the data and wrote the manuscript. C.L. conducted the literature review and wrote the manuscript. B.Z. refined the idea and revised the manuscript. All authors have read and approved the final manuscript.

Funding: This work was supported by the National Natural Science Foundation of China (Grant No. 71761012), the Humanities and Social Sciences Key Foundation of Education Department of Anhui Province (Grant No. SK2016A0971), and the Natural Science Foundation of Anhui Province (Grant No. 1808085MG224).

Acknowledgments: The authors are grateful for the support provided by the Risk Management and Financial Engineering Lab at the University of Florida, Gainesville, FL 32611, USA.

Conflicts of Interest: The authors declare no conflict of interest.

Appendix A

See Table A1.

Table A1. The definition of benchmark functions.

Function	Dim	Range	F_{min}	Type
$F1(x) = \sum_{i=1}^{n} x_i^2$	30,50,100	$[-100,100]$	0	Unimodal
$F2(x) = \sum_{i=1}^{n} |x_i| + \prod_{i=1}^{n} |x_i|$	30,50,100	$[-10,10]$	0	Unimodal
$F3(x) = \sum_{i=1}^{n} \left(\sum_{j=1}^{i} x_j\right)^2$	30,50,100	$[-100,100]$	0	Unimodal
$F4(x) = \max_i \{|x_i|, 1 \le i \le n\}$	30,50,100	$[-100,100]$	0	Unimodal
$F5(x) = \sum_{i=1}^{n-1} [100(x_{i+1} - x_i^2)^2 + (x_i - 1)^2]$	30,50,100	$[-30,30]$	0	Unimodal
$F6(x) = \sum_{i=1}^{n} ([x_i + 0.5])^2$	30,50,100	$[-100,100]$	0	Unimodal
$F7(x) = \sum_{i=1}^{n} i x_i^4 + random[0,1)$	30,50,100	$[-1.28,1.28]$	0	Unimodal
$F8(x) = \sum_{i=1}^{n} -x_i \sin\left(\sqrt{|x_i|}\right)$	30,50,100	$[-500,500]$	$-418.9829 \times n$	Multimodal
$F9(x) = \sum_{i=1}^{n} [x_i^2 - 10\cos(2\pi x_i) + 10]$	30,50,100	$[-5.12,5.12]$	0	Multimodal
$F10(x) = -20\exp\left(-0.2\sqrt{\frac{1}{n}\sum_{i=1}^{n} x_i^2}\right) - \exp\left(\frac{1}{n}\sum_{i=1}^{n}\cos(2\pi x_i)\right) + 20 + e$	30,50,100	$[-32,32]$	0	Multimodal
$F11(x) = \frac{1}{4000}\sum_{i=1}^{n} x_i^2 - \prod_{i=1}^{n}\cos\left(\frac{x_i}{\sqrt{i}}\right) + 1$	30,50,100	$[-600,600]$	0	Multimodal
$F12(x) = \frac{\pi}{n}\{10\sin^2(\pi y_1) + \sum_{i=1}^{n-1}(y_i - 1)^2[1 + 10\sin^2(\pi y_{i+1})] + (y_n + 1)^2\} + \sum_{i=1}^{n} u(x_i, 10, 100, 4)$ $y_i = 1 + \frac{x_i+1}{4}$ $u(x_i, a, k, m) = \begin{cases} k(x_i - a)^m, & x_i > a \\ 0, & -a \le x_i \le a \\ k(-x_i - a)^m, & x_i < -a \end{cases}$	30,50,100	$[-50,50]$	0	Multimodal
$F13(x) = 0.1\{\sin^2(3\pi x_1) + \sum_{i=1}^{n}(x_i - 1)^2[1 + \sin^2(3\pi x_i + 1)]\} + \sum_{i=1}^{n} u(x_i, 5, 100, 4)$	30,50,100	$[-50,50]$	0	Multimodal
$F14(x) = \left(\frac{1}{500} + \sum_{j=1}^{25}\frac{1}{j + \sum_{i=1}^{2}(x_i - a_{ij})^6}\right)^{-1}$	2	$[-65.536,65.536]$	1	Multimodal
$F15(x) = \sum_{i=1}^{11}\left[a_i - \frac{x_1(b_i^2 + b_i x_2)}{b_i^2 + b_i x_3 + x_4}\right]^2$	4	$[-5,5]$	0.00030	Multimodal
$F16(x) = 4x_1^2 - 2.1x_1^4 + \frac{1}{3}x_1^6 + x_1 x_2 - x2^2 + 4x_2^2$	2	$[-5,5]$	-1.0316	Multimodal
$F17(x) = (x_2 - \frac{5.1}{4\pi^2}x_1^2 + \frac{5}{\pi}x_1 - 6)^2 + 10(1 - \frac{1}{8\pi})\cos x_1 + 10$	2	$[-5,5]$	0.398	Multimodal
$F18(x) = [1 + (x_1 + x_2 + 1)^2(19 - 14x_1 + 3x_1^2 - 14x_2 + 6x_1 x_2 + 3x_2^2)] \times [30 + (2x_1 - 3x_2)(18 - 32x_1 + 12x_1^2 + 48x_2 - 36x_1 x_2 + 27x_2^2)]$	2	$[-5,5]$	3	Multimodal
$F19(x) = -\sum_{i=1}^{4} c_i \exp\left(-\sum_{j=1}^{3} a_{ij}(x_j - p_{ij})^2\right)$	3	$[1,3]$	-3.86	Multimodal
$F20(x) = -\sum_{i=1}^{4} c_i \exp\left(-\sum_{j=1}^{6} a_{ij}(x_j - p_{ij})^2\right)$	6	$[0,1]$	-3.32	Multimodal
$F21(x) = -\sum_{i=1}^{5}[(X - a_i)(X - a_i)^T + c_i]^{-1}$	4	$[0,10]$	-10.1532	Multimodal
$F22(x) = -\sum_{i=1}^{7}[(X - a_i)(X - a_i)^T + c_i]^{-1}$	4	$[0,10]$	-10.4028	Multimodal
$F23(x) = -\sum_{i=1}^{10}[(X - a_i)(X - a_i)^T + c_i]^{-1}$	4	$[0,10]$	-10.5363	Multimodal

References

1. Horst, R.; Pardalos, P.M. (Eds.) *Handbook of Global Optimization*; Springer: Medford, MA, USA, 1995.
2. Lera, D.; Sergeyev,Y.D. GOSH: Derivative-free global optimization using multi-dimensional space-filling curves. *J. Glob. Optim.* **2018**, *71*, 193–211. [CrossRef]
3. Sergeyev, Y.D.; Kvasov, D.E.; Mukhametzhanov, M.S. On the efficiency of nature-inspired metaheuristics in expensive global optimization with limited budget. *Sci. Rep.* **2018**, *8*, 453. [CrossRef]
4. Yang, X.S. Metaheuristic Optimization. *Scholarpedia* **2011**, *6*, 1–15. [CrossRef]
5. Goldfeld, S.M.; Quandt, R.E.; Trotter, H.F. Maximization by quadratic hill-climbing. *Econometrica* **1966**, *34*, 541–551. [CrossRef]
6. Abbasbandy, S. Improving Newton–Raphson method for nonlinear equations by modified Adomian decomposition method. *Appl. Math. Comput.* **2003**, *145*, 887–893. [CrossRef]
7. Jones, D.R.; Perttunen, C.D.; Stuckman, B.E. Lipschitzian optimization without the Lipschitz constant. *J. Optim. Theory Appl.* **1993**, *79*, 157–181. [CrossRef]
8. Lera, D.; Sergeyev, Y.D. An information global minimization algorithm using the local improvement technique. *J. Glob. Optim.* **2010**, *481*, 99–112. [CrossRef]
9. Sergeyev, Y.D.; Mukhametzhanov, M.S.; Kvasov, D.E.; Lera, D. Derivative-Free Local Tuning and Local Improvement Techniques Embedded in the Univariate Global Optimization. *J. Optim. Theory Appl.* **2016**, *171*, 186–208. [CrossRef]
10. Zhang, L.N.; Liu, L.Q.;Yang, X.S.; Dai, Y.T. A novel hybrid firefly algorithm for global optimization. *PLoS ONE* **2016**, *11*, e0163230. [CrossRef]
11. Whitley, D. A genetic algorithm tutorial. *Stat. Comput.* **1994**, *4*, 65–85. [CrossRef]
12. Kennedy, J.; Eberhart, R.C. Particle swarm optimization. In Proceedings of the 1995 IEEE International Conference on Neural Networks, Perth, Australia, 27 November–1 December 1995; pp. 1942–1948.
13. Dorigo, M.; Birattari, M.; Stutzle, T. Ant colony optimization. *IEEE Comput. Intell. Mag.* **2006**, *1*, 28–39. [CrossRef]
14. Storn, R.; Price, K.V. Differential evolution—A simple and efficient heuristic for global optimization over continuous spaces. *J. Glob. Optim.* **1997**, *11*, 341–359. [CrossRef]
15. Geem, Z.W.; Kim, J.H.; Loganathan, G.V. A New Heuristic Optimization Algorithm: Harmony Search. *Simulation* **2001**, *76*, 60–68. [CrossRef]
16. Passino, K.M. Biomimicry of bacterial foraging for distributed optimization and control. *IEEE Control Syst. Mag.* **2002**, *22*, 52–67
17. Hadad, O.B.; Afshar, A.; Marino, M.A. Honey Bees Mating Optimization (HBMO) Algorithm: A New Heuristic Approach for Water Resources Optimization. *Water Resour. Manag.* **2006**, *20*, 661–680. [CrossRef]
18. Karaboga, D.; Basturk, B. A powerful and efficient algorithm for numerical function optimization: artificial bee colony (ABC) algorithm. *J. Glob. Optim.* **2007**, *39*, 459–471. [CrossRef]
19. Simon, D. Biogeography-Based Optimization. *IEEE Trans. Evol. Comput.* **2008**, *12*, 702–713. [CrossRef]
20. Rashedi, E.; Nezamabadi-Pour, H.; Saryazdi, S. GSA: A gravitational search algorithm. *Inf. Sci.* **2009**, *179*, 2232–2248. [CrossRef]
21. Yang, X.S. Firefly algorithms for multimodal optimization. In *International Symposium on Stochastic Algorithms*; Springer: Berlin/Heidelberg, Germany, 2009; pp. 169–178.
22. Yang, X.S.; Deb S. Cuckoo Search via Lévy flights. In Proceedings of the 2009 World Congress on Nature & Biologically Inspired Computing (NaBIC), Coimbatore, India, 9–11 December 2009; pp.210–214.
23. Yang, X.S. A new metaheuristic bat-inspired algorithm. In *Nature Inspired Cooperative Strategies for Optimization (NICSO 2010)*; Springer: Berlin/Heidelberg, Germany, 2010; pp. 65–74.
24. Mirjalili, S.; Mirjalili, S.M.; Lewis, A. Grey wolf optimizer. *Adv. Eng. Softw.* **2014**, *69*, 46–61. [CrossRef]
25. Mirjalili, S. The ant lion optimizer. *Adv. Eng. Softw.* **2015**, *83*, 80–98. [CrossRef]
26. Mirjalili, S. Moth-flame optimization algorithm: A novel nature-inspired heuristic paradigm. *Knowl.-Based Syst.* **2015**, *89*, 228–249. [CrossRef]
27. Mirjalili, S. Dragonfly algorithm: A new meta-heuristic optimization technique for solving single-objective, discrete, and multi-objective problems. *Neural Comput. Appl.* **2016**, *27*, 1053–1073. [CrossRef]
28. Mirjalili, S.; Lewis, A. The whale optimization algorithm. *Adv. Eng. Softw.* **2016**, *95*, 51–67. [CrossRef]

29. Mirjalili, S.; Gandomi, A.H.; Mirjalili, S.Z.; Saremi S.; Faris H.; Mirjalili, S.M. Salp Swarm Algorithm: A bio-inspired optimizer for engineering design problems. *Adv. Eng. Softw.* **2017**, *114*, 163–191. [CrossRef]
30. Askarzadeh, A. A novel metaheuristic method for solving constrained engineering optimization problems: crow search algorithm. *Comput. Struct.* **2016**, *169*, 1–12. [CrossRef]
31. Połap, D. Polar bear optimization algorithm: Meta-heuristic with fast population movement and dynamic birth and death mechanism. *Symmetry* **2017**, *9*, 203. [CrossRef]
32. Cheraghalipour, A.; Hajiaghaei-Keshteli, M.; Paydar, M.M. Tree Growth Algorithm (TGA): A novel approach for solving optimization problems. *Eng. Appl. Artif. Intell.* **2018**, *72*, 393–414. [CrossRef]
33. AArora, S.; Singh, S. Butterfly optimization algorithm: A novel approach for global optimization. *Soft Comput.* **2019**, *23*, 715–734. [CrossRef]
34. Wolpert, D.H.; Macready, W.G. No Free Lunch Theorems for Optimization. *IEEE Trans. Evol. Comput.* **1997**, *1*, 67–82. [CrossRef]
35. Lai, X.S.; Zhang, M.Y. An Efficient Ensemble of GA and PSO for Real Function Optimization. In Proceedings of the 2009 2nd IEEE International Conference on Computer Science and Information Technology, Beijing, China, 8–11 August 2009; pp. 651–655.
36. Song, X.H.; Zhou, W.; Li, Q.; Zou, S.C.; Liang J. Hybrid particle swarm and ant colony optimization for Surface Wave Analysis. In Proceedings of the 2009 International Conference on Information Technology and Computer Science, Kiev, Ukraine, 25–26 July 2009; pp. 378–381.
37. Mirjalili, S.; Hashim, S.Z.M. A New Hybrid PSOGSA Algorithm for Function Optimization. In Proceedings of the 2010 International Conference on Computer and Information Application (2010 ICCIA), Tianjin, China, 3–5 December 2010; pp. 374–377.
38. Abdullah, A.; Deris, S.; Mohamad, M.S.; Hashim, S.Z.M. A New Hybrid Firefly Algorithm for Complex and Nonlinear Problem. In *Distributed Computing and Artificial Intelligence*; Springer: Berlin/Heidelberg, Germany, 2012; pp. 673–680.
39. Rizk-Allah, R.M.; Zaki, E.M.; El-Sawy, A.A. Hybridizing Ant Colony Optimization with Firefly Algorithm for Unconstrained Optimization Problems. *Appl. Math. Comput.* **2013**, *224*, 473–483. [CrossRef]
40. Rahmani, A.; Mirhassani, S.A. A Hybrid Firefly-Genetic Algorithm for the Capacitated Facility Location Problem. *Inf. Sci.* **2014**, *283*, 70–78. [CrossRef]
41. He, X.S.; Ding, W.J.; Yang, X.S. Bat algorithm based on simulated annealing and Gaussian perturbations. *Neural Comput. Appl.* **2013**, *25*, 459–468. [CrossRef]
42. Wang, G.G.; Gandomi, A.H.; Zhao, X.; Chu, H.C.E. Hybridizing harmony search algorithm with cuckoo search for global numerical optimization. *Soft Comput.* **2016**, *20*, 273–285. [CrossRef]
43. Seyedhosseini, S.M.; Esfahani, M.J.; Ghaffari M. A novel hybrid algorithm based on a harmony search and artificial bee colony for solving a portfolio optimization problem using a mean-semi variance approach. *J. Cent. South Univ.* **2016**, *23*, 181–188. [CrossRef]
44. Mafarja, M.M.; Mirjalili, S. Hybrid Whale Optimization Algorithm with simulated annealing for feature selection. *Neurocomputing* **2017**, *260*, 302–312. [CrossRef]
45. Rubinstein, R.Y. Optimization of Computer Simulation Models with Rare Events. *Eur. J. Oper. Res.* **1997**, *99*, 89–112. [CrossRef]
46. Rubinstein, R.Y. The Cross-Entropy Method for Combinatorial and Continuous Optimization. *Methodol. Comput. Appl. Probab.* **1999**, *1*, 127–190. [CrossRef]
47. Rubinstein, R.Y.; Kroese, D.P. *The Cross-Entropy Method: A Unified Approach to Combinatorial Optimization, Monte Carlo Simulation and Machine Learning*; Springer: New York, NY, USA, 2004.
48. Boer, P.T.; Kroese, D.P.; Mannor, S.; Rubinstein, R.Y. A Tutorial on the Cross-Entropy Method. *Ann. Oper. Res.* **2005**, *134*, 19–67. [CrossRef]
49. Kroese, D.P.; Portsky, S.; Rubinstein, R.Y. The Cross-Entropy Method for Continuous Multi-extremal Optimization. *Methodol. Comput. Appl. Probab.* **2006**, *8*, 383–407. [CrossRef]
50. Tang, R.; Fong, S.; Dey, N.; Wong, R.; Mohammed, S. Cross entropy method based hybridization of dynamic group optimization algorithm. *Entropy* **2017**, *19*, 533. [CrossRef]
51. Chepuri, K.; Homem-De-Mello T. Solving the vehicle routing problem with stochastic demands using the cross-entropy method. *Ann. Oper. Res.* **2005**, *134*, 153–181. [CrossRef]
52. Ho, S. L.; Yang, S. Multiobjective Optimization of Inverse Problems Using a Vector Cross Entropy Method. *IEEE Trans. Magnet.* **2012**, *48*, 247–250. [CrossRef]

53. Fang, C.; Kolisch, R.; Wang, L.; Mu, C. An estimation of distribution algorithm and new computational results for the stochastic resource-constrained project scheduling problem. *Flex. Serv. Manuf.* **2015**, *7*, 585–605. [CrossRef]

54. Peherstorfer, B.; Kramer, B.; Willcox, K. Multifidelity preconditioning of the cross-entropy method for rare event simulation and failure probability estimation. *SIAM/ASA J. Uncertain. Quantif.* **2018**, *6*, 737–761. [CrossRef]

55. Yang, X.S. Firefly Algorithm, Lévy Flights and Global Optimization. In *Research and Development in Intelligent Systems XXVI*; Springer: London, UK, 2009; pp. 209–218.

56. Yang, X.S. Multiobjective firefly algorithm for continuous optimization. *Eng. Comput.* **2013**, *29*, 175–184. [CrossRef]

57. Yang, X.S. Firefly algorithm, stochastic test functions and design optimization. *Int. J. Bio-Inspired Comput. Arch.* **2010**, *2*, 78–84. [CrossRef]

58. Marichelvam, M.K.; Prabaharan, T.; Yang, X.S. A discrete firefly algorithm for the multi-objective hybrid flowshop scheduling problems. *IEEE Trans. Evol. Comput.* **2014**, *28*, 301–305. [CrossRef]

59. Yang, X.S.; Hosseini, S.S.S.; Gandomi, A.H. Firefly algorithm for solving non-convex economic dispatch problems with valve loading effect. *Appl. Soft Comput.* **2012**, *12*, 1180–1186. [CrossRef]

60. Baykasoğlu, A.; Ozsoydan, F.B. Adaptive Firefly Algorithm with Chaos for Mechanical Design Optimization Problems. *Appl. Soft Comput.* **2015**, *36*, 152–164. [CrossRef]

61. Chandrasekaran, K.; Simon, S.P.; Padhy, N.P. Binary real coded firefly algorithm for solving unit commitment problem. *Inf. Sci.* **2013**, *249*, 67–84. [CrossRef]

62. Long, N.C.; Meesad, P.; Unger, H. A Highly Accurate Firefly Based Algorithm for Heart Disease Prediction. *Expert Syst. Appl.* **2015**, *42*, 8221–8231. [CrossRef]

63. Eiben, A.E.; Schipper, C.A. On Evolutionary Exploration and Exploitation. *Fund. Inform.* **1998**, *35*, 35–50.

64. Yao, X.; Liu, Y.; Lin, G.M. Evolutionary Programming Made Faster. *IEEE Trans. Evol. Comput.* **1999**, *3*, 82–102.

MDPI

St. Alban-Anlage 66

4052 Basel

Switzerland

Tel. +41 61 683 77 34

Fax +41 61 302 89 18

www.mdpi.com

Entropy Editorial Office

E-mail: entropy@mdpi.com

www.mdpi.com/journal/entropy

www.ingramcontent.com/pod-product-compliance
Lightning Source LLC
Chambersburg PA
CBHW051725210326
41597CB00032B/5613